Mathematics of Discrete Structures
for Computer Science

Gordon J. Pace

Mathematics of Discrete Structures for Computer Science

 Springer

Gordon J. Pace
Department of Computer Science,
    Faculty of Information and
    Communication Technology
University of Malta
Msida, Malta

ISBN 978-3-642-42988-0   ISBN 978-3-642-29840-0 (eBook)
DOI 10.1007/978-3-642-29840-0
Springer Heidelberg New York Dordrecht London

ACM Computing Classification (1998): F.4, F.2, G.2

Printed on acid-free paper

Springer is part of Springer Science+Business Media (www.springer.com)

*Dedicated to Claudia
and to my parents*

Foreword

Profound knowledge and skills in discrete mathematics are mandatory for a well educated computer scientist. Therefore a corresponding curriculum typically requires at least a basic course on this subject. This textbook, which is based on the lectures given by the author at the University of Malta, is a perfect companion for every student taking such a course.

When teaching a formal computer science topic one can either follow a so-called descriptive or algorithmic approach. Traditionally most textbooks on discrete mathematics follow the descriptive approach meaning that the properties of mathematical structures are listed first while the algorithms to compute corresponding results are of secondary interest. This textbook follows the algorithmic approach which seems to be better suited for computer science students: Mathematical structures are introduced and defined using algorithms which are then characterized by their properties in a second step.

The book covers the material that is essential knowledge for a computer science student. Starting with propositional and first order logic it discusses sets, relations and further discrete mathematical structures. It contains a chapter on type theory before introducing the natural numbers. In the final chapter it discusses how to reason about programs and gives the basics of computability theory.

Besides a thorough formal presentation the book always gives a convincing motivation for studying the corresponding structures and explains the main ideas using illustrations. A well chosen set of exercises rounds off each topic. After studying the presented material a computer science student will be well prepared for further, more specialized courses.

Lübeck, Germany

Martin Leucker

Preface

In computer science, as in other sciences, mathematics plays an important role. In fact, computer science has more in common with mathematics than with the traditional sciences. Consider physics, for example. Clearly, mathematics is right at the core of the subject, used to describe models which explain and predict the physical world. Without mathematics, most of modern physics would have been impossible to develop, since a purely qualitative analysis does not allow one to reason and deduce what logically follows from an observation or a hypothesised model. No manner of qualitative reasoning would have enabled Newton to formulate his law of gravity based on Kepler's claims about planetary motion. Computer science also exploits mathematical reasoning in a similar manner. Given a computer system, one can apply mathematical reasoning to prove that it will always calculate your paycheck correctly.

However, the link is even tighter than this. Computer science is a direct descendant of mathematics, and computers are nothing but the physical embodiment of mathematical systems developed long before the first modern computer was even conceived, let alone built.

In this book, we will be exploring the foundational mathematics which is necessary for the understanding of more advanced courses in computer science. Whether you are designing a digital circuit, a computer program or a new programming language, mathematics is needed to reason about the design—its correctness, robustness and dependability.

There are two distinct approaches used to present mathematical concepts and operators. In the first approach concepts and operators are defined in terms of properties which they satisfy. Based on these definitions, ways of computing the result of applying these operators are then developed and proved to be correct. In contrast, in computer science frequently one takes the opposite approach—one starts by defining ways of calculating the result of an operator, and then proves that the operator one knows how to compute satisfies a number of properties. The two approaches are, in the end, equivalent. However, given that this book is aimed at computer science students, the latter approach is usually adopted.

Finally, most sections are accompanied by exercises which are usually necessary as part of the learning process. Learning mathematics is surprisingly like learning to ride a bicycle. You must try to write formal expressions and proofs before you really understand the concepts you read about. Some of the concepts covered in this book may initially appear to be difficult. However, as you progress through the book and familiarise yourself with the concepts, you will hopefully start requiring less effort to follow the definitions and proofs, and begin to enjoy the beauty of how various concepts in mathematics can be built and reasoned about, based on just a small number of basic concepts.

Buenos Aires, Argentina   Gordon J. Pace

Acknowledgements

The original inspiration which led to this book dates back almost two decades, when I followed a course on mathematics for system specification lectured by Joe Stoy at the University of Oxford. His approach of building up the necessary mathematical machinery from the ground up, akin to building a software system, provided the spark which years later led to the design of the structure and content of a course in mathematics of discrete structures, and consequently to this book. Without Joe Stoy's spark, this book would never have been.

This book has been long in the making. Life kept getting in the way. Many students would have benefitted had the book been finished earlier. Instead, they were exposed to the intended content in my lectures—questions and blank faces exposing the parts of the book which I had to rethink. Thanks go out to those hundreds of students without whom the book would have been much poorer.

Thanks also go out to those people who used initial drafts of the book to deliver and support taught courses in mathematics of discrete structures: Christian Colombo, Adrian Francalanza, Ingram Bondin, Mark Bonnici, Andrew Calleja, Karlston Demanuele, Kevin Falzon, Andrew Gauci and Christine Vella. Their feedback was invaluable.

Finally, the shortest thanks go out to those whose support can never be fully acknowledged, no matter how long I write about it—Claudia, Dodie and Josie. Thank you.

Contents

Chapter 1
Why Mathematics?

How are programming and computing machines related to mathematics? Does one really need mathematics to understand computer science concepts?

1. In the sciences, mathematics is used to build a model of the observed reality. In physics, for example, scientists propose mathematical systems describing physical objects and phenomena pertaining to them. After accumulating data about planetary motion, Kepler formulated three mathematical laws which planetary motion satisfies. Using these, he could simulate planets moving around a sun, predict the actual position of planets and verify properties such as "Using these laws of planetary motion, there is no way in which Mars will collide with Earth in the next million years."

    Unlike these natural sciences, in computer science, we study artificial objects: computers, programs, programming languages, etc. Despite this there is no fundamental difference, in the sense that we would still like to have a model of these systems to reason mathematically about them. Will this program terminate? Does it give the right answer? Is this computer capable of performing this complex operation?

2. In programming, very frequently we have to represent complex structures. Imagine writing a program which has an internal representation of the map of the London Underground, chess game configurations or rules of English grammar. Describing these objects adequately requires the use of complex data structures. Fortunately, objects such as the ones mentioned have been very well studied by mathematicians. Or rather, no mathematician studied the London Underground map, but many have studied *graph theory,* using which it is straightforward to describe the map of the London Underground. The mathematical way of describing such systems can then be readily translated into a data structure to be used in our programs. What is even more interesting is that mathematicians have proved various things about these classes of objects, properties which would then also be true of our data structures, allowing us to analyse and optimise our programs.

3. Just as we can use mathematics to guide the design of data structures, computer designers and computer scientists use mathematics to build circuits and computers, and design programming languages. A circuit is nothing but a physical

G.J. Pace, *Mathematics of Discrete Structures for Computer Science,*
DOI 10.1007/978-3-642-29840-0_1, © Springer-Verlag Berlin Heidelberg 2012

instantiation of mathematical propositions, and a computer is nothing but a physical instance of a Turing machine (which is not a machine, but a mathematical model). Whereas in the natural sciences one usually uses mathematics to model reality, in computer science our objects of study are concrete instances of mathematical abstractions.

1.1 What Is Mathematics?

The layman usually associates mathematics with numbers. Numbers just happen to be one application of one particular field of mathematics. However, mathematics can be seen as the study of ideal, regular and formal structures. *Ideal* in the sense that it is detached from reality. No matter if in the real world no perfect circle exists, mathematicians in the field of geometry have been studying perfect circles for over two thousand years. *Regular* in the sense that the objects can be described in a compact way. Any circle on a plane can be described in terms of a point (the position of its centre) and a distance (its radius). Using this compact (and general) description, we can compare circles and study how they interact. *Formal* in the sense that the objects have a well-defined meaning and can be reasoned about in an unambiguous way. Once we agree on what a perfect circle is and the definition of the intersection of two circles, the deductions made cannot be disputed. If, based on the formal definitions, one proves that two particular circles do not intersect, the conclusion is final and cannot be further argued about.

The best way to understand what characteristics distinguish mathematical reasoning from other forms of reasoning is to take a brief look at the history of mathematics and see how mathematical thought was shaped and became what it is today.

1.2 A Historical Perspective

From the birth of the comparison of quantities and counting, it must have been inevitable that certain patterns were observed. If Paul has more sheep than Mary, and Mary has more sheep than Kevin, then it is inevitable that Paul has more sheep than Kevin, without any need to compare the flocks of Paul and Kevin directly. If Paul gives Mary some sheep, then Mary will still have more sheep than Kevin. It is typical human nature to try to identify patterns in observations, and in the same way that our ancestors observed the repetition and order of the seasons, they also observed patterns about quantities.

From the earliest numeric systems recorded on clay tablets, it is clear that the scribes were aware of properties about the numbers they were inscribing. For example, on a number of Babylonian tablets describing problem and solution, some sums were performed by adding the first quantity to the second, while in others, they added the second to the first. This implies that they were implicitly aware of the rule that $x + y = y + x$.

Numbers and geometry have various everyday applications: calculating areas for landlords to charge rent, working out profits and in architecture. The Ancient Greek civilisation, however, also gave great importance to abstract and mental pursuits, from which arose the study of numbers and other abstract notions for their own sake rather than simply as tools to calculate results. Sects such as the Pythagoreans gave an almost divine standing to the perfection of whole numbers. They identified classes of numbers: square numbers, perfect numbers, triangular numbers, prime numbers, etc., and asked questions about them: Is there an infinite number of prime numbers? Are there numbers which can be written as the sum of two square numbers in more than one way? Can the sum of two perfect numbers be itself perfect? They presented arguments which supported their claims. Unlike most philosophical claims which most fellow philosophers made, and which one could argue about for days and present arguments both for and against (Is the universe finite? Is there a basic building block out of which everything in the universe is made?), arguments about mathematical notions were irrefutable.

This gave rise to the question of whether such forms of argumentation can also be applied to other fields of philosophy and law. What was special about mathematical arguments that made them irrefutable? A mathematical argument always took the form of a sequence of statements, with each statement being either (i) a truth upon which the debating parties agreed, or (ii) built upon previous statements already proved and combined, once again in an agreed way. Therefore, one had to identify two types of shared truths to be able to build irrefutable arguments: a number of basic truths everyone agrees on, and a number of rules which dictate how one is allowed to conclude a new true statement from already known ones. For example, one basic truth may be that every circle has the same radius as itself, while a rule may say that, if two circles have the same radius, then they have the same area. If these truths are not agreed on, then one must seek even simpler truths upon which to agree.

Around 300 BC, Euclid constructed a mathematical system using this approach to reason about geometry. He proved various complex statements based on a small number of basic truths and rules. This mode of reasoning, usually called *formal* or *axiomatic reasoning*, is the basis of mathematics. Armed with a small number of obvious truths, the mathematician seeks out to prove complex properties of his or her objects of study.

But can these basic truths and rules be shown to be correct? One way of doing so is to give another group of rules and show the two to be equivalent. But that is just relegating the truth of one set of rules to another. What does it take to ensure that our basic truths and rules are correct? It means that they faithfully describe the objects they set out to describe. This is more of an empirical, or observational matter. What our proofs really do tell us, is "Give me a system which obeys these rules and this is what I can conclude about it ...". Nothing more, but more importantly, nothing less.

1.3 On the Superiority of Formal Reasoning

We have already argued that one of the main advantages of using formal reasoning is *incontrovertibility*—one can check an argument or proof for correctness quite easily. If the rules are adhered to, then the conclusions must be correct. This means that, as long as the basic truths remain undisputed, there can be no revisions of the mathematical conclusions.

Another advantage is that of *precision*. When we state something mathematically we are stating it in an unambiguous, exact manner. This is not so in English as in, for example, the following sentence: "Rugby is a sport played by men with oddly shaped balls". When one is designing a computer program, ambiguities and underlying assumptions in specifications can be very costly: "Yes, I just said that we wanted to be able to search the records, but why does your program search them by name? We always search by identity card numbers here!" In mathematics, any implicit assumptions can be discovered through analysis. If your sorting routine only works when the items to sort are positive numbers, it may lie hidden for years, until someone tries to sort a list of bank account balances. A mathematical analysis of the routine would have identified this hidden assumption immediately.

Abstraction is the act of ignoring irrelevant detail. Imagine someone is called in to check whether a sheep can go through all the doorways in a certain house. It is allowed to take measurements of the sheep but not physically push the sheep through the doors. Furthermore, justification of the conclusion is also required. One solution is to model the complex shape of the sheep mathematically using millions of points, and then build a computer program which rotates the sheep in different ways so as to assess whether it actually passes through all the doorways. A smarter solution, however, would be to physically check that the sheep fits inside a two metre by one metre by one metre box. Then, using pencil and paper, it can easily be shown that, since all the doorways are at least one metre wide and one metre high, the sheep can go through them all. What is so strong about this approach is that these mathematicians started by reasoning, 'if we know that the sheep will fit in such a box, then we can prove that it passes through all the doorways.' The act of using a box rather than the actual sheep gives a much simpler object to reason about, and is called *abstraction*.

One act of abstraction we are very familiar with is when we apply equalities. We take an equality $x = y$ and proceed to prove things by replacing x by y or y by x in a huge expression. This hides a very strong abstraction: x and y are equal no matter the context in which they occur. Whether x appears in a huge equation or a small one does not really matter.

We will be illustrating some applications of mathematics to computing in the coming section.

1.4 The Mathematics of Computing

Mathematics is a vast subject although, obviously, some topics are more important to computing than others. In this book we will be focussing on the mathematics of discrete structures—objects that vary in large steps, as opposed to small continuous changes. To mathematically model computers or programs, we can thus only use discrete structures made up of ones and zeros.[1]

Although the focus and aim of this book is to expose the mathematical foundations which are applied in various fields of computer science, the applications are exposed through examples throughout the text. As initial motivation, and a taste of where mathematics and computing overlap, we will look at a number of application areas.

1.4.1 The Analysis of Programs

Faced with the same specification to implement, different programmers may come up with different solutions. Consider the task of sorting a collection of numbers in ascending order. One person may propose the following algorithm:

> *Go through the whole list, choose the smallest, and put it in the first position. Then repeat the procedure with the other numbers until no numbers are left to sort.*

For example, faced with the sequence of numbers $\langle 2, 4, 1, 3 \rangle$, this approach will identify 1 as the smallest number and swap it with the number in the first position, resulting in $\langle 1, 2, 4, 3 \rangle$. The procedure is then reapplied to the sequence disregarding the first item $\langle 2, 4, 3 \rangle$. Eventually, one obtains the sorted list.

However, another person may come up with a different solution:

> *A list with no more than one number is already sorted. If the list is longer, remove the first item in the list, and separate the other numbers into two: those not larger than the first number, and those larger. These two collections of numbers are sorted (using this same procedure) and then just catenated, starting with the sorted list of smaller numbers, followed by the number which was the first item in the original list and finally the sorted list of larger numbers.*

If we consider sorting the sequence $\langle 2, 4, 1, 3 \rangle$, we would remove 2 (the first item in the list) and create two lists—those not larger than 2 and those larger than 2. This gives $\langle 1 \rangle$ and $\langle 4, 3 \rangle$. Using the same approach, these lists are sorted, giving $\langle 1 \rangle$ and $\langle 3, 4 \rangle$, and concatenated with the item 2 in between: $\langle 1, 2, 3, 4 \rangle$.

The two sorting procedures, although giving the same result, seem very different. Are they? By running the routines on some sample data, it is easy to see that they

---

[1]Some analogue devices can calculate values in a continuous manner. For example, one can build a device that, given two numbers encoded as voltages, outputs their sum as a voltage. This is an analogue device since its inputs and outputs can be varied in an infinitude of ways in arbitrarily small steps. Normal computers calculate things by first expressing them in terms of ones and zeros, thus allowing us to only change things in steps by changing ones to zeros and vice versa.

sort the numbers in different ways and take different times to sort the same list of numbers. But is one better than the other? If so, in what way is it better, and by what measure? Two important measures for computer programs are memory usage and execution time. We would like to be able to analyse these algorithms mathematically to be able to compare their relative efficiency. Wouldn't it be sufficient to test them together to see which runs faster? Although one could, such an analysis may not be easily generalised, and would not help us answer questions such as whether there are procedures which are even more efficient than these.

It can be mathematically proved that both algorithms work correctly. However, given a list of n numbers, on average the first algorithm (called *Selection Sort*) makes a number of comparisons proportional to n^2 to sort, while the second algorithm (called *Quicksort*) takes a number of steps proportional to $n \log n$. Since $\log n$ is smaller than n, the latter algorithm is preferred.

1.4.2 Formal Specification of Requirements

Every program is written to solve a problem. To judge whether or not a program works as intended, one must also know precisely what problem it intends to solve. Do the sorting routines above work? Well, it depends. If the task is to reverse a list of numbers, they certainly do not. However, if the specification is to sort the numbers in ascending order, they work correctly.

When software or hardware producers start on a project to design a system, the first step is to identify and document the requirements of the client or user. Most projects use informal notation (such as English), which means that ambiguity is inevitable, and formal analysis impossible. However, another school of thought proposes that mathematics be used to document the requirements. We have already discussed how effective mathematics is in describing things in a precise yet abstract fashion. We will see various examples of this later on in the book.

1.4.3 Reasoning About Programs

Once we have a specification of the requirements and the program itself, we would like to know whether or not the program works as expected. One way to do this is to feed the program with a large number of inputs and check that the outputs are as expected. However, to quote the computer scientist Edsger Dijkstra, "Testing shows the presence, not the absence of bugs". Even if it is usually not possible to test with every possible input, various techniques have been developed to intelligently choose values to test for or to quantify how much of the system's potential behaviour has been tested. But if testing can only detect the presence, not absence of bugs, can one ever be sure that a program is bug-free, or at least that it conforms to its specification?

Take an engineer trying to verify whether a particular bridge can take the weight of 50 cars without collapsing. One approach would be to test the statement by placing 50 cars in various configurations on the bridge: well distributed, all at one end, all at the other end, etc. Apart from being dangerous, this testing would only show us that, as long as the cars are in the positions we have tried, the bridge will not collapse. What a smart engineer would do, however, is to use the laws of physics to prove mathematically that no 50-car load distribution over the bridge would cause it to collapse. Similarly, since the 1960s, computer scientists have tried to develop analogous techniques for doing the same with computer programs.

The problem is that computer programs are not easily describable using the laws of physics.[2] To address this, computer scientists have developed and used various logics and mathematical tools to give meaning to computer programs. The translation of a program into a logic statement which describes its behaviour is called the *formal semantics* of the program. To check that a particular program satisfies a certain requirement, it suffices to translate the program into logic and prove that the requirements always follow from this logic description. Obviously, the requirements would also have to be written in a mathematical notation.

This is easier said than done. The proofs are usually very long, difficult and tedious to complete. However, when lives are on the line, it may still be desirable to verify critical sections of the code. Over recent decades, techniques have been developed to perform such verification in an automatic or semi-automatic manner.

1.4.4 The Limits of Computing

Is the computer omnipotent? Given a mathematical specification, can one always write a program to solve it? In the 1930s, Alan Turing proved that certain problems can never be solved using an algorithm. Surprisingly, it was not a difficult mathematical problem that he identified as incomputable, but one that most new programmers would think difficult, yet possible. Consider the problem of writing a program which decides if a given algorithm terminates. Obviously, we can write a program which can tell us whether particular programs terminate, however it will not work on all programs. For instance, if a program contains no loops or recursive calls, one can deduce that it will terminate. However, this does not help us answer the question for the sorting programs we saw earlier, both of which always terminate. Even simulating the algorithm is not sufficient. If the algorithm terminates, the program will correctly announce that it terminates, but it will never say anything about programs which do not terminate. Turing showed that, no matter how smart a programmer is, it is impossible to come up with a program which always works.

---

[2] Well, of course, if we model the transistors which make up the computer, the program would just be an instance of the transistors being set with particular voltages. However, we would be committing the cardinal sin of not abstracting away unnecessary detail, resulting in a precise description of the behaviour of a computer, but one too big to be of any practical use.

But couldn't Turing have been wrong? How could he have shown that something is impossible? Later on in this book, we will look at a proof of Turing's result, and you can judge for yourself. Interestingly, his proof, published in the 1930s, is still applicable to today's computing devices.

So, next time you are working with a compiler company, and you are asked to add an option to the compiler which reports whether the program being compiled terminates, just shrug your shoulders as you cite Turing's paper.

1.4.5 A Physical Embodiment of Mathematics

Computing was originally an offspring of mathematics. In the first half of the 20th century, mathematicians built all sorts of mathematical computation models. These were embodied into the first computers, and, conceptually, the computer has not changed much since then.

Mathematics is the reduction of domains into a finite description (the rules and basic truths) which can be used to construct complex truths. In a similar fashion, using a programming language is to give a finite description which can be used to solve complex problems. The similarity is not coincidental. Computers are a direct physical embodiment of mathematics. Computing and mathematics are intimately entwined, as the rest of the book will reveal.

1.5 Terminology

Finally, a short note about some terminology used in this (and other) books. When one makes a mathematical statement with no justification but which is believed to be true, it is called a *conjecture*. Once proved, it is no longer called a conjecture, but a theorem, lemma, proposition or corollary. The distinction between these names is not very important, but they are usually used to identify the importance of the result. Fundamental and general results are called *theorems*. In the process of proving a theorem, one sometimes needs to prove intermediate results which are interesting but are more of a stepping stone towards the theorem. These are called *lemmata* (singular *lemma*). When something follows almost immediately from a theorem, it is called a *corollary*. Finally, *propositions* are statements which follow directly from definitions. Apart from the term *conjecture,* the choice of which of the other terms to use may be rather arbitrary and not very important. No one will protest that a lemma should be a proposition or a theorem. As long as it carries a proof with the statement no one really minds. However, these terms give a structure to the mathematical results which may help their understanding.

It will all seem confusing at first, but as you read through this text you will get more familiar with these terms, and this primer is meant more for later reference than to be understood at this stage. When you read through the first few chapters and start wondering 'But why is this result called a theorem, whereas the other is called a lemma?' just refer back to this brief outline.

Chapter 2
Propositional Logic

One of the basic notions in mathematics, whether we are talking of geometry, arithmetic or any other area, is that of truth and falsity. Reasoning is what mathematics is all about, and even if we are discussing a topic such as geometry, we need a mathematical infrastructure to reason about the way the truth of a statement follows from other known truths. One area of mathematics dealing with these concepts is *propositional logic*.

2.1 What Is a Proposition?

A proposition is a statement which is either true or false. We may not know whether or not it is true, but it must have a definite value.

Example 'Mice eat snakes', 'The world is round', 'Elephants are pink' and 'Whenever it rains I always carry an umbrella' are all examples of propositions. On the other hand, 'Where am I?', 'Go to school', 'The dark side of the moon' are examples of non-propositions. The first is a question, and the second is an imperative, and are thus not statements. The third is a phrase and not a whole sentence. You would be baffled if I were to say: 'Tell me whether the following is true ... ' and then proceed to tell you 'Go to school'. With a proposition you may not know the answer, but you do know that it has to be either true or false. Examples of such propositions are 'Zejtun Red Stars play in the Maltese Premier league' and 'Jim Morrison died at the age of 28.'[1] Although you may not be able to answer 'true' or 'false' if asked about the truth of these propositions, you would be able to say 'I do not know, but it is either true or false'. ◇

---

[1] Assuming, of course, that you are not an ardent fan of The Doors.

G.J. Pace, *Mathematics of Discrete Structures for Computer Science*,
DOI 10.1007/978-3-642-29840-0_2, © Springer-Verlag Berlin Heidelberg 2012

Exercises

2.1 Which of the following are propositions:

    (a) If I am wrong, then I am an idiot.
    (b) Choose a number between 5 and 10.
    (c) Have I already arrived at the town of True?
    (d) 7 is the largest number.
    (e) Numbers are odd.

2.2 The Language of Discourse

Consider the propositions 'I am carrying an umbrella', 'It is raining' and 'I am drenched'. From these propositions we can construct more complex propositions such as 'I am carrying an umbrella and it is raining', 'if it is raining and I am carrying an umbrella, then I am drenched' and 'I am carrying an umbrella unless it is raining.' We will now define a number of operators which will allow us to combine propositions together. Why should we bother doing this? Consider the compound proposition 'It is raining and I am drenched'. Whatever the state of the weather and my clothes, I might as well have said 'I am drenched and it is raining'. Similarly, had I said 'I am carrying an umbrella and it is raining', I might as well have said 'it is raining and I am carrying an umbrella'. Obviously, I can generalise: for any propositions P and Q, we would like to show mathematically that saying 'P and Q' is the same as saying 'Q and P'. To reason mathematically, we should have a clear notion of what operators we will be using and their exact mathematical meaning.

Definition 2.1 *Given two propositions P and Q, the conjunction of P and Q is written as $P \wedge Q$. We read this as 'P and Q', and it is considered to be true whenever both P and Q are true.* ∎

Example We can write 'it is raining and I am carrying an umbrella' as 'it is raining' $\wedge$ 'I am carrying an umbrella'. If it is raining, and I am carrying a closed umbrella, this compound proposition is true, whereas if it is not raining, but I am carrying an open umbrella, it would be false. ◇

Definition 2.2 *Given two propositions P and Q, the disjunction of P and Q is written as $P \vee Q$. We read this as 'P or Q', and it is considered to be true whenever either P is true, or Q is true, or both.* ∎

Example We can write 'I am carrying an umbrella or I am drenched' as 'I am carrying an umbrella' $\vee$ 'I am drenched'. If I am carrying a closed umbrella in the rain, this is true. If I am under the shower, carrying an open umbrella, this is still true. It would be false if I am dry and carrying no umbrella. ◇

Definition 2.3 *Given two propositions P and Q, we write P ⇒ Q to mean 'If P is true, then so is Q', or 'P implies Q'. P ⇒ Q is considered to be true unless P is true and Q is false.* ∎

Example We can write 'If it is raining then I am drenched' as 'It is raining' ⇒ 'I am drenched'. If it is raining, but I am inside and thus not wet, this proposition is false. If it is raining and I am having a shower, then it is true. If it is not raining, then it is true whether or not I am wet.

As another example, consider a politician who makes the statement 'If the economy improves, we will build a new hospital'. This can be written as 'Economy improves' ⇒ 'A new hospital is built'. Does it correspond to our everyday use of the natural language statement? Let us consider the possible cases separately. If the economy improves and a new hospital is built, the politician has made a true statement. On the other hand, if the economy improves but no hospital is built, the politician clearly lied. Now let us look at the truth of the politician's statement if the economy does not improve. If the hospital is not built, everyone would agree that the politician did not make a false statement. What if a hospital is built anyway? One can argue that the politician made no commitment for the possible eventuality of the economy not improving. Therefore, even in this case, the statement was not a lie. If we compare this analysis to the meaning of implication, we see that the use of implication is justified. ◇

Definition 2.4 *Given two propositions P and Q, we take their bi-implication, written P ⇔ Q, to mean 'P if and only if Q'. This is considered to be true whenever P and Q are equivalent: either P and Q are both true, or P and Q are both false.* ∎

Example 'You will pass the exam if and only if you study and you work out the exercises' can be written formally as 'You will pass the exam' ⇔ ('you study' ∧ 'you work the exercises'). If this is a true proposition, not only does it mean that if you do not study or if you do not work out the exercises you will not pass the exams, but also that if you do study and work the exercises, you will pass the exam.

Another use of bi-implication can be seen when describing a logic gate. Consider the following conjunction gate which takes two input wires i_1 and i_2 and outputs their conjunction on wire o:

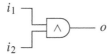

If we use a proposition for each wire (i_1, i_2 and o) which is true if that particular wire carries a high signal, we can describe the behaviour of the circuit as: the output wire is high if and only if both input wires are carrying a high signal. This can be written as: $o \Leftrightarrow (i_1 \wedge i_2)$. The behaviour of the other logic gates can be similarly described. ◇

Definition 2.5 *Given a proposition P, the* negation *of P is written as ¬P. We read this as 'not P', and it is considered to be true whenever P is not true (P is false).* ∎

Example 'If it is raining and I am not carrying an umbrella, then I will get wet or I will run quickly home' can be written as:

('it is raining' ∧ ¬ 'I am carrying an umbrella') ⇒
('I will get wet' ∨ 'I will run quickly home')   ◇

We call a sentence made up of these operators a propositional formula. It is called *a well-formed formula* if is syntactically correct. The propositional formulae given in the examples above are all well-formed while, for example, (∧ 'it is raining') is not. We will describe well-formed formulae using the following table:

wff ::= basic proposition
      | (wff)
      | *true*
      | *false*
      | ¬ wff
      | wff ∧ wff
      | wff ∨ wff
      | wff ⇒ wff
      | wff ⇔ wff

It means that a well-formed formula (wff) can take any of the forms separated by |. This describes the *syntax* of the language—it identifies the class of formulae which we will be able to reason about. Note that we have added two entries we have not mentioned: '*true*' and '*false*', which are the propositions which are always true and always false, respectively.

Example We will now translate English sentences into formal propositional logic formulae. Remember that natural languages tend to be informal and imprecise, so sometimes more than one translation exists, depending on how they are interpreted.

Fixing the car: I am not sure what is wrong with my car. It is either the carburetor or the distributor. My mechanic said, whenever there is a problem with the carburetor, the sparking plugs need changing. Another mechanic told me that the distributor and the sparking plugs are closely connected. If one finds problems in one, there are invariably problems with the other.

Let C be the proposition *'there is a problem with the carburetor'*; D be the proposition *'there is a problem with the distributor'*; and S be *'the sparking plugs need changing.'*

The statements about the way the problems are related can be expressed as:

- It is either the carburetor or the distributor: $C \vee D$.
- Whenever there is a problem with the carburetor, the sparking plugs need changing: $C \Rightarrow S$.

- The distributor and the sparking plugs are closely connected. If one finds problems in one, there are invariably problems with the other: $D \Leftrightarrow S$.

What do you think is wrong with my car?

Eating habits: Jack and Jill have just moved in together. They have discovered that their eating habits are quite different. For Jack to be happy, he insists that, if it is cold but he is not very hungry, he would want to eat soup. Jill, on the other hand, never eats soup when it is cold. Assuming that at their place on any one day only one dish is cooked, express these conditions formally.

Let S be the proposition for '*today, Jack and Jill will eat soup*'; H for '*Jack is hungry*'; and C for '*it is cold*'.

- Jack's constraint for happiness: $(C \wedge \neg H) \Rightarrow S$.
- Jill's constraint for happiness: $S \Rightarrow \neg C$.

There are different ways of writing Jill's constraint. For instance, another valid way of writing it is $C \Rightarrow \neg S$, or even $\neg(C \wedge S)$. Which to choose is largely a matter of interpretation of the English statement. Luckily, we will later develop mathematical tools to show that these three ways of writing her statement are equivalent.

Based on these known facts, do you think that they can live happily together?

Circuits: An engineer is trying to implement a multiplexer circuit which takes three input wires *sel*, *l* and *h*, and one output wire *out*. The behaviour of a multiplexer is very simple—when the selector wire *sel* is high, then the output wire *out* should be equal to the input *h*, while if *sel* is low, the output should be equal to the input *l*. We will use a propositional variable for each wire in the circuit—so *sel* will be a propositional variable which is true if the input *sel* carries a high signal. Similarly, we will have variables *out*, *l* and *h*.

Before designing a circuit to implement the multiplexer, the engineer lists a number of properties it should satisfy:

- If *sel* is high, then *out* should be equal to h: $sel \Rightarrow (out \Leftrightarrow h)$.
- If *sel* is low, then *out* should be equal to l: $\neg sel \Rightarrow (out \Leftrightarrow l)$.
- If l and h have the same value, then the output should be the same as either of them: $(l \Leftrightarrow h) \Rightarrow (out \Leftrightarrow h)$.

The engineer then proceeds to implement the multiplexer circuit using negation, conjunction and disjunction gates as shown below:

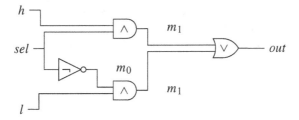

Note that we have named the intermediate wires between gates to be able to talk about their value. As we have already seen, the behaviour of logic gates can be described using the propositional logic operators. For instance, if a conjunction gate takes inputs i_1 and i_2 and has output o, its behaviour is described as $o \Leftrightarrow i_1 \wedge i_2$. The behaviour of the circuit designed by the engineer is thus equivalent to the conjunction of all the formulae describing the logic gates:

$$(out \Leftrightarrow m_1 \vee m_2)$$
$$\wedge \, (m_0 \Leftrightarrow \neg sel)$$
$$\wedge \, (m_1 \Leftrightarrow sel \wedge h)$$
$$\wedge \, (m_2 \Leftrightarrow m_0 \wedge l)$$

Do you think that the engineer's circuit really satisfies the properties identified earlier?   ◇

In arithmetic, to avoid overusing brackets, we give the operators a relative precedence. Thus, $3 \times 2 + 4$ is unambiguously interpreted as $(3 \times 2) + 4$. Mathematically, we say that multiplication has a higher precedence than addition—in an unbracketed expression, the multiplication is to be applied before addition. The operators with higher precedence can be seen as if surrounded by brackets. The following table shows the precedence levels of the propositional logic operators:

$$\neg \qquad \wedge \qquad \vee \qquad \Rightarrow \qquad \Leftrightarrow$$
$$\xrightarrow{}$$
$$\textit{decreasing precedence}$$

Therefore, $\neg A \Rightarrow B \wedge C \Leftrightarrow D$ effectively means $((\neg A) \Rightarrow (B \wedge C)) \Leftrightarrow D$. Although we can reduce brackets to a minimum, we usually use brackets to distinguish between $\wedge$ and $\vee$, and between $\Rightarrow$ and $\Leftrightarrow$. Therefore, we would usually write $A \vee (B \wedge C)$ even if $A \vee B \wedge C$ would do. Similarly, we write $A \Leftrightarrow (B \Rightarrow C)$ when $A \Leftrightarrow B \Rightarrow C$ would do.

Another issue is how to interpret an unbracketed expression with more than one use of the same operator, such as $A \wedge B \wedge C$. Should this be interpreted as $(A \wedge B) \wedge C$ or as $A \wedge (B \wedge C)$? Later on, we will prove that the two expressions are logically equivalent. However, syntactically, the expressions are different, and we thus have to decide which expression we mean when we leave out the brackets. Furthermore, the equivalence does not hold for all operators. Consider, for instance, subtraction. $(7 - 1) - 3$ is not equal to $7 - (1 - 3)$. Usually, when we write $7 - 1 - 3$, we mean $(7 - 1) - 3$. This is called *left associativity*. An operator $\oplus$ is said to be *right associative* when $x \oplus y \oplus z$ is to be interpreted as $x \oplus (y \oplus z)$. Conjunction and disjunction are left associative, while implication and bi-implication are right associative. Therefore, using precedence and associativity, we could write $A \wedge B \wedge C \Rightarrow D \Rightarrow E$ instead of $((A \wedge B) \wedge C) \Rightarrow (D \Rightarrow E)$.

Exercises

2.2 Write in mathematical notation:

(i) My loves include Sally or Jane. If I love Sally, I also love Jane.
(ii) Being a coward is the same as being a hero. If I were a hero, then either I can assess dangers well, or I am very foolish. In reply to the question 'Are you foolish?', I (truthfully) reply 'If that were so, then I am a coward.'

(iii) Swimming is fun unless it is not hot. If it is hot, it may be fun swimming, but playing tennis is certainly not.

(iv) In a circuit consisting of two negation gates connected in sequence, the output of the circuit is equal to its input.

2.3 Explain in English:

(i) B_1 and B_2 stand, respectively, for bulbs 1 and 2 being lit. S_1 and S_2 represent whether or not switches 1 and 2 are on or off. An engineer observes that $S_1 \wedge S_2 \Rightarrow B_1$, and $S_1 \Rightarrow B_2$. He also noticed that $B_2 \Rightarrow \neg S_2$ and that $B_1 \vee B_2$.

(ii) $T = $ I think, $A = $ I am. It has been said that $T \Rightarrow A$ and $A \vee \neg A$.

(iii) $A = $ I am thinking of an armadillo; $T = $ the animal I am thinking of has a tail; $F = $ the animal I am thinking of has feathers. I say: $T \vee \neg F \Rightarrow A$, $\neg(F \wedge T)$.

(iv) With the propositions $O = $ God is omnipotent and $C = $ God can create a rock She cannot lift, it is then true that $(O \Rightarrow C)$ and $(C \Rightarrow \neg O)$.

2.4 Add brackets to the following expressions:

(i) $A \wedge B \wedge C \Leftrightarrow A \wedge (B \wedge C)$.

(ii) $\neg A \vee B \Leftrightarrow A \Rightarrow B$.

(iii) $A \Rightarrow B \Rightarrow C \Leftrightarrow A \wedge B \Rightarrow C$.

2.5 Which of the formulae given in the previous exercise do you think are true no matter the values of A and B?

2.3 Model Theory

We have just defined what a formulae in propositional calculus should look like. We have also seen how such formulae correspond to English sentences, which can be either true or false. The next step is to build a series of mathematical tools to reason about these formulae. For instance, some statement can be shown to be true no matter what—if the weather report were to predict "Tomorrow, either it will rain or it will not," we can immediately conclude that the forecast is correct. We may also want to use the mathematical tools we will develop to show that two sentences are equivalent ("It is cold and it is raining" is intuitively equivalent to "It is raining and it is cold") or that one sentence follows from another ("A lion escaped from the zoo" follows from "No tiger escaped, and either a lion or a tiger, or both, escaped from the zoo"). Although these examples are short and easy to understand, in practice, when modelling complex systems or statements, one typically ends up with huge formulae to reason about. For example, if we were to translate a small electronic chip into a propositional formula, the result would have thousands of basic propositions and thousands of pages of formulae describing the behaviour of the chip.

2.3.1 Truth Tables: Propositional Operators

Since basic propositions can only be either true or false, we can list all possible values and use rules to generate the value of a compound formula. This approach would not work in another field of mathematics, such as numbers, where the number of values a basic variable can take ranges over an infinite collection.

2.3.1.1 Negation

Consider negation. We would like to draw up a table which precisely describes the value of $\neg P$ for any possible value of proposition P. The following table should satisfy our needs:

P	$\neg P$
true	false
false	true

The double vertical line indicates that what comes to the right is the answer. Before the double line, we list all possible values of the propositional variables. In this case, since we have one variable, we have two possibilities. Truth tables used to define the meaning of the propositional operators are *basic truth tables*. They are to be considered different from the truth tables of compound formulae which are *derived* from basic truth tables. We call such tables *derived truth tables*.

Example Using the basic truth table for negation, we can now draw up the derived truth table for the compound formula $\neg(\neg P)$:

P	$\neg P$	$\neg(\neg P)$
true	false	true
false	true	false

First of all, note that we have now used two double lines to separate the basic values from intermediate values we need to calculate the final result. To build the table we would start off with the left column filled (or the columns before the first double line when we have more than one variable), then proceed to use the basic truth tables to fill in the blanks, column by column. For example, to fill in the first entry of $\neg P$ column, we note that P is *true* (from the first column). Looking that up in the basic truth table, we find that the entry under $\neg P$ should be *false*. When we reach the last entry of the $\neg(\neg P)$, we note that $\neg P$ has the value *true* (from the second column), and thus the basic truth table decrees that the entry under $\neg(\neg P)$ will be *false*.   ◇

Exercises

2.6 Construct the truth table for $\neg(\neg(\neg P))$.

2.7 We noted that, since we have one variable, we will have two entries in the table. How many entries would we have with two variables? What about three? Can you generalise for n variables?

2.3.1.2 Conjunction

Let us now turn our attention to conjunction. Constructing the basic truth table for
$P \wedge Q$ is straightforward:

P	Q	$P \wedge Q$
true	true	true
true	false	false
false	true	false
false	false	false

We can now draw a number of more interesting derived truth tables.

Example Let us start by drawing the truth table for $\neg Q \wedge P$. First we observe that,
using the precedence rules, we can add brackets to get $(\neg Q) \wedge P$. We will have two
columns before the first double line (one for P, one for Q). The last column will
obviously have $\neg Q \wedge P$. Finally, we need to add a column $\neg Q$ (the only subformula
we need).

P	Q	$\neg Q$	$\neg Q \wedge P$
true	true	false	false
true	false	true	true
false	true	false	false
false	false	true	false

Consider filling one of the entries. The last entry in the rightmost column is to be
filled by the value of $\neg Q \wedge P$ when we know (from the previous columns) that the
value of $\neg Q$ is true and that of P is false. Consulting the second row of the basic
truth table for conjunction we know that this last entry has to take the value *false*.   ◇

Example Let us try another formula $P \wedge \neg(Q \wedge \neg R)$. Since we now have three variables, P, Q and R, we will have eight rows in the truth table. What about columns
for intermediate values? Well, we need $\neg(Q \wedge \neg R)$ to calculate the value of the
whole expression. To calculate this, we need to calculate the value of $Q \wedge \neg R$, for
which we need the value of $\neg R$.

P	Q	R	$\neg R$	$Q \wedge \neg R$	$\neg(Q \wedge \neg R)$	$P \wedge \neg(Q \wedge \neg R)$
true	true	true	false	false	true	true
true	true	false	true	true	false	false
true	false	true	false	false	true	true
true	false	false	true	false	true	true
false	true	true	false	false	true	false
false	true	false	true	true	false	false
false	false	true	false	false	true	false
false	false	false	true	false	true	false

◇

Exercises

2.8 Draw the truth tables for the following formulae:

(i) $\neg(\neg P \wedge \neg Q)$
(ii) $P \wedge Q \wedge R$ (Note: use associativity)
(iii) $Q \wedge P$

2.3.1.3 Disjunction

By now the pattern of how to draw up truth tables should be quite clear. It is sufficient to be given the basic truth table of a new operator, and we can derive truth tables of formulae using that operator. Let us look at the basic truth table of disjunction.

P	Q	$P \vee Q$
true	true	true
true	false	true
false	true	true
false	false	false

Note that, as we have already discussed, the or operator yields true, even if both its operands are true. You may argue that it would make more sense to have an exclusive-or meaning attached to the operator (where exactly one of the operands is true for the result to be true), but this is just a matter of choice of notation. Later on in this chapter, one of the exercises is to define such an operator and derive its properties.

Example Let us draw the truth table for $P \vee (Q \vee R)$:

P	Q	R	$Q \vee R$	$P \vee (Q \vee R)$
true	true	true	true	true
true	true	false	true	true
true	false	true	true	true
true	false	false	false	true
false	true	true	true	true
false	true	false	true	true
false	false	true	true	true
false	false	false	false	false

$\diamond$

Exercises

2.9 Draw the truth tables of the following formulae:

(i) $\neg(\neg P \vee \neg Q)$
(ii) $(P \vee Q) \vee R$

(iii) $P \vee P$
(iv) $\neg P \vee Q$

2.3.1.4 Bi-implication

Recall that the bi-implication, or if-and-only-if, operator results in true if the two operands have the same value.

P	Q	$P \Leftrightarrow Q$
true	true	true
true	false	false
false	true	false
false	false	true

Example Let us draw the truth table of $P \wedge Q \Leftrightarrow Q \wedge P$:

P	Q	$P \wedge Q$	$Q \wedge P$	$P \wedge Q \Leftrightarrow Q \wedge P$
true	true	true	true	true
true	false	false	false	true
false	true	false	false	true
false	false	false	false	true

◇

Example Here is another example: $P \Leftrightarrow P \vee Q$

P	Q	$P \vee Q$	$P \Leftrightarrow P \vee Q$
true	true	true	true
true	false	true	true
false	true	true	false
false	false	false	true

◇

Exercises

2.10 Draw the truth tables of the following formulae:

(i) $P \Leftrightarrow (Q \Leftrightarrow R)$
(ii) $(P \vee Q) \wedge (P \Leftrightarrow Q)$
(iii) $P \Leftrightarrow P$

2.3.1.5 Implication

Finally, let us turn our attention to implication. Recall that, if P is true, then Q must be true for $P \Rightarrow Q$ to be true. What about when P is false? We chose to take the view that when a politician says "If the economy improves, then we will build a new hospital," the statement is always true if the economy fails to improve. Therefore, if P is false, $P \Rightarrow Q$ is true, no matter the value of Q.[2]

P	Q	$P \Rightarrow Q$
true	true	true
true	false	false
false	true	true
false	false	true

Example Let us look at one final example to illustrate implication: $(P \Rightarrow Q) \Rightarrow R$:

P	Q	R	$P \Rightarrow Q$	$(P \Rightarrow Q) \Rightarrow R$
true	true	true	true	true
true	true	false	true	false
true	false	true	false	true
true	false	false	false	true
false	true	true	true	true
false	true	false	true	false
false	false	true	true	true
false	false	false	true	false

◇

Exercises

2.11   Draw the truth tables of the following formulae:

    (i) $(P \Leftrightarrow Q) \Rightarrow (P \Rightarrow Q)$

    (ii) $P \vee Q \Rightarrow P \wedge Q$

    (iii) $P \Rightarrow P$

    (iv) $P \Rightarrow (Q \Rightarrow R)$

    (v) $(P \Rightarrow Q) \wedge (Q \Rightarrow P)$

2.12   Four cards lie on a table, each of which, we are told, has a number written on one side and a letter on the other. The visible card faces show A, B, 4 and 8. We are asked to check whether the implication 'if a card has A on one side, then it has a 4 on the other' is true. Which cards would you have to turn over to check the truth of the statement?

2.3.1.6   Applications

Let us look at the examples we gave before at the end of Sect. 2.2, and draw truth tables for the formulae we came up with.

---

[2]Of course, if you are not happy with this interpretation, as we said in the case of disjunction, you are free to define another operator with your preferred semantics. However, if you choose the meaning of the new operator to be such that the politician's statement would be interpreted as a lie when a hospital is built despite the fact that the economy has not improved, you will discover that your operator is nothing other than bi-implication. Try it out to check.

Fixing the car: Recall that we set C to be the proposition '*there is a problem with the carburetor*', D the proposition '*there is a problem with the distributor*', and S '*the sparking plugs need changing.*'

The statements about the way the problems are related were expressed as follows:

- It is either the carburetor or the distributor: $C \lor D$.
- Whenever there is a problem with the carburetor, the sparking plugs need changing: $C \Rightarrow S$.
- The distributor and the sparking plugs are closely connected. If one finds problems in one, there are invariably problems with the other: $D \Leftrightarrow S$.

Now the whole statement was that all three expressions were true: $(C \lor D) \land (C \Rightarrow S) \land (D \Leftrightarrow S)$. Let us draw the truth table:

C	D	S	$C \lor D$	$C \Rightarrow S$	$D \Leftrightarrow S$	$(C \lor D) \land$ $(C \Rightarrow D)$	$(C \lor D) \land$ $(C \Rightarrow S) \land$ $(D \Leftrightarrow S)$
true	true	true	true	true	true	true	true
true	true	false	true	false	false	false	false
true	false	true	true	true	false	true	false
true	false	false	true	false	true	false	false
false	true	true	true	true	true	true	true
false	true	false	true	true	false	true	false
false	false	true	false	true	false	false	false
false	false	false	false	true	true	false	false

Now, we know that the statements are true, so the situation must refer to the first or fifth line, where we note in both cases, D and S are true. Therefore, there is a problem with the distributor, and the sparking plugs need changing.

Eating habits: We set S to be the proposition for '*today, Jack and Jill will eat soup*', H to be '*Jack is hungry*', and C to be '*it is cold*'.

- Jack's constraint: $(C \land \neg H) \Rightarrow S$.
- Jill's constraint: $S \Rightarrow \neg C$.

Again, let us draw the truth table of $((C \land \neg H) \Rightarrow S) \land (S \Rightarrow \neg C)$.

C	H	S	$(C \land \neg H) \Rightarrow S$	$S \Rightarrow \neg C$	$((C \land \neg H) \Rightarrow S) \land$ $(S \Rightarrow \neg C)$
true	true	true	true	false	false
true	true	false	true	true	true
true	false	true	true	false	false
true	false	false	false	true	false
false	true	true	true	true	true
false	true	false	true	true	true
false	false	true	true	true	true
false	false	false	true	true	true

It is worth noticing that, if it is cold and Jack is not very hungry, then there is no way to satisfy both constraints (rows 3 and 4). Unless Jack and Jill live in a country with a tropical climate or Jack never feels very hungry when it is cold, they will have to cook separate dishes.

Using a similar approach, we could draw the truth table for the circuit designed by the engineer, and the properties which the engineer expected it to satisfy. The only problem is that the circuit had seven propositional variables—three inputs, one output and three additional wires used to connect the logic gates. Since with every variable the number of entries in a truth table would double, the number of entries to describe all possible behaviours of the circuit would require $2^7 = 128$ entries.

2.3.2 Properties of Propositional Sentences

What we have done up to now is to be able to define the meaning of a sentence in terms of a table which we can use to check the value of a particular expression. We have made use of the truth tables informally to reason about situations expressed as propositional logic sentences. Now we need to define a number of concepts, using which we will be able to formulate better our conclusions.

2.3.2.1 Tautologies

Definition 2.6 *We say that a well-formed formula is a* tautology *if all the entries of the rightmost column in its truth table are true.*   ■

Informally, a tautology is something which is true no matter what the value of its constituent subformulae.

Example $P \vee \neg P$ is called the law of the excluded middle: there is no middle way; either something is true, or it is false.[3] We can show that this law holds by showing that it is a tautology.

P	$\neg P$	$P \vee \neg P$
true	false	true
false	true	true

Since the entries in the last column are all true, the statement is a tautology.   ◇

---

[3]Charles Dodgson, writing under the pseudonym Lewis Carroll, mentions this in *Alice Through the Looking Glass*:

> 'Everybody that hears me sing it–either it brings the tears into their eyes, or else–'
> 'Or else what?' said Alice, for the Knight had made a sudden pause.
> 'Or else it doesn't, you know.'

Example We can now also formalise what we were saying in the previous section. Consider the example with the car fixing problem. What we informally concluded from the given statements was that the distributor and the sparking plugs need changing, or to write it formally $D \wedge S$. Recall that the given information was expressed as: $(C \vee D) \wedge (C \Rightarrow D) \wedge (D \Leftrightarrow S)$. Our claim is that, if the given data is correct, then the distributor and the sparking plugs need changing:

$$((C \vee D) \wedge (C \Rightarrow D) \wedge (D \Leftrightarrow S)) \Rightarrow (D \wedge S).$$

It suffices to show that the above formula is a tautology, which we do by drawing up its truth table (we leave out some intermediate columns already given elsewhere):

C	D	S	$(C \vee D) \wedge$ $(C \Rightarrow D) \wedge$ $(D \Leftrightarrow S)$	$D \wedge S$	$((C \vee D) \wedge (C \Rightarrow D) \wedge (D \Leftrightarrow S))$ $\Rightarrow (D \wedge S)$
true	true	true	true	true	true
true	true	false	false	false	true
true	false	true	false	false	true
true	false	false	false	false	true
false	true	true	true	true	true
false	true	false	false	false	true
false	false	true	false	false	true
false	false	false	false	false	true

So the statement that the given information implies that we need to fix the distributor and change the plugs is true.   ◇

Exercises

2.13 Show that $(P \Rightarrow Q) \Leftrightarrow (\neg P \vee Q)$ is a tautology.

2.14 In Exercise 2.2, I gave some details about Sally and Jane. I love at least one of them, and if I love Sally, then I also love Jane. Show (using a tautology) that these constraints guarantee that I love Jane.

2.15 Using a tautology, show that two negation gates connected in sequence output the same value given as input.

2.16 Jack and Jill's dilemma can be expressed as 'If it is cold and Jack is not hungry, then at least one of them will be unhappy.' Express this formally, and show it to be a tautology.

2.3.2.2 Contradictions

A contradiction is the opposite of a tautology—an expression which is not true for any choice of values of the variables.

Definition 2.7 *We say that a well-formed formula E is a contradiction if all the entries in the rightmost column of its truth table are false.*   ∎

Example Intuitively, we know that P and $\neg P$ are opposite of each other. We can express this by showing that $P \wedge \neg P$ is a contradiction.

P	$\neg P$	$P \wedge \neg P$
true	false	false
false	true	false

Since the entries in the last column are all false, $P \wedge \neg P$ is a contradiction.   ◇

Example Using a tautology, we already saw that, when Jack is hungry and it is cold, Jack and Jill will be arguing about what to eat. We can also reach this conclusion by showing that, if $H \wedge C$ (Jack is hungry and it is cold) and both Jack and Jill's constraints hold, we end up with a contradiction.

C	H	S	$\neg H$	$\neg H \wedge C$	$((C \wedge \neg H) \Rightarrow S) \wedge$ $(S \Rightarrow \neg C)$	$(H \wedge C) \wedge$ $((C \wedge \neg H) \Rightarrow S) \wedge$ $(S \Rightarrow \neg C)$
true	true	true	false	false	false	false
true	true	false	false	false	true	false
true	false	true	true	true	false	false
true	false	false	true	true	false	false
false	true	true	false	false	true	false
false	true	false	false	false	true	false
false	false	true	false	true	true	false
false	false	false	false	true	true	false

This means that there is no way to satisfy their constraints under these circumstances.   ◇

Exercises

2.17 Show that $(P \Rightarrow P) \Rightarrow \neg(P \Rightarrow P)$ is a contradiction.

2.18 Show that $(P \Rightarrow \neg P) \Leftrightarrow P$ is a contradiction.

2.19 Jill lies 'If Jack eats fish then he eats meat.' Jack retorts that he doesn't eat meat. Can Jack be telling the truth?

2.20 If a formula E is a contradiction, can you construct a formula in terms of E which is a tautology? What about another contradiction?

2.3.2.3 Equivalence

Tautologies and contradictions tell us a lot about formulae. However, most interesting formulae are neither one nor the other. We can relate different formulae even if they are neither tautologies nor contradictions.

Definition 2.8 *Two propositional formulae E and F are said to be semantically equivalent if the values in their truth tables match. We write this as $E \models\mid F$.*

If they do not match on at least one row, we say that E and F are not semantically equivalent, and we write $E \not\models\mid F$.   ∎

It is easy to confuse semantic equivalence with bi-implication. It is true that saying that $E \models\!\!\!\models F$ is the same as saying that $E \Leftrightarrow F$ is a tautology. However, note that the symbol $\models\!\!\!\models$ is not a symbol which can appear in a well-formed propositional formula. When we talk about tautologies, semantic equivalences, etc., we are talking *about* propositions, and we are not writing other propositions.

This may take some time to sink in, but it is an important mathematical concept. For the moment, remember to look for symbols in the formula to decide whether it is a propositional formula or a statement about a propositional formula.

Example Semantic equivalence, in a certain sense, is equality between expressions, once we evaluate their symbols. For example, $P \Leftrightarrow Q$ is semantically equivalent to $P \Rightarrow Q$ and $Q \Rightarrow P$ (even the symbol used suggests this). If this is so, we should be able to show that $P \Leftrightarrow Q$ is semantically equivalent to $(P \Rightarrow Q) \wedge (Q \Rightarrow P)$:

P	Q	$P \Rightarrow Q$	$Q \Rightarrow P$	$P \Leftrightarrow Q$	$(P \Rightarrow Q) \wedge (Q \Rightarrow P)$
true	true	true	true	true	true
true	false	false	true	false	false
false	true	true	false	false	false
false	false	true	true	true	true

Since all the entries of the two expressions match, we can conclude that they are semantically equivalent: $P \Leftrightarrow Q \models\!\!\!\models (P \Rightarrow Q) \wedge (Q \Rightarrow P)$. ◇

Such equivalencies allow us to replace one expression with another (semantically equivalent one) within an expression to simplify it. We will talk more about this later, but for the moment it suffices to note that in this way we can reason about formulae by replacing semantically equivalent ones.

Example We will now show a number of properties of conjunction by constructing truth tables:

1. $P \wedge P \models\!\!\!\models P$

P	$P \wedge P$	P
true	true	true
false	false	false

2. $P \wedge Q \models\!\!\!\models Q \wedge P$

P	Q	$P \wedge Q$	$Q \wedge P$
true	true	true	true
true	false	false	false
false	true	false	false
false	false	false	false

3. $P \wedge (Q \wedge R) = \models (P \wedge Q) \wedge R$

P	Q	R	$P \wedge Q$	$Q \wedge R$	$P \wedge (Q \wedge R)$	$(P \wedge Q) \wedge R$
true	true	true	true	true	true	true
true	true	false	true	false	false	false
true	false	true	false	false	false	false
true	false	false	false	false	false	false
false	true	true	false	true	false	false
false	true	false	false	false	false	false
false	false	true	false	false	false	false
false	false	false	false	false	false	false

In all cases, we can deduce the semantic equivalence by inspecting the truth tables.

◇

This last example is an important one. When we define a new operator we would like to know whether these properties hold.

Definition 2.9 *A binary operator* $\oplus$ *is said to be* idempotent *if* $x \oplus x = x$, commutative *or* symmetric *if* $x \oplus y = y \oplus x$, *and* associative *if* $x \oplus (y \oplus z) = (x \oplus y) \oplus z$.

∎

Over propositional formulae, we can use semantic equivalence as our notion of equality.[4] In the previous example, we have thus shown that conjunction is idempotent, commutative and associative. Therefore, an unbracketed conjunction such as $P \wedge Q \wedge R$ gives the same resulting truth table, whether we take conjunction to be left or right associative.

Exercises
2.21 Show that *false* $\wedge$ $P = \models$ *false*, while *true* $\wedge$ $P = \models P$. Can you find similar equivalences for $\vee$?
2.22 Show that $\neg P \vee Q = \models P \Rightarrow Q$.
2.23 Show that $\neg(\neg P \wedge \neg Q) = \models P \vee Q$.
2.24 Show that $\neg(\neg P \vee \neg Q) = \models P \wedge Q$.
2.25 Show that $\Rightarrow$ is not commutative. Is it associative?
2.26 Show that $\vee$ is idempotent, commutative and associative.
2.27 $P \not\equiv Q$ is not equivalent to $P = \models \neg Q$. Give an example of two propositions P and Q to show that these two statements are not the same.

---

[4]Equalities should obey certain laws, which will be mentioned later on. For the moment, we can assume that semantic entailment satisfies these rules.

2.3.2.4 Semantic Entailment

While it is useful to have a notion of equivalence between propositional statements, in a logical argument the statements should follow from one another but not necessarily be equivalent. For instance, if at a certain point in a logical argument we know that, whenever at least one of two switches is on, then the bulb will be lit, and we also know that the first switch is on, we conclude that the bulb is on. Note that the conclusion is not equivalent to the knowledge we had before. However, for the argument to be valid, if the first statement is true, then so should the second statement. We call this notion *semantic entailment*.

Definition 2.10 *A well-formed propositional formula E is said to semantically entail formula F if every entry in the truth table of E which is true is also true in the truth table of F. We write this as* $E \models F$. *If E does not semantically entail F, we write* $E \not\models F$. ∎

It is easy to see that $E \models\mathrel{\models} F$ is the same as $E \models F$ and $F \models E$, hence the symbol. As noted before with semantic equivalence, it is easy to confuse semantic entailment with implication. In fact, saying that $E \models F$ is the same as saying that $E \Rightarrow F$ is a tautology.

Example We are told that a particular file is not owned by either Pam or Quentin. From this, it should follow that the file is not owned by Pam. The general rule from which this follows is: $\neg(P \vee Q) \models \neg P$. To check whether or not it holds in general, we draw the truth table for the two expressions.

P	Q	$P \vee Q$	$\neg(P \vee Q)$	$\neg P$
true	true	true	false	false
true	false	true	false	false
false	true	true	false	true
false	false	false	true	true

Since in every instance where $\neg(P \vee Q)$ is true, so is $\neg P$, it follows that $\neg(P \vee Q) \models \neg P$. ◇

Example Semantic entailment gives us another way of expressing our conclusions about the car example we gave earlier. To say that the car conditions we were given guarantee that the sparking plugs and distributor are to blame, we can simply write:

$$(C \vee D) \wedge (C \Rightarrow D) \wedge (D \Leftrightarrow S) \models D \wedge S$$

What about the carburetor? We can show that $(C \vee D) \wedge (C \Rightarrow D) \wedge (D \Leftrightarrow S) \not\models C$ but also $(C \vee D) \wedge (C \Rightarrow D) \wedge (D \Leftrightarrow S) \not\models \neg C$. Therefore, we cannot conclude anything about the carburetor from the statements we know.

In fact, we can show that $(C \vee D) \wedge (C \Rightarrow D) \wedge (D \Leftrightarrow S) \models\mathrel{\models} D \wedge S$, meaning that the three statements can be replaced by one saying that the sparking plugs and the distributor need changing. This statement says nothing about the carburetor—it may be working fine, or not working at all.   ◇

Example If we were to draw the truth table for the multiplexer circuit, we would then want to check whether the circuit behaviour guarantees the properties:

$$\left(\begin{array}{cc} & (out \Leftrightarrow m_1 \vee m_2) \\ \wedge & (m_0 \Leftrightarrow \neg sel) \\ \wedge & (m_1 \Leftrightarrow sel \wedge h) \\ \wedge & (m_2 \Leftrightarrow m_0 \wedge l) \end{array}\right) \models \left(\begin{array}{cc} & (sel \Rightarrow (out \Leftrightarrow h)) \\ \wedge & (\neg sel \Rightarrow (out \Leftrightarrow h)) \\ \wedge & ((l \Leftrightarrow h) \Rightarrow (out \Leftrightarrow h)) \end{array}\right)$$

Unfortunately, as we have already seen, this would require drawing a truth table with 128 entries to confirm.   ◇

   Recall that a tautology is a proposition which is true no matter what the value of the propositional variables in the expression. Therefore, a tautology E should satisfy $true \models E$—since $true$ always holds, all the entries of E should be true. We sometimes abbreviate this to $\models E$.

Exercises

2.28 Show that the following semantic entailments hold:

   (i) $(P \Rightarrow Q) \wedge P \models Q$
   (ii) $P \wedge Q \models P$
   (iii) $(P \vee Q) \wedge \neg P \models Q$

2.29 Write a computer program to check whether the multiplexer circuit is correct, by generating and looking at all 128 possibilities.

2.30 Show that $\models P \vee (\neg P \wedge \neg Q) \vee Q$.

2.31 $P \not\models Q$ is not equivalent to $P \models \neg Q$. Give an example to show that these two statements are not the same.

2.3.3 Conclusion

This concludes this section about model theory for propositional logic. Reasoning about propositions using truth tables is rather straightforward, if laborious at times. However, these notions give us means of reasoning about statements in a precise and unambiguous manner.

Exercises

2.32 The exclusive-or operator (usually referred to simply as *xor*) is a binary operator which is true if exactly one of its operands is true. Thus, "Paul loves men xor Paul loves women" is true if Paul is purely heterosexual or homosexual. It would be false if Paul is asexual or bisexual. The symbol we will use for xor is ⋈.

   (i) Draw the basic truth table for the $\bowtie$ operator.
  (ii) Draw the derived truth table for $(P \bowtie Q) \vee Q$.
 (iii) Show that $P \bowtie P$ is a contradiction, while $P \bowtie \neg P$ is a tautology.
  (iv) Show that $\bowtie$ is commutative and associative.
   (v) Show that $P \bowtie Q \models P \vee Q$.
  (vi) Give a well-formed formula using P, Q, $\vee$, $\wedge$ and $\neg$ which is equivalent
       to $P \bowtie Q$. Show that they are semantically equivalent.

2.4 Proof Theory

Propositional model theory is just one way of reasoning about propositions. In model theory, we explicitly build every possible model of the system (using truth tables or similar techniques) and painstakingly verify our claims for every case. We have already noted that the number of entries in a truth table with n variables is 2^n. This made the manual checking of the multiplexer example with 128 entries in the truth table impractical. But it rapidly gets worse—with 10 variables, we will have over 1,000 rows to fill and with 20 variables it is already over 1,000,000. Even using a computer program to check all rows quickly becomes unfeasible as the number of variables increases. The sheer magnitude of these tables means that maybe there are more effective techniques, which at least in certain cases, reduce the work required to show that a proposition is a tautology or equivalent to another. Furthermore, when we move on to numbers and mathematical systems which have an infinite number of possible values, performing an exhaustive, case-by-case analysis is impossible. The definitions used in model theory gave us the meaning of the operators—we will now look at how we can build a proof theory to reason about propositions.

2.4.1 What Is a Proof?

The concept of proof is possibly the most fundamental notion in mathematics. Everyone is familiar with the concept of a proof from an everyday viewpoint—whether it is from lawsuits or normal conversational argumentation. In mathematics, a proof is a way of justifying a mathematical statement based on a number of basic truths, using acceptable rules to move from one true statement to the next.

   We thus have two important elements in any mathematical system in which we want to write proofs:

Basic truths: Basic truths, or *axioms* as they are usually called, are statements in
   the mathematical system under analysis which we accept to be true without justification. For example, in the case of propositional logic, we may choose to take
   $P \vee \neg P$ to be one of the basic truths, meaning that $P \vee \neg P$ needs no justification
   to be proved. When defining the odd numbers, we can have an axiom that 1 is an
   odd number. One may ask why, if we are using such an approach, do we not define

all true statements as axioms. In the first place, we would like our axioms to be *basic* truths. The proposition $((P \Rightarrow \neg Q) \wedge Q) \Rightarrow \neg P$ is true; however, not everyone would readily accept it as a basic truth which needs no justification to believe. Secondly, we want to be able to list all basic truths in a finite manner. Unfortunately, there exist an infinite number of true statements, and we limit ourselves to start with only a finite collection of basic truths to avoid unnecessary complexity.

Acceptable rules: *Rules of inference* tell us how statements follow from one another. Just as in the case of axioms, we will accept their validity as is. In the case of propositional logic, for example, we may use a rule saying that knowledge of proposition P follows directly from knowledge of proposition $P \wedge Q$. If we were talking about odd numbers, we would say that, from the fact that n is an even number, we can conclude that $n + 2$ is also an odd number. Again, here one could adopt complex rules which allow us to do everything in one step. However, this defeats the purpose of the exercise. We will define a simple format of rules which are allowed, and all rules are to be expressed in terms of this format. Furthermore, it is considered good mathematics to have a small number of rules whose truth is self-evident.

The notion of *self-evident* is not very precise. A discussion as to what should be self-evident will ensue later on in this chapter.

We call a statement with an accompanying proof justifying it a *theorem*. To avoid long-winded proofs, we will also accept steps in a proof which are the application of a theorem proved earlier.

Definition 2.11 *A proof that a statement Y follows from statement X in a mathematical system S is a finite sequence of statements finishing with Y, such that every line of the proof is either* (i) *statement X (called the hypothesis), or* (ii) *an axiom of S, or* (iii) *follows from previous lines in the proof and a rule of inference of S, or* (iv) *follows from previous lines in the proof and a theorem already proved in S.* ∎

2.4.2 Example Axioms and Rules of Inference

Let us take a look at the notation we will use before we start giving the axioms and rules of inference of propositional logic. Let us take a look at a system in which we can prove that a number is odd. Instead of using the normal representation of numbers, we will use a number of blobs to stand for numbers—to represent a number n, we will write n blobs in line. For example, 3 would be written as ●●●, 1 as ● and 7 as ●●●●●●●. Furthermore, we will use the notation $\mathcal{O}(n)$ to mean that n is an odd number.

To represent a logic in which we can prove numbers to be odd, it suffices to use one axiom and one rule of inference:

Axiom: The only axiom of the system will assert that 1 is an odd number. We write this as follows:

$$\frac{}{\mathcal{O}(\bullet)}$$

Sometimes this is also written (more concisely) as $\vdash \mathcal{O}(\bullet)$. Since we need to refer to this axiom in our proofs, we will name this axiom *1-odd*.

Rule of inference: The rule of inference we will now need states that, if we know that a string of n blobs is an odd number, then adding another two blobs to n will still give us an odd number:

$$\frac{\mathcal{O}(n)}{\mathcal{O}(\bullet\bullet n)}$$

The rule says that, if the top formula can be proved, then the formula below follows directly (in one step). We will name this rule *odd-n + 2*. Note that the formulae in the rule use a variable to stand for any number of blobs (of course, the constraint allows only an odd number of blobs, but we know that only because we have an intuitive understanding of $\mathcal{O}$). The important thing is that n is the same collection of blobs below and above the inference line. We will also use this in our rules for propositional logic. In general, rules can have more than one formula above the line. These formulae are called the antecedents of the rule of inference.

Let us now proceed to prove that 5 is an odd number using these rules. Recall that a formal proof consists of a sequence of lines, each containing (i) a well-formed formula, and (ii) justification in terms of either an axiom name, the name of a rule of inference and the lines (earlier in the proof) where we can find the antecedents, or the name of a theorem we proved earlier. Here is what the proof would look like:

> 1. $\mathcal{O}(\bullet)$ (axiom *1-odd*)
> 2. $\mathcal{O}(\bullet \bullet \bullet)$ (rule *odd-n + 2* on line 1)
> 3. $\mathcal{O}(\bullet \bullet \bullet \bullet \bullet)$ (rule *odd-n + 2* on line 2)

Notice the three columns: the first is just the line number; the second contains a well-formed formula; and the third contains the axiom used, or rule applied and line numbers on which it is applied.

We have therefore just proved that $\mathcal{O}(\bullet\bullet\bullet\bullet\bullet)$ based on just the axioms and rules of inference. This is called a *theorem*.

Sometimes, proofs are of statements of the form "From formulae $E_1, E_2, \ldots, E_n$, it follows that F". In this case, the Es are called the hypotheses and can be used in the proof. Here is a simple example: From $\mathcal{O}(\bullet\bullet)$, it follows that $\mathcal{O}(\bullet\bullet\bullet\bullet\bullet\bullet)$.

> 1. $\mathcal{O}(\bullet\bullet)$ (hypothesis)
> 2. $\mathcal{O}(\bullet \bullet \bullet \bullet)$ (rule *odd-n + 2* on line 1)
> 3. $\mathcal{O}(\bullet \bullet \bullet \bullet \bullet \bullet)$ (rule *odd-n + 2* on line 2)

"But wait," you may argue, "we have just proved that 6 is an odd number. There is something going wrong!" Not really. We have only proved that if 2 is an odd number, then we can prove that 6 is also an odd number. Anything follows from a false statement. For instance from $1 = 2$, it should follow that I am the Pope.[5]

Definition 2.12 *If from a number of well-formed formulae* $\mathcal{X}$, *we can prove a well-formed formula* Y, *we say that* Y *is provable from* X, *written as* $\mathcal{X} \vdash Y$. $\mathcal{X}$ *are called the* hypotheses *and* Y *the* conclusion. $\mathcal{X} \vdash Y$ *is said to be a theorem.*

If no hypotheses are needed (the proof uses only the axioms and rules of inference) we write $\vdash Y$, *and we simply say that* Y *is a* theorem *of the system.*

We say that X *and* Y *are equivalent, written* $X \dashv\vdash Y$, *if both* $X \vdash Y$ *and* $Y \vdash X$ *hold.* ■

The above two proofs thus enable us to write that $\vdash \mathcal{O}(\bullet\bullet\bullet\bullet\bullet)$ and that $\mathcal{O}(\bullet\bullet) \vdash \mathcal{O}(\bullet\bullet\bullet\bullet\bullet\bullet)$. Let us now give the rules for propositional logic.

2.4.2.1 Conjunction

Note that we will need rules for all the operators in the logic. Let us start by looking at conjunction. We will need rules to remove conjunctions appearing in the hypotheses, and others to produce conjunction to appear in the conclusion. The rules are rather straightforward. Let us start with the introduction rule:

$$\frac{A, \ B}{A \wedge B}$$

This says that, if we can prove A and also B, then it follows in one additional step that $A \wedge B$. We call this the $\wedge$-*introduction* rule.

What if we have a conjunction which we would like to break into its constituent parts? Well, from $A \wedge B$, it follows both that A and also B. We write this as two separate rules:

$$\frac{A \wedge B}{A} \qquad \frac{A \wedge B}{B}$$

We call these rules $\wedge$-*elimination 1* and $\wedge$-*elimination 2*, respectively. Note that this is not the only way to axiomatise conjunction. Other approaches exist, which you may find in other books. The important thing is that we are identifying one set of axioms and rules of inference and sticking to them as our basic truths.

Example Let us start by giving a simple example of a proof using only these rules of inference. We would like to prove that conjunction is commutative: $P \wedge Q \vdash Q \wedge P$.

---

[5]Here is the proof: Either I am the Pope, or I am not. If I am, then fine, we have proved what was required. If I am not, then the Pope and I are two persons. But $1 = 2$, and therefore we are one person. Hence, I am the Pope.

The proof is quite straightforward:

1. $P \wedge Q$ (hypothesis)
2. P ($\wedge$-elimination 1 on line 1)
3. Q ($\wedge$-elimination 2 on line 1)
4. $Q \wedge P$ ($\wedge$-introduction on lines 3 and 2)

$\diamond$

Example Here is another proof of associativity of conjunction shown in one direction: $P \wedge (Q \wedge R) \vdash (P \wedge Q) \wedge R$.

1. $P \wedge (Q \wedge R)$ (hypothesis)
2. P ($\wedge$-elimination 1 on line 1)
3. $Q \wedge R$ ($\wedge$-elimination 2 on line 1)
4. Q ($\wedge$-elimination 1 on line 3)
5. R ($\wedge$-elimination 2 on line 3)
6. $P \wedge Q$ ($\wedge$-introduction on lines 2 and 4)
7. $(P \wedge Q) \wedge R$ ($\wedge$-introduction on lines 6 and 5)

$\diamond$

One important rule of applying the rules of inference is that the antecedents must match the whole formula in the quoted line. For instance, from the formula $(P \wedge Q) \wedge R$, we cannot directly conclude that $R \wedge R$ using the second rule of $\wedge$-elimination on the subexpression $P \wedge Q$. The known formula is a conjunction of $(P \wedge Q)$ and R, and thus, the elimination rules can only conclude these results. To prove Q requires applying the elimination rules twice. This may seem extremely pedantic, but as we introduce new operators, applying rules of inference on subexpressions can invalidate proofs.

Exercises
2.33 Prove the following:

(i) $(P \wedge Q) \wedge R \vdash P \wedge (Q \wedge R)$
(ii) $P \wedge P \vdash P$
(iii) $P \vdash P \wedge P$

2.4.2.2 Implication

Let us move on to the next operator: implication. To eliminate an implication $A \Rightarrow B$, we first have to prove the left-hand side of the implication A, from which we can then conclude the right-hand side B:

$$\frac{A \Rightarrow B, \ A}{B}$$

We will call this rule $\Rightarrow$-*elimination*, although the classical name which you will find in most logic books is *modus ponens*.[6]

Example A simple rule involving implication: $(A \wedge (A \Rightarrow B)) \wedge (B \Rightarrow C) \vdash C$

1.	$(A \wedge (A \Rightarrow B)) \wedge (B \Rightarrow C)$	(hypothesis)
2.	$A \wedge (A \Rightarrow B)$	($\wedge$-elimination 1 on line 1)
3.	A	($\wedge$-elimination 1 on line 2)
4.	$A \Rightarrow B$	($\wedge$-elimination 2 on line 2)
5.	$B \Rightarrow C$	($\wedge$-elimination 2 on line 1)
6.	B	($\Rightarrow$-elimination on lines 4 and 3)
7.	C	($\Rightarrow$-elimination on lines 5 and 6)

$\diamond$

What about introducing an implication? What knowledge would allow us to conclude that $A \Rightarrow B$? Looking at the meaning of implication—if A holds then B must also hold—provides a way of approaching implication introduction. If we manage to write a proof which has A as an additional hypothesis, and which concludes proposition B, then we can cite the *whole proof* as evidence of $A \Rightarrow B$.

Let us look at the introduction rule $\Rightarrow$-*introduction* for implication:

$$\frac{S,\ A \vdash B}{S \vdash A \Rightarrow B}$$

This rule is more complex than the ones we have seen, so let us look at it step by step. Unlike the conjunction rules we have seen, the statement that needs to be proved is not a single well-formed formula, but a theorem. Ignoring the S for the moment, the rule says that, if we write a formal proof to show that $A \vdash B$, then we can conclude that $A \Rightarrow B$. The addition of S is to be able to enrich the rule in the following manner: if we have already proved a number of statements S, from which we would like to conclude that $A \Rightarrow B$, then we can also use statements from S to prove B from A—in other words, we require a proof of S, $A \vdash B$ to be able to conclude $A \Rightarrow B$ when we already know S.

How do we write such a proof? Well, from the rule it seems like we would have to write a separate small proof, and then apply it in the main proof to conclude $A \Rightarrow B$. That would not be very practical, so we incorporate subproofs into our proofs directly. The proof will be written as part of the main proof but marked with

---

[6]*Modus ponens* is Latin for 'the method that affirms'.

a separator to show it is a subproof and should thus be treated as such. This is the structure of such a proof:

$$\vdots$$

8.	A	(subhypothesis)
$\vdots$		
12.	B	(?)
13.	$A \Rightarrow B$	($\Rightarrow$-introduction on subproof 8–12)

$$\vdots$$

As you can see, lines 8 to 12 are marked as a separate subproof with the hypothesis (line 8) marked as a subhypothesis to avoid confusion with the hypotheses of the main proof. The propositions to be found on lines 1 to 6 are what we were calling $\mathcal{S}$, and can therefore be used within the subproof in lines 8 to 12. Lines 8 to 12 should be seen as a separate proof and can only be referred to as a whole to conclude that $A \Rightarrow B$ using implication introduction, as we did on line 13. This means that, beyond line 12, we are not allowed to refer to individual lines that appear within the subproof.

Definition 2.13 A subproof *is a proof appearing within another proof or subproof. It has a scope ranging over a number of lines, and can be* discharged *using implication introduction. No individual lines within the scope of a subproof can be used outside its scope.*   ■

Example The best way to understand this business about subproofs is to see a couple of examples. Let us start by proving that implication is transitive: $(P \Rightarrow Q) \wedge (Q \Rightarrow R) \vdash P \Rightarrow R$.

The best way to understand a proof is to build it from scratch, rather than read the full proof from top to bottom. We will see how this proof is constructed. We start by copying the hypothesis and conclusion of the proof, and apply conjunction elimination to break down the hypothesis into its constituent parts:

1.	$(P \Rightarrow Q) \wedge (Q \Rightarrow R)$	(hypothesis)
2.	$P \Rightarrow Q$	($\wedge$-elimination 1 on line 1)
3.	$Q \Rightarrow R$	($\wedge$-elimination 2 on line 1)
	$\vdots$	
7.	$P \Rightarrow R$	(?)

Since we need to conclude the implication $P \Rightarrow R$, we need to open a subproof with subhypothesis P and concluding with R:

1.	$(P \Rightarrow Q) \wedge (Q \Rightarrow R)$	(hypothesis)
2.	$P \Rightarrow Q$	($\wedge$-elimination 1 on line 1)
3.	$Q \Rightarrow R$	($\wedge$-elimination 2 on line 1)
4.	P	(subhypothesis)
	$\vdots$	
6.	R	(?)
7.	$P \Rightarrow R$	($\Rightarrow$-introduction on subproof 4–6)

The proof can now be completed using implication elimination:

1.	$(P \Rightarrow Q) \wedge (Q \Rightarrow R)$	(hypothesis)
2.	$P \Rightarrow Q$	($\wedge$-elimination 1 on line 1)
3.	$Q \Rightarrow R$	($\wedge$-elimination 2 on line 1)
4.	P	(subhypothesis)
5.	Q	($\Rightarrow$-elimination on lines 2 and 4)
6.	R	($\Rightarrow$-elimination on lines 3 and 5)
7.	$P \Rightarrow R$	($\Rightarrow$-introduction on subproof 4–6)

Now that was easy, wasn't it? Note that lines 5 and 6 use information which was known before the subproof was opened, namely lines 2 and 3.   ◇

Example Rather surprisingly, $P \Rightarrow (Q \Rightarrow R) \dashv\vdash P \wedge Q \Rightarrow R$. Let us prove just one direction of the equivalence here: $P \Rightarrow (Q \Rightarrow R) \vdash P \wedge Q \Rightarrow R$, leaving the other as an exercise.

1.	$P \Rightarrow (Q \Rightarrow R)$	(hypothesis)
2.	$P \wedge Q$	(subhypothesis)
3.	P	($\wedge$-elimination 1 on line 2)
4.	$Q \Rightarrow R$	($\Rightarrow$-elimination on lines 1 and 3)
5.	Q	($\wedge$-elimination 2 on line 2)
6.	R	($\Rightarrow$-elimination on lines 4 and 5)
7.	$P \wedge Q \Rightarrow R$	($\Rightarrow$-introduction on subproof 2–6)

Build the proof yourself to understand it better.   ◇

Example What if we were asked to prove something very similar: $P \Rightarrow (Q \Rightarrow R) \vdash Q \wedge P \Rightarrow R$. The proof should be easy if we use the previous theorem and the theorem of commutativity of conjunction.

1.	$P \Rightarrow (Q \Rightarrow R)$	(hypothesis)
2.	$P \wedge Q \Rightarrow R$	(theorem $P \Rightarrow (Q \Rightarrow R) \vdash P \wedge Q \Rightarrow R$ on line 1)
3.	$Q \wedge P$	(subhypothesis)
4.	$P \wedge Q$	(theorem $A \wedge B \vdash B \wedge A$ on line 3)
5.	R	($\Rightarrow$-elimination on lines 2 and 4)
6.	$Q \wedge P \Rightarrow R$	($\Rightarrow$-introduction on subproof 3–5)

Note that every time we use a theorem we must clearly state what we are applying, and make sure that it has already been proved elsewhere.   ◇

Example Subproofs can appear within other subproofs. As an example of a proof which uses nested subproofs, consider a formal proof of: $A \Rightarrow C \vdash A \Rightarrow (B \Rightarrow C)$. We start by quoting the hypothesis and adding the conclusion as the last line of the proof:

$$
\begin{array}{lll}
1. & A \Rightarrow C & \text{(hypothesis)} \\
& \vdots & \\
6. & A \Rightarrow (B \Rightarrow C) & \text{(?)}
\end{array}
$$

Since the last line is an implication, we use implication introduction, introducing a subproof:

$$
\begin{array}{lll}
1. & A \Rightarrow C & \text{(hypothesis)} \\
\hline
2. & A & \text{(subhypothesis)} \\
& \vdots & \\
5. & B \Rightarrow C & \text{(?)} \\
\hline
6. & A \Rightarrow (B \Rightarrow C) & (\Rightarrow\text{-introduction on subproof 2–5)}
\end{array}
$$

Once again, the line at the bottom is an implication, thus forcing us to open a further subproof to be able to prove the bottom line using implication introduction.

Using implication introduction, we obtain:

$$
\begin{array}{lll}
1. & A \Rightarrow C & \text{(hypothesis)} \\
\hline
2. & A & \text{(subhypothesis)} \\
3. & B & \text{(subhypothesis)} \\
& \vdots & \\
4. & C & \text{(?)} \\
5. & B \Rightarrow C & (\Rightarrow\text{-introduction on subproof 3–4)} \\
6. & A \Rightarrow (B \Rightarrow C) & (\Rightarrow\text{-introduction on subproof 2–5)}
\end{array}
$$

The proof can now be concluded using implication elimination:

$$
\begin{array}{lll}
1. & A \Rightarrow C & \text{(hypothesis)} \\
\hline
2. & A & \text{(subhypothesis)} \\
3. & B & \text{(subhypothesis)} \\
4. & C & (\Rightarrow\text{-elimination on lines 1 and 2)} \\
5. & B \Rightarrow C & (\Rightarrow\text{-introduction on subproof 3–4)} \\
6. & A \Rightarrow (B \Rightarrow C) & (\Rightarrow\text{-introduction on subproof 2–5)}
\end{array}
$$

◇

Exercises

2.34 Prove the other direction of the equivalence in the last example.

2.35 Prove that $(P \Rightarrow Q) \Rightarrow R \vdash P \Rightarrow (Q \Rightarrow R)$.

2.36 Using previous results, prove that $(P \Rightarrow Q) \Rightarrow R \vdash P \wedge Q \Rightarrow R$.

2.37 Prove that $P \Rightarrow Q \wedge R \vdash (P \Rightarrow Q) \wedge (P \Rightarrow R)$.

2.4.2.3 Bi-implication

Bi-implication has very simple rules. We can either break up a bi-implication into two implications, or combine two implications into a bi-implication:

$$\frac{A \Rightarrow B, \;\; B \Rightarrow A}{A \Leftrightarrow B} \qquad \frac{A \Leftrightarrow B}{A \Rightarrow B} \qquad \frac{A \Leftrightarrow B}{B \Rightarrow A}$$

We call these rules $\Leftrightarrow$-*introduction*, $\Leftrightarrow$-*elimination 1* and $\Leftrightarrow$-*elimination 2*, respectively.

Example Let us prove that $P \wedge Q \vdash P \Leftrightarrow Q$

1.	$P \wedge Q$	(hypothesis)
2.	P	(subhypothesis)
3.	Q	($\wedge$-elimination 2 on line 1)
4.	$P \Rightarrow Q$	($\Rightarrow$-introduction on subproof 2–3)
5.	Q	(subhypothesis)
6.	P	($\wedge$-elimination 1 on line 1)
7.	$Q \Rightarrow P$	($\Rightarrow$-introduction on subproof 5–6)
8.	$P \Leftrightarrow Q$	($\Leftrightarrow$-introduction on lines 4 and 7)

$\diamond$

Exercises

2.38 Prove the following:

    (i) $A \Leftrightarrow B \vdash B \Leftrightarrow A$

    (ii) $\vdash (A \wedge B) \wedge C \Leftrightarrow A \wedge (B \wedge C)$

    (iii) $\vdash A \wedge B \Leftrightarrow B \wedge A$

    (iv) $\vdash A \Leftrightarrow A$

2.4.2.4 Disjunction

Let us move on to disjunction. Introducing a disjunction is straightforward:

$$\frac{A}{A \vee B} \qquad \frac{B}{A \vee B}$$

We call these rules $\vee$-*introduction 1* and $\vee$-*introduction 2*, respectively.

Example To prove that disjunction follows from conjunction: $P \wedge Q \vdash P \vee Q$, a short proof does the job:

$$
\begin{array}{lll}
1. & P \wedge Q & \text{(hypothesis)} \\
2. & P & (\wedge\text{-elimination 1 on line 1}) \\
3. & P \vee Q & (\vee\text{-introduction 1 on line 2})
\end{array}
$$

◇

What about eliminating a disjunction? If we know that $P \vee Q$, can we conclude something which does not involve a disjunction? At first sight, not much. However, consider a scenario in which we know that (i) if a car is rusty then it is cheap, and (ii) if a car is old, then it is cheap. Now, if I own a car which is either old or rusty (possibly both), I should be able to conclude that my car is cheap. Therefore, if we know that $P \Rightarrow R$ and $Q \Rightarrow R$, then the proposition R must follow from $P \vee Q$.

This is the elimination rule $\vee$-*elimination* for disjunction:

$$
\frac{A \Rightarrow C, \; B \Rightarrow C, \; A \vee B}{C}
$$

Consider a partially finished proof, in which we have to conclude C from a statement $A \vee B$:

$$
\begin{array}{lll}
 & \vdots & \\
9. & A \vee B & (\dots) \\
 & \vdots & \\
23. & C & (?) \\
 & \vdots &
\end{array}
$$

To be able to apply disjunction elimination, we need two additional results, namely that $A \Rightarrow C$ and $B \Rightarrow C$:

$$
\begin{array}{lll}
 & \vdots & \\
9. & A \vee B & (\dots) \\
 & \vdots & \\
21. & A \Rightarrow C & (?) \\
22. & B \Rightarrow C & (?) \\
23. & C & (\vee\text{-elimination on lines 9, 21 and 22}) \\
 & \vdots &
\end{array}
$$

This leaves two implications to be proved. To prove them, we need to open a sub-proof for each:

$$\vdots$$

9.	$A \vee B$	
10.	A	(subhypothesis)
	$\vdots$	
15.	C	(?)
16.	B	(subhypothesis)
	$\vdots$	
20.	C	(?)
21.	$A \Rightarrow C$	($\Rightarrow$-introduction on subproof 10–15)
22.	$B \Rightarrow C$	($\Rightarrow$-introduction on subproof 16–20)
23.	C	($\vee$-elimination on lines 9, 21 and 22)

$$\vdots$$

This structure of proofs in which we need to eliminate disjunction is thus rather uniform, as we will see in the coming examples.

Example Let us start by proving that disjunction is commutative:

$$P \vee Q \vdash Q \vee P$$

1.	$P \vee Q$	(hypothesis)
2.	P	(subhypothesis)
3.	$Q \vee P$	($\vee$-introduction 2 on line 2)
4.	Q	(subhypothesis)
5.	$Q \vee P$	($\vee$-introduction 1 on line 4)
6.	$P \Rightarrow (Q \vee P)$	($\Rightarrow$-introduction on subproof 2–3)
7.	$Q \Rightarrow (Q \vee P)$	($\Rightarrow$-introduction on subproof 4–5)
8.	$Q \vee P$	($\vee$-elimination on lines 1, 6 and 7)

The proof follows the structure we have seen for proofs which eliminate a disjunction, with the remaining parts being rather straightforward applications of disjunction introduction.   ◇

Example An important law of propositional logic is that of distributivity of conjunction over disjunction: $P \wedge (Q \vee R) \dashv\vdash (P \wedge Q) \vee (P \wedge R)$. Let us look at one direction of this equivalence: $P \wedge (Q \vee R) \vdash (P \wedge Q) \vee (P \wedge R)$.

The proof of this direction of the law is the following:

1.	$P \wedge (Q \vee R)$	(hypothesis)
2.	P	($\wedge$-elimination 1 on line 1)
3.	$Q \vee R$	($\wedge$-elimination 2 on line 1)
4.	Q	(subhypothesis)
5.	$P \wedge Q$	($\wedge$-introduction on lines 2 and 4)
6.	$(P \wedge Q) \vee (P \wedge R)$	($\vee$-introduction 1 on line 5)
7.	R	(subhypothesis)
8.	$P \wedge R$	($\wedge$-introduction on lines 2 and 7)
9.	$(P \wedge Q) \vee (P \wedge R)$	($\vee$-introduction 2 on line 8)
10.	$Q \Rightarrow ((P \wedge Q) \vee (P \wedge R))$	($\Rightarrow$-introduction on subproof 4–6)
11.	$R \Rightarrow ((P \wedge Q) \vee (P \wedge R))$	($\Rightarrow$-introduction on subproof 7–9)
12.	$(P \wedge Q) \vee (P \wedge R)$	($\vee$-elimination on lines 3, 10 and 11)

◇

Example Let us now look at a slightly more complex example: associativity. Only one direction is proved, $P \vee (Q \vee R) \vdash (P \vee Q) \vee R$.

1.	$P \vee (Q \vee R)$	(hypothesis)
2.	P	(subhypothesis)
3.	$P \vee Q$	($\vee$-introduction 1 on line 2)
4.	$(P \vee Q) \vee R$	($\vee$-introduction 1 on line 3)
5.	$Q \vee R$	(subhypothesis)
6.	Q	(subhypothesis)
7.	$P \vee Q$	($\vee$-introduction 2 on line 6)
8.	$(P \vee Q) \vee R$	($\vee$-introduction 1 on line 7)
9.	R	(subhypothesis)
10.	$(P \vee Q) \vee R$	($\vee$-introduction 2 on line 9)
11.	$Q \Rightarrow ((P \vee Q) \vee R)$	($\Rightarrow$-introduction on subproof 6–8)
12.	$R \Rightarrow ((P \vee Q) \vee R)$	($\Rightarrow$-introduction on subproof 9–10)
13.	$(P \vee Q) \vee R$	($\vee$-elimination on lines 5, 11 and 12)
14.	$P \Rightarrow ((P \vee Q) \vee R)$	($\Rightarrow$-introduction on subproof 2–4)
15.	$(Q \vee R) \Rightarrow ((P \vee Q) \vee R)$	($\Rightarrow$-introduction on subproof 5–13)
16.	$(P \vee Q) \vee R$	($\vee$-elimination on lines 1, 14 and 15)

Although it is easy to check that it is correct, this proof is practically impossible to understand if you try to read it from top to bottom. Instead, build the proof gradually, using the rules and creating subproofs as necessary, to ensure that you understand it.

◇

Exercises

2.39 Give a different three-line proof that disjunction follows from conjunction from the one given in the example.

2.40 Complete the proof of distributivity of conjunction over disjunction.
2.41 Prove that disjunction is distributive over conjunction: $P \vee (Q \wedge R) \dashv\vdash (P \vee Q) \wedge (P \vee R)$.
2.42 Complete the proof that disjunction is associative.
2.43 Prove the other distributivity law that $P \vee (Q \wedge R) \dashv\vdash (P \vee Q) \wedge (P \vee R)$.

2.4.2.5 Negation

It is not immediately obvious what rules of inference to use for negation. Let us start with negation elimination. Eliminating a single negation is not simple to do; however, we can eliminate negations in pairs, concluding A from $\neg(\neg A)$:

$$\frac{\neg\neg A}{A}$$

As for the introduction of negation, how can we prove that something is false: $\neg A$. A technique which philosophers developed to be able to deal with this is that of *reductio ad absurdum* or *proof by contradiction*.

Example Consider the following scenario. An operating system ensures two properties: (i) system files cannot be owned by a normal user, and (ii) a user may only delete files owned by him- or herself. One would expect that, if the operating system satisfies these two properties, then normal users may not delete system files. An informal argument would be as follows: If a normal user may delete a system file then, by rule (ii), the user must own that file. However, by rule (i), the user does not own it. Therefore, from the statement that a normal user can delete a system file, we concluded two contradictory things—that the user owns the file, and that the user does not own it. This means that it is impossible for the statement that a normal user may not delete a system file to be true. This is a proof by contradiction.   ◇

We will embody this reasoning in a rule of inference: if a statement A implies two contradictory statements ($A \Rightarrow B$ and $A \Rightarrow \neg B$) then A cannot be true, and we can conclude $\neg A$:

$$\frac{A \Rightarrow B, \ A \Rightarrow \neg B}{\neg A}$$

This rule ($\neg$-*introduction*) allows us to prove negations. The biggest challenge is that B appears only in the antecedents of the rule. Therefore, if we need to prove that $\neg A$, we have to guess what is the appropriate B that will allow us to use the rule.

Let us see some applications of these rules of inference.

Example The first example is a rather straightforward application of the rule of inference. If both a statement P and its inverse hold, then we should be able to prove anything: $P \wedge \neg P \vdash Q$. As usual, we start by writing the desired proof structure:

$$
\begin{array}{lll}
1. & P \wedge \neg P & \text{(hypothesis)} \\
& \vdots & \\
9. & Q & \text{(?)}
\end{array}
$$

Since there is nothing we can do with the conclusion (and breaking down the hypothesis does not help), we add negations to be able to use negation introduction:

$$
\begin{array}{lll}
1. & P \wedge \neg P & \text{(hypothesis)} \\
& \vdots & \\
8. & \neg\neg Q & \text{(?)} \\
9. & Q & \text{($\neg$-elimination on line 8)}
\end{array}
$$

Now, the result we have to prove, $\neg\neg Q$, is a negation, which means that we can use negation introduction, by proving that $\neg Q$ implies some statement and its opposite:

$$
\begin{array}{lll}
1. & P \wedge \neg P & \text{(hypothesis)} \\
& \vdots & \\
6. & \neg Q \Rightarrow ? & \text{(?)} \\
7. & \neg Q \Rightarrow \neg ? & \text{(?)} \\
8. & \neg\neg Q & \text{($\neg$-introduction on lines 6 and 7)} \\
9. & Q & \text{($\neg$-elimination on line 8)}
\end{array}
$$

What statement to choose is the key to the proof. Looking at the hypothesis, we realise, however, that both P and $\neg P$ are known, and would thus be good candidates for the statements following from $\neg Q$:

$$
\begin{array}{lll}
1. & P \wedge \neg P & \text{(hypothesis)} \\
& \vdots & \\
6. & \neg Q \Rightarrow P & \text{(?)} \\
7. & \neg Q \Rightarrow \neg P & \text{(?)} \\
8. & \neg\neg Q & \text{($\neg$-introduction on lines 6 and 7)} \\
9. & Q & \text{($\neg$-elimination on line 8)}
\end{array}
$$

The proof can now be easily completed:

1.	$P \wedge \neg P$	(hypothesis)
2.	$\neg Q$	(subhypothesis)
3.	P	($\wedge$-elimination 1 on line 1)
4.	$\neg Q$	(subhypothesis)
5.	$\neg P$	($\wedge$-elimination 2 on line 1)
6.	$\neg Q \Rightarrow P$	($\Rightarrow$-introduction on subproof 2–3)
7.	$\neg Q \Rightarrow \neg P$	($\Rightarrow$-introduction on subproof 4–5)
8.	$\neg\neg Q$	($\neg$-introduction on lines 6 and 7)
9.	Q	($\neg$-elimination on line 8)

$\diamond$

Example When talking about model theory, we mentioned the rule of the excluded middle: $\vdash P \vee \neg P$. We can now prove it using our rules of inference.

1.	$\neg(P \vee \neg P)$	(subhypothesis)
2.	P	(subhypothesis)
3.	$P \vee \neg P$	($\vee$-introduction 1 on line 2)
4.	P	(subhypothesis)
5.	$\neg(P \vee \neg P)$	(copy line 1)
6.	$P \Rightarrow (P \vee \neg P)$	($\Rightarrow$-introduction on subproof 2–3)
7.	$P \Rightarrow \neg(P \vee \neg P)$	($\Rightarrow$-introduction on subproof 4–5)
8.	$\neg P$	($\neg$-introduction on lines 6 and 7)
9.	$P \vee \neg P$	($\vee$-introduction 2 on line 9)
10.	$\neg(P \vee \neg P)$	(subhypothesis)
11.	$\neg(P \vee \neg P) \Rightarrow (P \vee \neg P)$	($\Rightarrow$-introduction on subproof 1–9)
12.	$\neg(P \vee \neg P) \Rightarrow \neg(P \vee \neg P)$	($\Rightarrow$-introduction on subproof 10–10)
13.	$\neg\neg(P \vee \neg P)$	($\neg$-introduction on lines 11 and 12)
14.	$P \vee \neg P$	($\neg$-elimination on line 13)

The proof is not straightforward, and it helps to reconstruct it step by step to understand exactly how it works. $\diamond$

These last two theorems we have proved, $P \wedge \neg P \vdash Q$ and $\vdash P \vee \neg P$, sometimes allow us to prove results in a more direct way than using the rules of inference for negation directly. Consider the following proof:

Example Implication can be expressed in terms of negation and disjunction—$P \Rightarrow Q$ is equivalent to $\neg P \vee Q$. The following proof shows one direction, namely that $P \Rightarrow Q \vdash \neg P \vee Q$.

The key to this proof is that, when we open the hypothesis and conclusion, nothing much can be done, so we apply the law of the excluded middle to be able to conclude that $P \vee \neg P$, after which we can use disjunction elimination to complete the proof.

1.	$P \Rightarrow Q$	(hypothesis)
2.	$P \vee \neg P$	(theorem $\vdash A \vee \neg A$)
3.	P	(subhypothesis)
4.	Q	($\Rightarrow$-elimination on lines 1 and 3)
5.	$\neg P \vee Q$	($\vee$-introduction 2 on line 4)
6.	$\neg P$	(subhypothesis)
7.	$\neg P \vee Q$	($\vee$-introduction 1 on line 6)
8.	$P \Rightarrow (\neg P \vee Q)$	($\Rightarrow$-introduction on subproof 3–5)
9.	$\neg P \Rightarrow (\neg P \vee Q)$	($\Rightarrow$-introduction on subproof 6–7)
10.	$\neg P \vee Q$	($\vee$-elimination on lines 2, 8 and 9)

The proof of $\neg P \vee Q \vdash P \Rightarrow Q$ is left as an exercise.   ◇

Example Here are a couple of lemmata that we will need for the theorem that follows. The first says that a proposition which holds follows from anything: $P \vdash Q \Rightarrow P$:

1.	Q	(subhypothesis)
2.	P	(hypothesis)
3.	$Q \Rightarrow P$	($\Rightarrow$-introduction on subproof 1–2)

The second says that anything follows from a false statement: $P \vdash \neg P \Rightarrow Q$:

1.	P	(hypothesis)
2.	$\neg P$	(subhypothesis)
3.	$P \wedge \neg P$	($\neg$-introduction on lines 1 and 2)
4.	Q	(theorem $A \wedge \neg A \vdash B$ on line 3)
5.	$\neg P \Rightarrow Q$	($\Rightarrow$-introduction on subproof 2–4)

◇

Example The contrapositive law of implication says that $P \Rightarrow Q \dashv\vdash \neg Q \Rightarrow \neg P$. Let us prove this in one direction $P \Rightarrow Q \vdash \neg Q \Rightarrow \neg P$.

1.	$P \Rightarrow Q$	(hypothesis)
2.	$\neg(\neg Q \Rightarrow \neg P)$	(subhypothesis)
3.	P	(subhypothesis)
4.	Q	($\Rightarrow$-elimination on lines 1 and 3)
5.	$\neg Q \Rightarrow \neg P$	(lemma $A \vdash \neg A \Rightarrow B$ on line 4)
6.	P	(subhypothesis)
7.	$\neg(\neg Q \Rightarrow \neg P)$	(copy line 2)
8.	$\neg P$	($\neg$-introduction on lines 3–5 and 6–7)
9.	$\neg Q \Rightarrow \neg P$	(lemma $A \vdash B \Rightarrow A$ on line 8)
10.	$\neg(\neg Q \Rightarrow \neg P)$	(subhypothesis)
11.	$\neg\neg(\neg Q \Rightarrow \neg P)$	($\neg$-introduction on lines 2–9 and 10–10)
12.	$\neg Q \Rightarrow \neg P$	($\neg$-elimination on line 11)

◇

As you can see from these examples, proofs which involve negation are the most challenging ones we have encountered. When there seems to be no way forward in a proof, an approach which frequently works is to use the rule of the excluded middle: $\vdash P \vee \neg P$, and then use disjunction elimination to break the proof into two parts.

Exercises

2.44 Negation elimination allows us to conclude P from $\neg\neg P$. Prove the opposite direction: $P \vdash \neg\neg P$.

2.45 Prove that $\neg P \wedge (P \vee Q) \vdash Q$.

2.46 Prove that $\neg P \vdash \neg(P \wedge Q)$.

2.47 Prove that $\neg(P \vee Q) \vdash \neg P$ and that $\neg(P \vee Q) \vdash \neg Q$.

2.48 Prove the opposite direction of the contrapositive law.

2.49 *Modus tollens*, meaning 'the method that denies,' says that $(P \Rightarrow Q) \wedge \neg Q \vdash \neg P$. Prove it.

2.50 One way of decomposing bi-implication in terms of other operators is to rewrite $P \Leftrightarrow Q$ as $(P \wedge Q) \vee (\neg P \wedge \neg Q)$. Prove that $P \Leftrightarrow Q \vdash (P \wedge Q) \vee (\neg P \wedge \neg Q)$ and $(P \wedge Q) \vee (\neg P \wedge \neg Q) \vdash P \Leftrightarrow Q$.

2.51 Prove that $\neg P \vee Q \vdash P \Rightarrow Q$.

2.52 De Morgan's laws show how conjunction can be expressed in terms of disjunction and negation, and similarly, how disjunction can be expressed in terms of conjunction and negation. Prove them:

  (i) $P \wedge Q \dashv\vdash \neg(\neg P \vee \neg Q)$
  (ii) $P \vee Q \dashv\vdash \neg(\neg P \wedge \neg Q)$

2.4.2.6 Truth and Falsity

The only remaining terms in the language of propositional calculus which we do not know how to handle are *true* and *false*.

The well-formed formula *true* is easy to introduce, because it always holds:

$$\frac{}{true}$$

true-introduction is the only axiom we will be using. You may have observed that the others were all rules of inference.

What about eliminating *true*? Since nothing follows from just *true* (apart from *true* itself), we need no such rule of inference.

As for *false*, we can introduce it only when we reach a contradiction (by proving that P and $\neg P$ both hold). But we have proved that $P \wedge \neg P \vdash Q$ (for any Q), and thus we can already introduce *false* without the need of a new inference rule. What about the elimination of *false*? Using the dictum that anything follows from falsity, we express *false-elimination* as follows:

$$\frac{false}{P}$$

Example The proof that $\vdash P \vee true$ turns out to be straightforward:

> 1. *true*   (*true*-introduction)
> 2. $P \vee true$ ($\vee$-introduction 2 on line 1)

◇

Exercises

2.53 Prove the following:

> (i) $P \vee true$ is equivalent to *true*
> (ii) $P \wedge true$ is equivalent to P
> (iii) $P \vee false$ is equivalent to P
> (iv) $P \wedge false$ is equivalent to *false*

2.4.3 Why Proofs?

One may be put off using this approach to prove things due to its complexity. However, as we have noted earlier an enumerative way of showing things is not possible for most domains. Furthermore, axioms and rules of inference allow us, or rather force us, to make explicit our underlying assumptions. Proofs can also be much shorter than the exponentially sized truth tables. For example, we can prove that $P \wedge (Q_1 \wedge \cdots \wedge Q_{1000}) \vdash P$ in just one line, rather than by drawing a table with 2^{1000} rows.

The proof approach is applicable to all areas of mathematics, from geometry and topology to propositional calculus and arithmetic. It was mentioned earlier, that this is sometimes called 'syntactic reasoning', and the reason is that we can apply the axioms and rules of inference without bothering to understand the real meaning of the symbols. You may have found yourself reading a proof, and understanding each and every line, without really understanding how each line contributes towards the theorem. This is an advantage, since it means that once we agree on the axioms and rules of inference, once someone produces a proof, there is no disagreeing about it. Each step can be easily checked (even by a computer), since it must match one of a small number of axioms or rules of inference. On the other hand, the level of detail one has to go through is usually daunting, and even mathematicians rarely prove theorems at this level. We will discuss this later in this chapter, when we will compare different types of proofs.

2.5 Comparing Truth Tables and Proofs

So we now have two different mathematical descriptions of propositional calculus: one model based, the other proof based. In the first we encoded the whole meaning of the symbols in the mathematical system, while in the second we tried to mimic the way we would reason about formulae. The obvious question is now how the two systems compare. Is every semantic entailment a theorem? If $P \models Q$, can we always find a proof of $P \vdash Q$? In other words, can every true statement (in terms of truth tables) be proved? Vice versa, is every theorem a semantic entailment? In other words, is everything that can be proved true?

2.5.1 Completeness

Definition 2.14 *A mathematical system S is said to be* complete *with respect to another mathematical system S', if every truth in S' is a truth in S.*   ■

In our case, we ask whether for every P and Q such that $P \models Q$, it is also the case that $P \vdash Q$. Propositional calculus turns out to be complete with respect to model theory. The proof of completeness is beyond the scope of the book, however its implications are not. If we are trying in vain to prove that $P \vdash Q$, we can use a truth table, and if we can show that $P \models Q$, it then follows by completeness that $P \vdash Q$.

2.5.2 Soundness

Definition 2.15 *A mathematical system S is said to be* sound *with respect to another mathematical system S', if every truth in S is also a truth in S'.*   ■

With propositional logic, the question is whether every proof yields a semantic entailment: if $P \vdash Q$, does it follow that $P \models Q$? Again, propositional calculus turns out to be sound, and we can thus avoid working out huge truth tables by using proofs. The proof of soundness is also beyond the scope of the book.

Note that saying that S is complete with respect to S' is equivalent to saying that S' is sound with respect to S. Therefore, it follows that $\models$ and $\vdash$ are both sound and complete with respect to each other.

2.6 More About Propositional Logic

2.6.1 Boolean Algebra

You have probably been using the word algebra in mathematics for quite a few years. But have you ever wondered what an algebra really is? In mathematics, an algebra is a collection of identities (equalities). Any operators which satisfy these identities are called an instance of the algebra.

George Boole came up with what is today called a Boolean algebra, when he observed some laws of sets and realised that they are also satisfied by logic sentences and numbers. He chose a number of such laws and went on to prove things about the objects which satisfy the laws. Any class of objects which satisfies the basic laws automatically satisfied his results.

A Boolean algebra is a collection of objects X, together with binary operators $\odot$ and $\oplus$, unary operator $\bar{\ }$ (a bar over the operand) and a notion of equality $=$, which satisfy the following laws:

Commutativity: Both $\odot$ and $\oplus$ are commutative.

$$x \odot y = y \odot x$$
$$x \oplus y = y \oplus x$$

Associativity: Both $\odot$ and $\oplus$ are associative.

$$x \odot (y \odot z) = (x \odot y) \odot z$$
$$x \oplus (y \oplus z) = (x \oplus y) \oplus z$$

Identities: There are two special values of S, namely 1 and 0, which act as identities of $\odot$ and $\oplus$ respectively:

$$x \odot 1 = x$$
$$x \oplus 0 = x$$

Distributivity: $\odot$ and $\oplus$ distribute over each other:

$$x \odot (y \oplus z) = (x \odot y) \oplus (x \odot z)$$
$$x \oplus (y \odot z) = (x \oplus y) \odot (x \oplus z)$$

Complement: The unary operator interacts with $\odot$ and $\oplus$ as a complement operator:

$$x \odot \bar{x} = 0$$
$$x \oplus \bar{x} = 1$$

What operators satisfy these laws? At first numbers with addition and multiplication seem to fit the bill, but some laws fail (for example the distributivity laws). One solution is to take S to be just the numbers 0 and 1, $\odot$ to be multiplication, $\oplus$ to be addition modulo 2,[7] and the unary operator $\bar{x}$ to be defined to be $1 - x$. All the identities of Boolean algebra are satisfied by these operators.

Another way of satisfying these laws is to take 1 to be *true*, 0 to be *false*, $\odot$ to be $\wedge$, $\oplus$ to be $\vee$ and $\bar{x}$ to be $\neg x$. If we take equality to be either $=\!\!\models$ or $\dashv\vdash$,

---

[7]In which we have only the numbers 0 and 1, and taking $1 \oplus 1 = 1$.

propositional logic turns out to be an instance of Boolean algebra. Various other interesting domains turn out to be Boolean algebrae.

What is interesting about these laws, is that whatever we prove based only on the laws of the algebra, can be concluded for propositional logic, addition modulo 2, or any other instance of a Boolean algebra. Various interesting laws can be derived from the basic identities of the algebra.

The following table gives a number of these laws.

Idempotency laws:	$x \oplus x = x$
	$x \odot x = x$
Dominance laws:	$x \oplus 1 = 1$
	$x \odot 0 = 0$
De Morgan's laws:	$x \oplus y = \overline{\overline{x} \odot \overline{y}}$
	$x \odot y = \overline{\overline{x} \oplus \overline{y}}$
Absorption laws:	$x \oplus (x \odot y) = x$
	$x \odot (x \oplus y) = x$
Double complement:	$\overline{\overline{x}} = x$

Just to give an idea of how these are proved, we will show the proof of one of the idempotency laws:

$$x$$
$$= \quad \text{identities law } (x = x \odot 1)$$
$$x \odot 1$$
$$= \quad \text{complement law } (x \oplus \overline{x} = 1)$$
$$x \odot (x \oplus \overline{x})$$
$$= \quad \text{distributivity law } (x \odot (y \oplus z) = (x \odot y) \oplus (x \odot z))$$
$$(x \odot x) \oplus (x \odot \overline{x})$$
$$= \quad \text{complement law } (x \odot \overline{x} = 0)$$
$$(x \odot x) \oplus 0$$
$$= \quad \text{identities law } (x \oplus 0 = x)$$
$$x \odot x$$

Note the different style of proof. This is called an algebraic proof, where we replace a term by an equal term until we reach the desired result.

Since propositional logic satisfies the Boolean algebra basic laws, we immediately know that anything that follows from these laws is also satisfied by propositional logic.

Exercises

2.54 Prove the other idempotency law.

2.55 Prove the dominance laws.

2.6.2 Doing with Less

For convenience, we have defined our logic in terms of a rather large number of operators. Can we do with less? We have already seen how some operators can be expressed in terms of simpler ones. The equivalences below show how we can make do with just negation and conjunction or negation and disjunction.

$$P \wedge Q \dashv\vdash \neg(\neg P \vee \neg Q)$$

$$P \vee Q \dashv\vdash \neg(\neg P \wedge \neg Q)$$

$$P \Rightarrow Q \dashv\vdash \neg P \vee Q$$

$$P \Leftrightarrow Q \dashv\vdash (P \Rightarrow Q) \wedge (Q \Rightarrow P)$$

$$true \dashv\vdash P \vee \neg P$$

$$false \dashv\vdash P \wedge \neg P$$

The last four laws show how all the operators can be expressed in terms of just conjunction, disjunction and negation. The first line then shows how one can convert conjunction to disjunction and negation, thus reducing any formula into these two operators. Alternatively, one may apply the law given in the second line to convert the operators into just conjunction and negation.

Another related question you may ask is whether someone can define a new interesting operator which cannot be expressed in terms of our operators. It turns out not to be possible. The operators are universal in the sense that, for any truth table, we can express the last column in terms of our operators.

Exercises

2.56 A propositional formula is said to be a *literal* if it consists of either just a propositional variable, or the negation of a propositional variable. It is said to be a *clause* if it is the disjunction of a number of literals (possibly just one). Finally, it is said to be in *conjunctive normal form* if it is a conjunction of a number of clauses (again, possibly just one). For example, $\neg P$ and Q are literals, while $(\neg P \vee Q \vee \neg R)$ is a clause and $(\neg P \vee Q \vee \neg R) \wedge (P \vee Q) \wedge \neg Q$ is in conjunctive normal form. It can be shown that all propositional formulae can be transformed into conjunctive normal form. Using the results we proved in this chapter, transform the formula $P \Rightarrow (\neg Q \wedge R)$ into an equivalent formula in conjunctive normal form.

2.6.3 Doing with Even Less

Can we do with even less? Consider the nor operator (written as $\triangledown$). $P \triangledown Q$ is true only if neither P nor Q are true (both are false).

P	Q	$P \triangledown Q$
true	*true*	*false*
true	*false*	*false*
false	*true*	*false*
false	*false*	*true*

Now look at the following truth table for $P \triangledown P$:

P	$P \triangledown P$
true	*false*
false	*true*

Therefore, we can write $\neg P$ as $P \triangledown P$.

As its name suggests, nor behaves much like $\neg$ and $\vee$: $P \triangledown Q = \neg(P \vee Q)$. It thus follows that $P \vee Q = \neg(P \triangledown Q)$. But we already know how to write $\neg$ in terms of $\triangledown$. We can thus express $\vee$ in terms of $\triangledown$.

But we have shown earlier that all our operators can be expressed in terms of $\neg$ and $\vee$. We can thus express all expressions in terms of $\triangledown$.

Exercises

2.57 Write an expression equivalent to $P \vee Q$ using only $\triangledown$. Verify it is correct using truth tables.

2.58 Nor is not the only universal operator. Another operator is the *nand* operator written as $\triangle$. $P \triangle Q$ is false only when both P and Q are true. Give the basic truth table of nand and use the same kind of arguments we used with nor to show that nand is a universal operator.

2.6.4 What if the Axioms Are Wrong?

You may be asking why we should believe the axioms and rules of inference. Of course, the fact that we have shown them to be sound and complete guarantees that they are correct with respect to the results obtainable using truth tables. If we accept truth tables to be the 'truth', our rules of inference are correct.

What about systems where we are not sure of their soundness? Well, let us look at propositional logic. We were calling the symbols and, or, not etc. However, the conclusions are true for all systems which obey the axioms and rules of inference. In fact, the short section about Boolean algebra should have convinced you that the results hold for addition and multiplication modulo 2. Therefore, we should see our results as holding not just on propositions, but also any system which happens to satisfy the axioms.

Over 2,000 years ago, Euclid gave an axiomatised geometry. Amongst his axioms was a rather complex one which people tried for 2,000 years to prove in terms of the other axioms. In the 1700s, mathematicians showed that, if we take the opposite of the axiom to be true, not only can no contradiction be reached, but the resulting system models a different kind of geometry which has a physical (albeit non-Euclidian) instance. In other words, there is nothing holy about axioms. Their opposite may equally well describe another phenomenon.

One such case in propositional logic is intuitionistic logic, where the law of the excluded middle is no longer a theorem of the system. The result is a logic, not

unlike the one we have explored, but where all proofs are constructive (hence sometimes called *constructive logic*) and no proofs by contradiction are possible. It turns
out to be a better logic to reason about certain systems.

2.7 Some Notes

Now that we have explored an area of mathematics in depth, we can look back at
the subject as a whole and raise some issues which were not evident during the
exposition, or which are peripheral to the area.

2.7.1 Notation and Symbols

We started this chapter by presenting the language of propositional logic—its syntax, from which we can derive meaningful sentences in the language, or well-formed
formulae as we have called them. One frequently overlooked aspect of mathematics
is the notation and symbols used. Indeed, notation in itself is not a prerequisite to
good mathematics.[8] However, mathematicians all agree that elegance plays a strong
role in mathematics. Einstein is said to have noted that, given the choice between
two mathematical models, the simpler, more elegant one turns out to be the correct
one. Although elegance is usually applied to the axioms and proofs, notation can
also be elegant or not.

Designers have an important rule of thumb when it comes to designing functional
(as opposed to decorative) devices: *the user should never have to stop and think how
to use a device.* Have you ever came across a door which you were unsure whether
it should be pushed or pulled? Or does it slide sideways? One solution is to put up
a sign saying 'Turn clockwise, then push'; however, the best solution is to have the
device suggesting its use: a flat bar as a handle immediately indicates that you have
to push. You wouldn't even stop to think what you should do. The same should hold
with mathematical notation. Here are a few things to keep in mind when defining
new operators:

- A symmetric symbol should be used for commutative functions or symmetric
  relations. This indicates to the reader a property of the operator. Good examples
  of this rule are the symbols $+$ and $\times$. A bad example of an operator which gives
  the visual illusion of symmetry is subtraction.

---

[8]The celebrated mathematician John Conway was once attending a presentation where the mathematician noted, in passing, that he had not yet managed to prove one particular statement which he
believed to be true. At the end of the presentation, Conway stood up and sketched out the proof of
the statement. 'Yes, your notation is much better, which allows you to derive the proof easily' was
the presenter's first comment, to which Conway is said to have tersely reprimanded 'Mathematics
is about notions, not notations.'

- In operators where the order of the operands is important, the use of an asymmetric symbol can be useful. Not only does it allow the user to remember better how the operands are applied, but also allows him or her to write them the other way round if more convenient. Consider the less-than operator. Not only does $x < y$ help us remember which is the smaller, but it also allows us to write $y > x$ (reversing the operator symbol) as a syntactic equivalent of $x < y$. In programming, the assignment symbol $:=$ is much more appropriate than $=$. Indeed, in some circumstances one might prefer to write $\texttt{sin(x)+cos(y)} \; =: \; \texttt{z;}$ instead of $\texttt{z} \; := \; \texttt{sin(x)+cos(y);}$ because it would make the meaning of a program more evident to the reader of the code. Alas, mainstream programming languages do not support this.[9]
- Operators which are related, duals or inverses of each other, should be graphically related. In propositional logic, conjunction and disjunction are duals (see De Morgan's laws) and are thus visually related: $\land$ and $\lor$. We will later see that set intersection and union are related to conjunction and disjunction (respectively), and will use the symbols $\cap$ and $\cup$ to represent them. The similarity is not coincidental. It indicates a similarity between the operators, and suggests that they obey similar laws.
- We say that a symbol is *overloaded* when it is used to represent more than one operation. As we have mentioned in Sect. 2.6.1 about Boolean algebra, some texts use $+$ for disjunction and $\cdot$ for conjunction. At first, one may find this confusing. However, once the similarity of the laws obeyed by disjunction and conjunction with the ones obeyed by addition (counting up to 1) and multiplication becomes evident, we realise that the choice of symbols is appropriate since it hints at these similarities. It is not surprising that $x \times (y + z) = x \times y + x \times z$ is true in both interpretations of the operator.
- However clear and intuitive your symbols are, always tell the reader how expressions which use your symbol are read.

2.7.2 Proof Elegance

Some proofs are more elegant than others. It is not a matter of simplicity or comprehensibility, but one of style, even if reaching a conclusion using simpler arguments and theorems is usually a prevalent element in elegant proofs. Such a proof usually depends on a twist or insight which, although at first may seem tangential to the result, leads to the desired conclusion in an unexpected, yet more direct manner. You will know one when you see one. For example, proofs by exhaustive case analysis rarely fall under this category.

---

[9]Most programming languages are notoriously bad when it comes to notation and meaning. You may have always believed that $+$ is commutative in C or Java, but think again. If you write a function $\texttt{incX()}$ which increments the value of a global variable $\texttt{x}$ and returns the value 0, the value of $\texttt{incX()+x}$ is now clearly different from $\texttt{x+incX()}$.

Paul Erdös, one of the most influential mathematicians of this past century, had more than 1,000 papers published under his name, each of which contained numerous proofs of varying complexity. He frequently talked about the Book—a book with a perfect proof on each of its (infinite number of) pages possessed by God. Mathematicians were allowed one glance of a page before being born, which they strive throughout their life to reproduce. These were what he called 'proofs from the Book'.

It takes time to learn how to write an elegant proof. You may still be wondering why, given the soundness and completeness results we gave, we do not simply prove everything by constructing the truth table. Well, firstly, the size of the truth table of an expression grows exponentially with the number of variables. Propositional formulae are used to model the behaviour of circuits, which can run into thousands of variables. Despite the size of the truth table, a proof of the formula may be much shorter. Secondly, from the next chapter we will start reasoning about infinite domain systems. Proving that every number divisible by four is also divisible by two is quite easy using axioms and rules of inference, while showing its truth by testing it on all multiples of four is impossible. Thirdly, it is a matter of elegance. Never go for brute force when more elegant approaches exist.

2.7.3 How to Come up with a Proof

Proofs are not always easy to come up with, being mostly difficult because there seems to be no obvious pattern, method or approach that will always work. So is there one? Can we write a computer program which outputs whether a given conjecture is true or not?

For propositional logic, we have shown that the axioms and rules are complete and sound. We can thus try out all combinations of the truth table. As noted before, however, this is of exponential size, and thus very inefficient. Is there a better way? Interestingly enough, this question turns out to be a very difficult one. It was posed in 1970, and no one has yet managed to show whether it is possible to write a program which does not require exponential time to solve the problem.

For more complex systems, the situation is even worse. Following the work of Kurt Gödel and Alan Turing in the 1930s, we know that (i) a mathematical system which can handle arithmetic (addition and multiplication) cannot be both sound and complete, and (ii) it is not possible to write an algorithm which, given a statement in such a system, tells us whether or not it is true.

So to conclude, sorry, but there is no surefire easy method of coming up with a proof.

2.7.4 Not All Proofs Are Equal

We have already mentioned that the proofs we gave in this chapter are extremely detailed and tedious to write. Writing every proof in such a manner is not much fun, and in fact, most proofs in mathematics and computer science are given in a less detailed fashion. Proofs written in this style are called *formal proofs,* with every single argument going back to the axioms or rules of inference. We will be proving things formally in this and the following chapter. What is particularly enticing about these proofs is that, since they leave no details out, they are easy to check, even by machine.

Less fastidious are *rigorous proofs,* where the proof is given in an exact mathematical manner, but not going back to the axioms. Most of the proofs in this book starting from Chap. 4 are rigorous. Rigorous does not mean leaving out parts of the proof, or stating that parts are obvious, but still proving everything, possibly taking several obvious steps at a time to make the proof easier to write and comprehend.

At the bottom level are *informal,* or *hand-waving* proofs. These can be useful to convey the general idea of how the actual proof works. Informal arguments will only be used to given an intuition of material beyond the scope of this book, and should not be considered as proofs.

Chapter 3
Predicate Calculus

3.1 The Limits of Propositional Logic

Propositional logic is not the only logic mathematicians use. Although at first sight it may seem expressive enough to formulate most, if not all, statements, once we try applying it in more complex domains, we find that there are limits to what can be expressed in propositional logic. Consider the following statement:

"Every car has four wheels."

Clearly, this is either true or false. How can we express it as a proposition? One possibility is to use one propositional variable to express the whole statement: W. But now, someone tells us that *"A Ford is a car"*, and we would like to conclude that *"A Ford has four wheels"*. Clearly, just naming the statement W does not work. Another solution would be that of having a pair of propositions for every object o: C_o and F_o. C_o would tell us whether o is a car, and F_o would tell us whether it has four wheels. The above statement would now become

$$(C_o \Rightarrow F_o) \wedge (C_p \Rightarrow F_p) \wedge (C_q \Rightarrow F_q) \wedge \cdots$$

Now, if someone tells us that *"The Eiffel Tower does not have four wheels"*, we would be able to formulate this as $\neg F_e$ and thus (using the contrapositive of $C_e \Rightarrow F_e$) be able to conclude that $\neg C_e$. *"The Eiffel Tower is not a car."* This still has problems. One problem is that we need an infinite number of expressions to write this proposition, and all our definitions for propositional logic dealt with finite formulae. Not only is the formula infinite, but it also uses an infinite number of basic propositions, meaning that a truth table would have an infinite number of columns, meaning that semantic entailment, tautologies, etc. are impossible to judge.

Predicate logic extends propositional logic by adding predicates and quantifiers. A predicate is a way of expressing properties of objects. For example, $car(x)$ states that x is a car. *car* is the predicate. $mother(x, y)$ states that x is the mother of y. *mother* is the predicate. Predicates need not be written in this way. One predicate

G.J. Pace, *Mathematics of Discrete Structures for Computer Science*,
DOI 10.1007/978-3-642-29840-0_3, © Springer-Verlag Berlin Heidelberg 2012

we are all familiar with is $\leq$, which is usually written in infix[1] notation: $x \leq y$. Quantifiers allow us to qualify the use of variables as parameters of such predicates.

3.2 The Language of Discourse

The language of predicate logic is very similar to that of propositional logic, but with the addition of two operators.

Universal quantification: This is a way of saying that all objects of a certain type satisfy a particular property. For example, to write *"All cars have four wheels"*, we write $\forall c : \mathsf{CARS} \cdot fourWheels(c)$. $\forall$, read *'for all'*, is called the *quantifier*. c is a variable standing for a single object of type CARS. We will see the significance and importance of types later. The $\cdot$ symbol is just used to separate the quantifier part from the actual predicate.

To say that *"All natural numbers[2] are greater than 100"* we write $\forall n : \mathbb{N} \cdot n > 100$. *"Every number is either even or greater than zero"*: $\forall n : \mathbb{N} \cdot even(n) \vee n > 0$.

The statement *"Not everyone is unhappy"* can be expressed as $\neg \forall p : \mathsf{PERSON} \cdot unhappy(p)$.

For a final, more complex example, consider *"Every natural number is either greater than 0, or it is less than every other natural number.*

$$\forall n : \mathbb{N} \cdot n > 0 \vee \forall m : \mathbb{N} \cdot m \neq n \Rightarrow n < m$$

Note that unless explicit bracketing is used, the quantifier applies over the rest of the expression.

Existential quantification: When we want to state that there exists at least one object of a certain type satisfies a certain property, we use existential quantification. *"There really are yellow cars"* would be written as $\exists c : \mathsf{CARS} \cdot yellow(c)$. $\exists$, read *'there exists'*, is used in a similar manner as $\forall$.

"There is at least one natural number which is both even and prime": $\exists n : \mathbb{N} \cdot even(n) \wedge prime(n)$. *"Some people are odd"* would be written as $\exists p : \mathsf{PERSON} \cdot odd(p)$. *"There is no one who can beat me"*: $\neg \exists p : \mathsf{PERSON} \cdot beats(p, me)$.

One final example: *"There is a person who can beat anyone at chess"* could be written as $\exists p : \mathsf{PERSON} \cdot \forall q : \mathsf{PERSON} \cdot p \neq q \Rightarrow chessbeats(p, q)$.

After this informal exposition, and examples, we can write the syntax of well-formed predicates in a similar manner as we did with propositions.

---

[1]Infix describes the way an operator is written—*between* its operands. $\wedge$ and $\vee$ are infix operators, for instance. Prefix operators are written *before* the operands, as in the case of $\neg$ and *car*. Finally postfix operators appear *after* the operands. The squaring operator in mathematics is an example of such an operator.

[2]The natural numbers are the numbers used for counting—whole numbers starting from zero. We usually use $\mathbb{N}$ to stand for these numbers, which will be discussed in more detail later on in the book.

predicate	::=	basic-predicate
	|	(predicate)
	|	*true*
	|	*false*
	|	¬ predicate
	|	predicate ∧ predicate
	|	predicate ∨ predicate
	|	predicate ⇒ predicate
	|	predicate ⇔ predicate
	|	∀ variable:type · predicate
	|	∃ variable:type · predicate

We will keep the same precedence and associativity as before. The bracketing related to the quantifiers (∀ and ∃) is taken to extend till the end of the formula unless closed by brackets. Therefore, $∃x : X · P ∧ ∀y : Y · Q$ is implicitly bracketed as $∃x : X · (P ∧ (∀y : Y · Q))$. We use brackets if a different bracketing is desired, such as: $(∃x : X · P) ∧ (∀y : Y · Q)$.

We will sometimes write $∀x : X, y : Y · P$ as a shorthand notation for $∀x : X · ∀y : Y · P$. When the two types X and Y are the same, we may further shorten it to: $∀x, y : X · P$. The same notation is also used with existential quantification.

Finally, one important observation is that we can only have a variable after a quantifier, and not a predicate. $∃P : \mathsf{COMPARISON} · P(3, 2)$ is not allowed.

Example Let us try to express properties of a multi-user operating system. We will assume we have a type USER of all valid usernames and a type RIGHT of all rights a user may have. The predicate *activated*(u) is true if and only if u is a username which is activated on the system. The predicate *admin*(u) (where u is a USER) will mean that u is an administrator of the system, while *normal*(u) means that user u is a normal user. Finally, the predicate *hasRight*(u, r) is true exactly when user u has right r. Let us look at a number of properties which we can express using predicate logic. Keep in mind that these properties can be written in different ways:

1. There is at least one activated administrator:

$$∃u : \mathsf{USER} · activated(u) ∧ admin(u)$$

2. Every activated user is either an administrator or a normal user:

$$∀u : \mathsf{USER} · activated(u) ⇒ (admin(u) ∨ normal(u))$$

3. No user is both an administrator and a normal user:

$$¬∃u : \mathsf{USER} · admin(u) ∧ normal(u)$$

4. Every administrator has the right *CreateUser*:

$$∀u : \mathsf{USER} · admin(u) ⇒ hasRight(u, CreateUser)$$

5. Normal users do not have the right *CreateUser*:

$$\forall u : \text{USER} \cdot normal(u) \Rightarrow \neg hasRight(u, CreateUser)$$

6. At least one administrator has all rights:

$$\exists u : \text{USER} \cdot admin(u) \wedge \forall r : \text{RIGHT} \cdot hasRight(u, r)$$

7. All normal users have the same rights:

$$\forall u_1 : \text{USER}, \ u_2 : \text{USER}, \ r : \text{RIGHT} \cdot$$

$$(normal(u_1) \wedge normal(u_2)) \Rightarrow (hasRight(u_1, r) \Leftrightarrow hasRight(u_2, r)) \quad \diamond$$

Exercises

3.1 Translate into English:

    (i) $\forall p : \text{PERSON} \cdot \forall q : \text{PERSON} \cdot loves(p, q)$
    (ii) $\forall p : \text{PERSON} \cdot \exists q : \text{PERSON} \cdot loves(p, q)$
    (iii) $\exists p : \text{PERSON} \cdot \forall q : \text{PERSON} \cdot loves(p, q)$
    (iv) $\exists p : \text{PERSON} \cdot \exists q : \text{PERSON} \cdot loves(p, q)$

3.2 Translate into predicate logic syntax:

    (i) All men are mortal. Socrates is a man. Hence, Socrates is mortal.
    (ii) No politician is honest. The same holds for lawyers and priests. I know someone who is either a politician or a priest but not a lawyer, even though he knows all the lawyers.
    (iii) There is an animal which has more legs than any other animal and which no animal can beat at singing Michael Jackson songs. There is also an animal which has 10,000 legs.

3.3 Which of the following do you think are true:

    (i) $\exists n : \mathbb{N} \cdot n^2 = 2$
    (ii) $(\exists p : \text{PERSON} \cdot chinese(p)) \wedge (\exists p : \text{PERSON} \cdot \neg chinese(p))$
    (iii) $\exists n : \mathbb{N} \cdot \forall m : \mathbb{N} \cdot n > m$
    (iv) $\forall m : \mathbb{N} \cdot \exists n : \mathbb{N} \cdot n > m$

3.4 In this question, we will look at arrays of numbers. Given an array A, we will assume we have a way of knowing the length of the array: $size(A)$. Similarly, $item(A, n)$ gives us the item in array A at position n (with $0 \leq n < size(A)$). We will define a number of operators based on these functions and predicate logic. For example, to define the predicate $hasZero(A)$ which is true if the number zero is an item in array A, we would write:

$$hasZero(A) \stackrel{\text{df}}{=} \exists n : \mathbb{N} \cdot 0 \leq n \wedge n < size(A) \wedge item(A, n) = 0$$

Similarly, *positive*(*A*) can be defined to be true if and only if all items in the array are positive:

$$positive(A) \stackrel{\mathrm{df}}{=} \forall n : \mathbb{N} \cdot (0 \leq n \wedge n < size(A)) \Rightarrow item(A, n) > 0$$

Define the following predicates:

(a) Define *empty*(*A*), which is true only when *A* is the empty array (contains no items).
(b) Define *zero*(*A*), which is true only when all items in *A* are zero.
(c) Define *numbers*(*A*), which is true only when in array *A*, the item at position *n* is in fact the number *n*.
(d) Define *constant*(*A*), which is true only when all items in *A* are equal to each other.
(e) Define *seq*(*A*), which is true if and only if *A* contains two sequential items which are the same.
(f) Define *elt*(*A*, *x*), which is true if *x* appears as an item in array *A*.
(g) Define *cap*(*A*, *n*), which is true if and only if *n* is greater than or equal to all items in array *A*.
(h) Define *max*(*A*, *n*), which is true if and only if *n* is the largest item in array *A*.
(i) Define *equal*(*A*, *B*), which is true if and only if arrays *A* and *B* are equal (same length, same items, same order).
(j) Define *sorted*(*A*), which is true only when the items in *A* are in sorted order.

3.5 $\exists_1 x : X \cdot P$ means *there exists exactly one x of type X satisfying P*. Express the same property in terms of the quantifiers and equality.

3.3 The Use of Variables

The use of quantifiers in predicate logic brings in the use of variables. An expression with variables which do not appear in a quantifier is ambiguous. Consider the statement *parent*(*x*, *John*). What does this statement really mean? When is it true? On the other hand, if we add a quantifier above the use of the variable, as in $\forall p : \mathsf{PERSON} \cdot parent(x, John)$, the meaning of the formula becomes clearer: 'Everyone is John's mother.' Quantifiers act very much like variable declarations in programming, in that they make the use of a variable legitimate in their body. Variables which have not been declared in a quantifier are called *free* instances of that variable. Ones which have been declared are called *bound* instances of the variable.

As a further complication, different uses of the same variable may be related to different quantifiers, as in the following sentence:

$$(\exists x : \mathsf{PERSON} \cdot maltese(x)) \wedge (\exists x : \mathbb{N} \cdot odd(x))$$

Clearly, the first x refers to the x : PERSON, while the second to x : $\mathbb{N}$. Even though they carry the same name, they are not in any way related. In fact, they do not even have the same type. The formula following the quantifier and variable is called the *scope* of the quantifier. This means that one can have a variable appearing both free and bound in the same formula:

$$maltese(x) \wedge (\exists x : \mathbb{N} \cdot odd(x))$$

As a final consideration, we may have quantifiers within quantifiers using the same variable name. Consider the sentence:

$$\exists x : \mathbb{N} \cdot ((\forall x : \text{PERSON} \cdot maltese(x)) \wedge odd(x))$$

The situation is now a bit different. The second existential quantification appears within the first. However, it uses the same variable name. The second instance of x appears only in the context of the first quantifier, and thus clearly refers to the existential quantification over x : $\mathbb{N}$. What about the first? It is taken to be bound to the closest quantifier, that is to the universal quantification over x : PERSON. We will now formalise these concepts.

3.3.1 Free Variables

The use of a variable is said to be *free* in a an expression P if it is undeclared—there is no quantifier with that variable around its use.

Definition 3.1 *We say that x occurs free in P, written $free(x, P)$, if and only if x appears free in formula P. It is defined recursively over the structure of the formula as follows:*

- *Variable x does not appear free in the basic formulae true and false:*

$$free(x, true) \stackrel{\text{df}}{=} false$$

$$free(x, false) \stackrel{\text{df}}{=} false$$

- *Variable x does not appear free in a formula which starts with a quantifier over variable x itself:*

$$free(x, \exists x : X \cdot P) \stackrel{\text{df}}{=} false$$

$$free(x, \forall x : X \cdot P) \stackrel{\text{df}}{=} false$$

- *If a formula starts with a quantifier over a variable y different from x ($y \neq x$), then x occurs free in that formula if and only if it occurs free in the body of the* *quantifier*

$$free(x, \exists y : X \cdot P) \stackrel{\mathrm{df}}{=} free(x, P)$$

$$free(x, \forall y : X \cdot P) \stackrel{\mathrm{df}}{=} free(x, P)$$

- *If a formula consists of a propositional logic operator then x occurs free in the top formula if and only if it occurs free in at least one of the subformulae:*

$$free(x, \neg P) \stackrel{\mathrm{df}}{=} free(x, P)$$

$$free(x, P \wedge Q) \stackrel{\mathrm{df}}{=} free(x, P) \ or \ free(x, Q)$$

$$free(x, P \vee Q) \stackrel{\mathrm{df}}{=} free(x, P) \ or \ free(x, Q)$$

$$free(x, P \Rightarrow Q) \stackrel{\mathrm{df}}{=} free(x, P) \ or \ free(x, Q)$$

$$free(x, P \Leftrightarrow Q) \stackrel{\mathrm{df}}{=} free(x, P) \ or \ free(x, Q)$$

- *Finally, x occurs free in a predicate $p(x_1, x_2, \ldots, x_n)$ if and only if x is one of the variables appearing as a parameter of p:*

$$free(x, p(x_1, x_2, \ldots, x_n)) \stackrel{\mathrm{df}}{=} x = x_i \ for \ some \ value \ of \ i$$

We will write $x \setminus P$ to signify that x does not occur free in P, and is defined to be the negation of free(x, P). ■

The use of the symbol $\stackrel{\mathrm{df}}{=}$, indicates that the use of the operator on the left is defined to be the term on the right. It acts as the replacement of an expression in a particular form into another, without the need for further calculation. In proofs one can go from the left-hand side to the right-hand side expression or vice versa.

Example Consider the following formula P:

$$P \stackrel{\mathrm{df}}{=} printerjob(p) \wedge (\forall p : \mathsf{PROCESS} \cdot \exists u : \mathsf{USER} \cdot owner(p, u))$$

Is p free in P? We can calculate this by applying the definition on the expression:

$$free(p, P)$$
$$= free(p, printerjob(p)) \ or$$
$$\qquad free(p, \forall p : \mathsf{PROCESS} \cdot \exists u : \mathsf{USER} \cdot owner(p, u))$$
$$= true \ or \ false$$
$$= true$$

Variable p thus turns out to be free in P. What about variable u?

$$free(u, P)$$

$$= free(u, printerjob(p)) \text{ or}$$

$$\qquad free(u, \forall p : \mathsf{PROCESS} \cdot \exists u : \mathsf{USER} \cdot owner(p, u))$$

$$= false \text{ or } free(u, \exists u : \mathsf{USER} \cdot owner(p, u))$$

$$= false \text{ or } false$$

$$= false$$

Therefore, we can conclude that $u \setminus P$.   ◇

Exercises

3.6 Recall that a variable occurs bound in an expression if there an instance of the variable which is not free. Define $bound(x, P)$ which is true if and only if x occurs bound in expression P.

3.7 Give an example of an expression P in which variable x occurs both free and bound. Confirm that both $bound(x, P)$ and $free(x, P)$ evaluate to true.

3.3.2 Substitution of Variables

Consider the expression $\forall u : \mathsf{USER} \cdot admin(u) \Rightarrow access(u, passwd)$, which says that all administrator-level users can access the password file. If *john* and *root* are both users, we expect the property to hold for both:

$$admin(john) \Rightarrow access(john, passwd)$$

$$admin(root) \Rightarrow access(root, passwd)$$

Now consider the expression:

$$\forall u : \mathsf{USER} \cdot admin(u) \vee \neg \forall u : \mathsf{USER} \cdot access(u, passwd))$$

This says that, for every user u, either u is an administrator, or not every user can access the password file. Although we would probably want to write such a formula in a different form, it is a well-formed formula, and we would expect the top-level universal quantifier to be applicable to all users, including *john* and *root*:

$$admin(john) \vee \neg \forall u : \mathsf{USER} \cdot access(u, passwd)$$

$$admin(root) \vee \neg \forall u : \mathsf{USER} \cdot access(u, passwd)$$

Note that the other instance of u is not replaced by *john* or *root*, because it is bound to a different quantifier than the one which we are opening. Another way of seeing it is that we are replacing all free instances of u in the body of the quantifier with the value *john* or *root*. In general, from $\forall x : X \cdot P$, we expect to be able to conclude that P holds replacing all free instances of x in P with any value in type T. To be able to reason about quantifiers, we thus need to formalise the notion of substitution of all free instances of a variable by a particular term:

Definition 3.2 *Given a well-formed formula P, variable x and term t, the substitution of t for x in P, written as $P[x \leftarrow t]$, is defined to be the same as P but replacing all free instances of x by t. Formally, this is defined as follows:*

- *Since variable x does not appear free in true and false they remain the same after substitution:*

$$true[x \leftarrow t] \stackrel{\mathrm{df}}{=} true$$

$$false[x \leftarrow t] \stackrel{\mathrm{df}}{=} false$$

- *Similarly, substituting free instances of x by t in a formula starting with a quantifier over variable x itself leaves the formula unchanged:*

$$(\exists x : X \cdot P)[x \leftarrow t] \stackrel{\mathrm{df}}{=} \exists x : X \cdot P$$

$$(\forall x : X \cdot P)[x \leftarrow t] \stackrel{\mathrm{df}}{=} \forall x : X \cdot P$$

- *If a formula starts with a quantifier over a variable y different from x ($y \neq x$), then x will be substituted by t in the subformula:*

$$(\exists y : X \cdot P)[x \leftarrow t] \stackrel{\mathrm{df}}{=} \exists y : Y \cdot P[x \leftarrow t]$$

$$(\forall y : X \cdot P)[x \leftarrow t] \stackrel{\mathrm{df}}{=} \forall y : Y \cdot P[x \leftarrow t]$$

- *If a formula consists of a propositional logic operator then we must apply the substitution to all the subformulae:*

$$(\neg P)[x \leftarrow t] \stackrel{\mathrm{df}}{=} \neg(P[x \leftarrow t])$$

$$(P \wedge Q)[x \leftarrow t] \stackrel{\mathrm{df}}{=} P[x \leftarrow t] \wedge Q[x \leftarrow t]$$

$$(P \vee Q)[x \leftarrow t] \stackrel{\mathrm{df}}{=} P[x \leftarrow t] \vee Q[x \leftarrow t]$$

$$(P \Rightarrow Q)[x \leftarrow t] \stackrel{\mathrm{df}}{=} P[x \leftarrow t] \Rightarrow Q[x \leftarrow t]$$

$$(P \Leftrightarrow Q)[x \leftarrow t] \stackrel{\mathrm{df}}{=} P[x \leftarrow t] \Leftrightarrow Q[x \leftarrow t]$$

- *Finally, substituting term t for the free instances of x in a predicate $p(x_1, x_2, \ldots, x_n)$ will replace all variables appearing in the parameter list and matching x with t:*

$$(p(x_1, x_2, \ldots, x_n))[x \leftarrow t] \stackrel{\mathrm{df}}{=}$$
$$p(x_1', x_2', \ldots, x_n') \text{ where } x_i' = t \text{ if } x_i = x, \text{ otherwise } x_i' = x_i \quad \blacksquare$$

Example Consider the formula $admin(u) \Rightarrow \forall f : \text{FILE} \cdot canWrite(u, f)$. We can now calculate the formula but with all free instances of u replaced by *john*:

$$(admin(u) \Rightarrow \forall f : \text{FILE} \cdot canWrite(u, f))[u \leftarrow john]$$

$$= (admin(u))[u \leftarrow john] \Rightarrow (\forall f : \text{FILE} \cdot canWrite(u, f))[u \leftarrow john]$$

$$= admin(john) \Rightarrow \forall f : \text{FILE} \cdot (canWrite(u, f))[u \leftarrow john]$$

$$= admin(john) \Rightarrow \forall f : \text{FILE} \cdot canWrite(john, f)$$

On the other hand, since f is not free in the formula, we can show that substituting free instances of f by *passwd* will leave the formula unchanged:

$$(admin(u) \Rightarrow \forall f : \text{FILE} \cdot canWrite(u, f))[f \leftarrow passwd]$$

$$= (admin(u))[f \leftarrow passwd] \Rightarrow (\forall f : \text{FILE} \cdot canWrite(u, f))[f \leftarrow passwd]$$

$$= admin(u) \Rightarrow \forall f : \text{FILE} \cdot canWrite(u, f)$$ ◇

One of the applications of substitution is that we can rename variables in a formula: $\forall x : X \cdot property(x)$ can be renamed to $\forall y : X \cdot property(y)$ which, as you may notice, can be achieved by renaming free instances of x in the body of the quantifier by y: $\forall y : X \cdot (property(x)[x \leftarrow y])$. Renaming of variables in a quantifier is called alpha-conversion, but we have to be careful not to cause name clashes—renaming a variable to a name which is already used. Consider $\forall n : \mathbb{N} \cdot n < m$. If we rename n to m, we get $\forall m : \mathbb{N} \cdot m < m$, which has a different meaning from the original formula. For alpha-conversion to work as expected, we require that the new variable does not occur free in the formula.

Definition 3.3 *Given a formula P of the form $\forall x : X \cdot Q$ (or $\exists x : X \cdot Q$), and a variable y such that $y \setminus P$, the alpha-conversion of P from x to y is defined to be $\forall y : X \cdot (Q[x \leftarrow y])$ (respectively $\exists y : X \cdot (Q[x \leftarrow y])$).* ■

One important law which substitution satisfies is that $P[x \leftarrow x] = P$. We do not yet have the tools to be able to prove it, but it should be clear that it holds just by looking at the definition of substitution, and noting that, if $x = t$, then the left-hand and right-hand sides of the definition clauses are the same.

Law   Substitution of a variable by itself leaves the formula unchanged: $P[x \leftarrow x] = P$.

Exercises

3.8 Show that $P[x \leftarrow y][y \leftarrow x] = P$ is not always true. Give a condition which guarantees its truth.

3.9 Do you think that $P[x \leftarrow y][y \leftarrow z] = P[x \leftarrow z]$? Justify your answer.

3.10 What about $P[x \leftarrow s][y \leftarrow t]$ and $P[y \leftarrow t][x \leftarrow s]$?

3.4 Axiomatisation of Predicate Logic

Reasoning about predicates is an extension to the rules we used for propositional logic. The extra rules dictate the behaviour of the quantifiers.

3.4.1 Universal Quantification

We have already seen how substitution can help us reason about universal quantification. From the formula $\forall p : \mathsf{PROCESS} \cdot priority_\le(p,\ kernel)$, which says that all processes should have a priority which does not exceed that of the *kernel*, we should be able to conclude that the printer queue process *pqueue* has a priority which is not higher than that of the kernel: $priority_\le(pqueue,\ kernel)$. We can produce such specialised formulae by replacing the free instances of p in the body of the universal quantifier by the particular value.

The rule of inference to perform such reasoning is called *specialisation* or $\forall$-*elimination* and is the following:

$$\frac{\forall x : X \cdot A}{A[x \leftarrow t]}\ t : X$$

Note the expression on the right side of the rule—we call this a side-condition, and it must be checked that it holds if the rule is to be correctly applied. In this case, we must make sure that the term we are replacing for x is of the correct type.

One way of applying the specialisation rule is to take a variable as a term. For instance, from $\forall n : \mathbb{N} \cdot n \ge 0$, we can conclude that $(n \ge 0)[n \leftarrow m]$ or $m \ge 0$. As we will see in the proofs, frequently we will just replace x by itself. We allow such free instances of variables in expressions in intermediate parts of the proof.[3] Note that, if no assumption is made about m, we can generalise once again to $\forall m : \mathbb{N} \cdot m \ge 0$. This will be used as the rule of inference to introduce universal quantification:

$$\frac{A}{\forall x : X \cdot A}\ x : X,\ x \setminus \text{open (sub)hypotheses of } A$$

This rule is called *generalisation* or $\forall$-*introduction*. Note that we now have two side-conditions: (i) that x is of the right type, and (ii) we have no standing assumptions (hypotheses or subhypotheses) which refer to x. We will see an example to justify the latter requirement later on.

---

[3]For this approach to be valid, we must limit ourselves to non-empty types. Otherwise, this approach will be invalid.

Example Let us prove that the order of two consecutive universal quantifiers does not matter: $\forall x : X \cdot \forall y : U \cdot P \vdash \forall y : U \cdot \forall x : X \cdot P$.

1.	$\forall x : X \cdot \forall y : Y \cdot P$	(hypothesis)
2.	$(\forall y : Y \cdot P)[x \leftarrow x]$	($\forall$-elimination on line 1)
3.	$\forall y : Y \cdot P$	(law $P[x \leftarrow x] = P$ on line 2)
4.	$P[y \leftarrow y]$	($\forall$-elimination on line 3)
5.	P	(law $P[x \leftarrow x] = P$ on line 4)
6.	$\forall x : X \cdot P$	($\forall$-introduction on line 5)
7.	$\forall y : Y \cdot \forall x : X \cdot P$	($\forall$-introduction on line 5)

Note that the side-conditions are checked as we go along and do not appear in our proofs. Technically they should also be there, but we will just ensure that they are correct. In this case, clearly, x and y are of the correct type since they arise from quantification over $x : X$ and $y : U$.   ◇

Example Universal quantification distributes inside implications in the following manner:

$$\forall x : X \cdot P \Rightarrow Q \vdash (\forall x : X \cdot P) \Rightarrow (\forall x : X \cdot Q)$$

1.	$\forall x : X \cdot P \Rightarrow Q$	(hypothesis)
2.	$(P \Rightarrow Q)[x \leftarrow x]$	($\forall$-elimination on line 1)
3.	$P \Rightarrow Q$	(law $P[x \leftarrow x] = P$ on line 2)
4.	$\forall x : X \cdot P$	(subhypothesis)
5.	$P[x \leftarrow x]$	($\forall$-elimination on line 4)
6.	P	(law $P[x \leftarrow x] = P$ on line 5)
7.	Q	($\Rightarrow$-elimination on lines 3 and 6)
8.	$\forall x : X \cdot Q$	($\forall$-introduction on line 7)
9.	$\forall x : X \cdot P) \Rightarrow (\forall x : X \cdot Q)$	($\Rightarrow$-introduction on subproof 4–8)

It is important that we verify that x is not free in the open subhypotheses in line 8. Since the only ones start with a quantifier over x, we know that x is not free within these formulae.   ◇

Let us reconsider the side-condition that the variable we generalise over in $\forall$-introduction does not appear free in any open hypotheses or subhypotheses.

Here is an example of a proof which ignores this constraint, and proves a false property—all natural numbers are 42 or larger using nothing but the known fact that $43 \geq 42$:

1.	$n \geq 42$	(subhypothesis)
2.	$\forall n : \mathbb{N} \cdot n \geq 42$	($\forall$-introduction on line 1)
3.	$n \geq 42 \Rightarrow \forall n : \mathbb{N} \cdot n \geq 42$	($\Rightarrow$-introduction on subproof 1–2)
4.	$\forall n : \mathbb{N} \cdot n \geq 42 \Rightarrow \forall n : \mathbb{N} \cdot n \geq 42$	($\forall$-introduction on line 3)
5.	$(n \geq 42 \Rightarrow$ $\forall n : \mathbb{N} \cdot n \geq 42)[n \leftarrow 43]$	($\forall$-elimination on line 4)
6.	$43 \geq 42 \Rightarrow \forall n : \mathbb{N} \cdot n \geq 42$	(application of substitution)
7.	$43 \geq 42$	(known fact)
8.	$\forall n : \mathbb{N} \cdot n \geq 42$	($\Rightarrow$-elimination on lines 6 and 7)

The mistake lies in line 2, where we generalise over variable n, despite the fact that it occurs free in the subhypothesis in line 1.

Exercises

3.11 Prove that $\forall x : X \cdot P \wedge Q \dashv\vdash (\forall x : X \cdot P) \wedge (\forall x : X \cdot Q)$.

3.12 Prove that alpha-conversion leaves the meaning of a formula unchanged. Given that $y \neq x$ and $y \setminus P$, prove that $\forall x : X \cdot P \dashv\vdash \forall y : X \cdot (P[x \leftarrow y])$.

3.13 Prove that $(\forall x : X \cdot P) \vee (\forall x : X \cdot Q) \vdash \forall x : X \cdot P \vee Q$.

3.14 Prove that $\forall x : X \cdot P \Leftrightarrow Q \vdash (\forall x : X \cdot P) \Leftrightarrow (\forall x : X \cdot Q)$.

3.15 Give informal examples to show that the statements in the last two exercises are not equivalences.

3.4.2 Existential Quantification

Universal quantification was rather straightforward to axiomatise. We will now turn to existential quantification. Existential quantification can be introduced by showing a value for which the property holds. For instance, if we want to show that $\exists n : \mathbb{N} \cdot n > 42$, it is sufficient to prove that 43 (which is a natural number) satisfies the property. In this case, we can prove that $(n > 42)[n \leftarrow 43]$—there is a value for n, in this case 43, which satisfies the body of the existential quantifier. This satisfies the proof obligation (or requirement) to reach the desired conclusion.

The introduction rule for existential quantification ($\exists$-*introduction*) is shown below:

$$\frac{A[x \leftarrow t]}{\exists x : X \cdot A} \; t : X$$

Example Let us prove that, if a property holds for all instances of a type, then there must be one value of that type which satisfies the property: $\forall x : X \cdot P \vdash \exists x : X \cdot P$.

1.	$\forall x : X \cdot P$	(hypothesis)
2.	$P[x \leftarrow x]$	($\forall$-elimination on line 1)
3.	$\exists x : X \cdot P$	($\exists$-introduction on line 2)

Here is where we note that the assumption of non-empty types is necessary. If we allowed X to be an empty type, the universal property would be trivially true, while the existential property false. $\diamond$

Elimination of existential quantification is more intricate. It is worth, however, noting the similarity between the existential quantifier and disjunction. The formula $\exists x : X \cdot P$ is conceptually equivalent to the disjunction: $(P[x \leftarrow x_1] \vee P[x \leftarrow x_2] \vee \cdots)$. Although the latter formula is not well formed (since it is infinite), we will use it to understand how to obtain an existential quantifier elimination rule. If our goal is to prove that an expression Q holds, we can imagine the application of a generalised disjunction rule:

$$P[x \leftarrow x_1] \vee P[x \leftarrow x_2] \vee \cdots$$
$$P[x \leftarrow x_1] \Rightarrow Q$$
$$P[x \leftarrow x_2] \Rightarrow Q$$
$$\vdots$$
$$\rule{7cm}{0.4pt}$$
$$Q$$

The first line is equivalent to $\exists x : X \cdot P$, while the remaining lines are equivalent to saying that all values of x satisfy $P \Rightarrow Q$: $\forall x : X \cdot P \Rightarrow Q$. This is the intuition behind the existential elimination rule ($\exists$-*elimination*):

$$\frac{\exists x : X \cdot P, \ \forall x : X \cdot P \Rightarrow Q}{Q} \ x \setminus Q$$

Note the side-condition, which insists that x is not free in Q. This is required, since in the informal version of the rule, we replaced x by particular values only in P, and not in Q. Later on we will see a concrete example to show that ignoring this side-condition can lead to false conclusions.

How is the rule applied? Once we know an existential quantification of the form $\exists x : X \cdot P$ to hold, we take predicate Q to be the property we would like to conclude (hopefully with no free instances of x). If we manage to prove that $\forall x : X \cdot P \Rightarrow Q$, we can then deduce Q.

Visually, imagine we have reached the following stage in a proof, where we still have to complete the proof between lines 7 and 15:

$$\vdots$$

7.	$\exists x : X \cdot P$	(...)
	?	
15.	Q	(?)

$$\vdots$$

At this stage, the next step is to add a new proof obligation—to prove the universal quantification to obtain:

$$\vdots$$

7. $\exists x : X \cdot P$ (...)
   ?
14. $\forall x : X \cdot P \Rightarrow Q$ (?)
15. Q ($\exists$-elimination on lines 7 and 14)

$$\vdots$$

Obviously, one should check that x does not occur free in formula Q. The following examples should make the application of this rule clearer.

Example We start by showing that existential quantification distributes through disjunction. Here is one direction of the proof:

$$\exists x : X \cdot P \vee Q \vdash (\exists x : X \cdot P) \vee (\exists x : X \cdot Q)$$

Starting from setting the proof up, with the know hypothesis, and the sought-after conclusion, we obtain

1. $\exists x : X \cdot P \vee Q$ (hypothesis)

$$\vdots$$

16. $(\exists x : X \cdot P) \vee (\exists x : X \cdot Q)$ (?)

Since the first line is an existential quantification formula, we need to eliminate it, but for this we need to prove that, for all values of the variable, the body of the existential quantification implies the conclusion we seek:

1. $\exists x : X \cdot P \vee Q$ (hypothesis)

$$\vdots$$

15. $\forall x : X \cdot (P \vee Q) \Rightarrow$
        $((\exists x : X \cdot P) \vee (\exists x : X \cdot Q))$ (?)
16. $(\exists x : X \cdot P) \vee (\exists x : X \cdot Q)$ ($\exists$-elimination on lines 15 and 1)

This can be obtained by applying $\forall$-introduction, which results in an implication:

1.	$\exists x : X \cdot P \vee Q$	(hypothesis)
2.	$P \vee Q$	(subhypothesis)
	$\vdots$	
13.	$(\exists x : X \cdot P) \vee (\exists x : X \cdot Q)$	(?)

14. $(P \vee Q) \Rightarrow$
        $((\exists x : X \cdot P) \vee (\exists x : X \cdot Q))$ ($\Rightarrow$-introduction on subproof 2–13)
15. $\forall x : X \cdot (P \vee Q) \Rightarrow$
        $((\exists x : X \cdot P) \vee (\exists x : X \cdot Q))$ ($\forall$-introduction on line 14)
16. $(\exists x : X \cdot P) \vee (\exists x : X \cdot Q)$ ($\exists$-elimination on lines 15 and 1)

The proof can now be completed by eliminating the disjunction on line 2:

1.	$\exists x : X \cdot P \vee Q$	(hypothesis)
2.	$P \vee Q$	(subhypothesis)
3.	P	(subhypothesis)
4.	$P[x \leftarrow x]$	(law $P[x \leftarrow x] = P$ on line 3)
5.	$\exists x : X \cdot P$	($\exists$-introduction on line 4)
6.	$(\exists x : X \cdot P) \vee (\exists x : X \cdot Q)$	($\vee$-introduction 1 on line 5)
7.	Q	(subhypothesis)
8.	$Q[x \leftarrow x]$	(law $P[x \leftarrow x] = P$ on line 7)
9.	$\exists x : X \cdot Q$	($\exists$-introduction on line 8)
10.	$(\exists x : X \cdot P) \vee (\exists x : X \cdot Q)$	($\vee$-introduction 2 on line 9)
11.	$P \Rightarrow$ $((\exists x : X \cdot P) \vee (\exists x : X \cdot Q))$	($\Rightarrow$-introduction on subproof 3–6)
12.	$Q \Rightarrow$ $((\exists x : X \cdot P) \vee (\exists x : X \cdot Q))$	($\Rightarrow$-introduction on subproof 7–10)
13.	$(\exists x : X \cdot P) \vee (\exists x : X \cdot Q)$	($\vee$-elimination on lines 2, 11 and 12)
14.	$(P \vee Q) \Rightarrow$ $((\exists x : X \cdot P) \vee (\exists x : X \cdot Q))$	($\Rightarrow$-introduction on subproof 2–13)
15.	$\forall x : X \cdot ((P \vee Q) \Rightarrow$ $((\exists x : X \cdot P) \vee (\exists x : X \cdot Q))$	($\forall$-introduction on line 14)
16.	$(\exists x : X \cdot P) \vee (\exists x : X \cdot Q)$	($\exists$-elimination on lines 15 and 1)

The other direction of the distributivity law is left as an exercise.   ◇

Example As in the case of universal quantification, the order of two existential quantifiers in sequence may be switched without changing the meaning of the formula:

$$\exists x : X \cdot \exists y : Y \cdot P \vdash \exists y : Y \cdot \exists x : X \cdot P$$

Since the hypothesis is existentially quantified, we have to start by eliminating the quantifier:

1.	$\exists x : X \cdot \exists y : Y \cdot P$	(hypothesis)
	$\vdots$	
12.	$\forall x : X \cdot (\exists y : Y \cdot P) \Rightarrow$ $(\exists y : Y \cdot \exists x : X \cdot P)$	(?)
13.	$\exists y : Y \cdot \exists x : X \cdot P$	($\exists$-elimination on lines 13 and 1)

Continuing the proof by introducing the universal quantifier of line 12, and the implication, we end up with the following proof fragment:

1.	$\exists x : X \cdot \exists y : Y \cdot P$	(hypothesis)
2.	$\exists y : Y \cdot P$	(subhypothesis)
	$\vdots$	
10.	$\exists y : Y \cdot \exists x : X \cdot P$	(?)
11.	$(\exists y : Y \cdot P) \Rightarrow$ $(\exists y : Y \cdot \exists x : X \cdot P)$	($\Rightarrow$-introduction on subproof 2–10)
12.	$\forall x : X \cdot (\exists y : Y \cdot P) \Rightarrow$ $(\exists y : Y \cdot \exists x : X \cdot P)$	($\forall$-introduction on line 11)
13.	$\exists y : Y \cdot \exists x : X \cdot P$	($\exists$-elimination on lines 13 and 1)

Eliminating the existentially quantifier at the top part of the proof:

1.	$\exists x : X \cdot \exists y : Y \cdot P$	(hypothesis)
2.	$\exists y : Y \cdot P$	(subhypothesis)
	$\vdots$	
9.	$\forall y : Y \cdot$ $P \Rightarrow (\exists y : Y \cdot \exists x : X \cdot P)$	(?)
10.	$\exists y : Y \cdot \exists x : X \cdot P$	($\exists$-elimination on lines 9 and 2)
11.	$(\exists y : Y \cdot P) \Rightarrow$ $(\exists y : Y \cdot \exists x : X \cdot P)$	($\Rightarrow$-introduction on subproof 2–10)
12.	$\forall x : X \cdot (\exists y : Y \cdot P) \Rightarrow$ $(\exists y : Y \cdot \exists x : X \cdot P)$	($\forall$-introduction on line 11)
13.	$\exists y : Y \cdot \exists x : X \cdot P$	($\exists$-elimination on lines 13 and 1)

The proof can now be completed:

1.	$\exists x : X \cdot \exists y : Y \cdot P$	(hypothesis)
2.	$\exists y : Y \cdot P$	(subhypothesis)
3.	P	(subhypothesis)
4.	$P[x \leftarrow x]$	(law $P[x \leftarrow x] = P$ on line 3)
5.	$\exists x : X \cdot P$	($\exists$-introduction on line 4)
6.	$(\exists x : X \cdot P)[y \leftarrow y]$	(law $P[x \leftarrow x] = P$ on line 5)
7.	$\exists y : Y \cdot \exists x : X \cdot P$	($\exists$-introduction on line 6)
8.	$P \Rightarrow (\exists y : Y \cdot \exists x : X \cdot P)$	($\Rightarrow$-introduction on subproof 3–7)
9.	$\forall y : Y \cdot$ $P \Rightarrow (\exists y : Y \cdot \exists x : X \cdot P)$	($\forall$-introduction on line 8)
10.	$\exists y : Y \cdot \exists x : X \cdot P$	($\exists$-elimination on lines 9 and 2)
11.	$(\exists y : Y \cdot P) \Rightarrow$ $(\exists y : Y \cdot \exists x : X \cdot P)$	($\Rightarrow$-introduction on subproof 2–10)
12.	$\forall x : X \cdot (\exists y : Y \cdot P) \Rightarrow$ $(\exists y : Y \cdot \exists x : X \cdot P)$	($\forall$-introduction on line 11)
13.	$\exists y : Y \cdot \exists x : X \cdot P$	($\exists$-elimination on lines 13 and 1)

$\diamond$

Example Consider the following two sentences: (i) Given any sorting routine, it cannot be more efficient than Quicksort; and (ii) There is no sorting routine which is more efficient than Quicksort. With some thought, it should be intuitively clear that the two statements are equivalent. If we were to write them formally, using the predicate *betterThanQSort(p)* to mean that sorting routine p is more efficient than Quicksort, we would get:

(i) $\forall p : \text{SORTER} \cdot \neg betterThanQSort(p)$
(ii) $\neg \exists p : \text{SORTER} \cdot betterThanQSort(p)$

In fact, we can show that in general: $\forall x : X \cdot \neg P$ is equivalent to $\neg \exists x : X \cdot P$. This will follow from two important equivalences relating the two quantifiers: (i) $\forall x : X \cdot P$ is equivalent to $\neg \exists x : X \cdot \neg P$, and (ii) $\exists x : X \cdot P$ is equivalent to $\neg \forall x : X \cdot \neg P$. These laws are interesting, since they show that having only one quantifier is sufficient.

We will prove one of these laws in one direction:

$$\forall x : X \cdot P \vdash \neg \exists x : X \cdot \neg P$$

The proof of this statement is the following:

1.	$\forall x : X \cdot P$	(hypothesis)
2.	$\exists x : X \cdot \neg P$	(subhypothesis)
3.	$\neg P$	(subhypothesis)
4.	P	($\forall$-elimination on line 1)
5.	$P \wedge \neg P$	($\wedge$-introduction on lines 4 and 3)
6.	$\neg(\exists x : X \cdot \neg P)$	(theorem $P \wedge \neg P \vdash Q$ on line 5)
7.	$\neg P \Rightarrow \neg(\exists x : X \cdot \neg P)$	($\Rightarrow$-introduction on subproof 3–6)
8.	$\forall x : X \cdot$ $\neg P \Rightarrow \neg(\exists x : X \cdot \neg P)$	($\forall$-introduction on line 7, $x\exists \setminus x : X \cdot \neg P$)
9.	$\neg(\exists x : X \cdot \neg P)$	($\exists$-elimination on lines 2 and 8)
10.	$\exists x : X \cdot \neg P$	(subhypothesis)
11.	$(\exists x : X \cdot \neg P) \Rightarrow$ $(\exists x : X \cdot \neg P)$	($\Rightarrow$-introduction on subproof 10–10)
12.	$(\exists x : X \cdot \neg P) \Rightarrow$ $\neg(\exists x : X \cdot \neg P)$	($\Rightarrow$-introduction on subproof 2–9)
13.	$\neg(\exists x : X \cdot \neg P)$	($\neg$-introduction on lines 11 and 12)

⬦

Since the use of existential elimination introduces a universal quantifier and an implication within, the pattern of use is always very similar:

$$\vdots$$

7.	$\exists x : X \cdot P$	(...)

8.	P	(subhypothesis)

$$\vdots$$

12.	Q	(?)

13.	$P \Rightarrow Q$	($\Rightarrow$-introduction on subproof 8–12)
14.	$\forall x : X \cdot P \Rightarrow Q$	($\forall$-introduction on line 13)
15.	Q	($\exists$-elimination on lines 7 and 14)

$$\vdots$$

One important thing to ensure when applying existential elimination is that the variable does not occur free in the conclusion. Consider the following wrong proof:

1.	$\exists x : X \cdot P$	(hypothesis)

2.	P	(subhypothesis)
3.	P	(copy line 2)

4.	$P \Rightarrow P$	($\Rightarrow$-introduction on subproof 2–3)
5.	$\forall x : X \cdot (P \Rightarrow P)$	($\forall$-introduction on line 4)
6.	P	($\exists$-elimination on lines 1 and 5)
7.	$\forall X : X \cdot P$	($\forall$-introduction on line 6)

Clearly, this should not be the case. Otherwise, the existence of a single male person in the world implies that all persons are male, and the existence of vanilla-flavoured ice cream would imply that all ice cream is vanilla flavoured. What is wrong with the proof? Look carefully at line 6, and you will realise that we have no guarantee that x does not occur free in P, and therefore the existential elimination rule is not applicable.

Exercises

3.16 Prove the following:

(i) $\exists x : X \cdot P \vee Q \dashv\vdash (\exists x : X \cdot P) \vee (\exists x : X \cdot Q)$

(ii) $\exists x : X \cdot P \wedge Q \vdash (\exists x : X \cdot P) \wedge (\exists x : X \cdot Q)$

Give an informal counterexample to show that (ii) is not an equivalence.

3.17 Prove the following:

(i) $\exists x : X \cdot \forall y : Y \cdot P \vdash \forall y : Y \cdot \exists x : X \cdot P$

(ii) $\exists x : X \cdot P \vdash \neg \forall x : X \cdot \neg P$

Give a counterexample to show that (i) is not an equivalence.

3.18 Prove that alpha-conversion of existentially quantified formulae leaves the meaning of the formula intact—provided that $y \neq x$ and $y \setminus P$:

$$\exists x : X \cdot P \dashv\vdash \exists y : X \cdot (P[x \leftarrow y])$$

3.19 Consider the following alternative proof that, from an existentially quantified
      formula, we can derive the same but universally quantified one:

1.	$\exists x : X \cdot P$	(hypothesis)
2.	P	(subhypothesis)
3.	$P[x \leftarrow x]$	(law $P[x \leftarrow x] = P$ on line 2)
4.	$\forall x : X \cdot P$	($\forall$-elimination on line 3)
5.	$P \Rightarrow \forall x : X \cdot P$	($\Rightarrow$-introduction on subproof 2–4)
6.	$\forall x : X \cdot P \Rightarrow (\forall x : X \cdot P)$	($\forall$-introduction on line 5)
7.	$\forall x : X \cdot P$	($\exists$-elimination on lines 1 and 6)

What is wrong with the proof this time?

3.4.3 Existential Quantification and Equality

Since we are now also using variables, and terms, we will be using equality of terms
in our proofs. We will be formalising our notion of equivalence later on in the book,
but for the moment, it is sufficient to note that, rules of equality which are important
for us at this stage are (i) reflexivity: a term is always equal to itself; and (ii) substitutivity: if two terms are equivalent, we should be able to freely interchange them.
From these rules, an important result we will be extensively using in the coming
chapters is the *one-point rule*, which says that, provided that x does not occur free
in term t: $\exists x : X \cdot P \wedge x = t \dashv\vdash P[x \leftarrow t]$.

We will be seeing applications of this rule in the coming chapters.

3.5 Beyond Predicate Calculus

This concludes our overview of predicate calculus. You may wonder whether *everything* can be expressed in an effective manner using predicate calculus. In fact,
various other logics have been developed to handle better other situations.

For instance, fuzzy logic enables approximate truth values and reasoning e.g. the
fuzzy proposition *SystemLoadHigh* may range from completely false to completely
true, including the values in between, depending on the load the system is encounter-
ing. These fuzzy propositions would still be combined with the logical operators to
write formulae such as *SystemLoadHigh* $\wedge$ *TemperatureHigh* to express compound
fuzzy propositions, which themselves may lie along the whole range from being
completely false to being completely false.

A class of logics which are also of particular interest to computer science are
modal logics, which can handle a wide variety of models in which the truth of propositions and predicates may be qualified in different ways. One particular application
is that of representing beliefs (apart from whether P is true, also whether or not I
believe P). Such logics are of particular interest when dealing with the design of

agents with limited knowledge, leading to the setting up of beliefs which may be revised as more information is discovered. You may think that this is rather pedantic, but look at the following situation to see why it may sometimes be necessary. Consider a person who mistakenly believes that the city of Copenhagen is in Finland. Now, even though any atlas will confirm that Copenhagen is the capital city of Denmark, and this person believes that Copenhagen is in Finland, it is certainly not the case that he or she believes that the capital city of Denmark is in Finland. We cannot apply the rule of substitutivity to replace 'Copenhagen' by 'the capital city of Finland' in the second statement, despite the two being known to be equal. This shows that the indiscriminate combination of truths and beliefs can lead to incorrect conclusions—while the equality of 'Copenhagen' and 'the capital city of Finland' lies within our knowledge, the statement that 'the capital city of Denmark is in Finland' appears as a belief.

Another application of modal logics (or rather a particular subclass of modal logics called *temporal logics*) is to describe systems which change over time. These have been extensively used over these past years in the verification of computer systems, to specify properties such as 'after a user makes a request, eventually the system will acknowledge it', or 'no two users may be accessing the same file at the same time'. These logics are beyond the scope of this book, but can be reasoned about in a similar manner as the logics we see here.

Chapter 4
Sets

Now that we can reason about predicates, we can use logic to construct more complex mathematical systems. An important concept in mathematics and computer science is that of a collection of objects, or a *set*. In this chapter we will formally define sets and prove a number of properties about sets and set operators in terms of predicate logic.

4.1 What Are Sets?

A set is a collection of distinct objects but of a similar type. Different types of collections are required for different applications—answering the questions of (i) whether one is interested in keeping track of multiple instances of the same object, and (ii) whether the order of the items is important. There are four ways of answering these two yes-or-no questions, and all four are useful for different applications.

Printer queue: Consider the situation where one has to keep track of the files sent to be printed. Clearly order is important—it is reasonable to expect that printouts appear in the order they were printed. Since one may send the same file multiple times, one also has to keep track of copies of the same object. One may want to query which file appears at a particular location in the queue, or how many times a particular file appears in the queue. To model a printer queue a collection in which both order and repetition can be reasoned about is required.

Employees: Keeping track of employees on the payroll of a company requires a different type of collection. In such a situation, no implicit order is important, and neither would one require to have the same employee appear more than once in the collection. The sort of query one would need in this scenario is whether a particular employee is on the payroll. In contrast to the printer queue example, in such a situation, neither order nor repetition is important.

Stock inventory: Keeping multiple instances of the same employee is never necessary, but in the case of a stock inventory—a list of the items currently held in

G.J. Pace, *Mathematics of Discrete Structures for Computer Science*,
DOI 10.1007/978-3-642-29840-0_4, © Springer-Verlag Berlin Heidelberg 2012

stock—it is desirable to be able to keep track of multiple copies of the same object. One would not simply want to know whether a particular item is in stock, but also the number of copies of that item held in stock. As in the case of the printer queue, keeping track of repetition is important, but as in the case of the employee records, order is not.

Queue in a bank: Now consider a queue of persons waiting in a bank. Order is important—just try jumping to the head of a long queue! On the other hand, no person may be at two distinct positions in the queue. This is an example where we would require to use a mathematical model for collections with order but no repetition. Queries one may want to make in this case may be asking at which position a particular person is in the queue, and who lies at a particular position of the queue.

As one can deduce from these examples, all four types of collections are useful to model different situations. In fact, mathematicians use all four types of collections for different purposes. In this chapter, we will look at sets, which are collections in which neither order nor repetition is important. Later on, we will build the mathematical tools to reason about multisets (no order, but repetition is possible), sequences (both order and repetition are important) and injective sequences (ordered but with no repetition).

We have already discussed the notion of *types* in predicate calculus, when we insisted that every variable must be given a particular type. We will show the importance of types later in this chapter. For the moment, it is sufficient to accept types as conceptual collections which are mutually disjoint (no two types contain objects in common). Sets will be built as selected subcollections of types.

4.2 Queries About Sets

After reading about the axiomatic approach to building mathematical models to reason about propositions and predicates, one would be tempted to start modelling sets by giving axioms and rules of inference. Although possible, in this book we take a different approach—we try to reduce reasoning about sets to statements in predicate and propositional logic, which we already know how to handle formally.

Predicate logic deals with statements which reduce to true or false. In contrast, sets are collections of objects and are thus neither true nor false. How can one reduce questions about sets into predicates? Rather than defining a set directly as a predicate, the trick is to reduce *queries* about sets into predicates.

The most basic query is whether a particular object can be found in a set. Although we do not yet know how to describe a set, we introduce this notion of whether an object is an element of a set:

Definition 4.1 *We say that an object $x : X$ is an element of set S of objects of type X, written $x \in S$, to mean that x appears in S. We will normally write $x \notin S$ to stand for $\neg(x \in S)$.* ∎

To avoid confusion, we will use upper-case letters for sets, and lower-case letters for elements of sets.

Example If C is the set of all independent countries, we would be able to say that *Malta* $\in C \wedge$ *India* $\in C$. If we are not sure whether Scotland and Wales are considered as independent countries, but we know that if Wales is independent, then so is Scotland, we would write: *Wales* $\in C \Rightarrow$ *Scotland* $\in C$.

If Z is the set of all honest lawyers, and we can deduce that Z contains no elements, we can write:

$$\forall x : \mathsf{PERSON} \cdot x \notin Z$$

We can decompose this by defining the set of honest people H and the set of lawyers L, and write:

$$\neg \exists x : \mathsf{PERSON} \cdot x \in L \wedge x \in H \qquad \diamond$$

4.3 Comparing Sets

We will start by defining ways of comparing sets for inclusion and equality.

4.3.1 Subsets

Very frequently, we would like to say that one set is contained in another—or equivalently that every element of one set can be found in the other set.

Consider the statement: 'Every car owner is a polluter'. If we have two sets *CarOwners* and *Polluters*, this can be expressed using predicate calculus as follows:

$$\forall p : \mathsf{PERSON} \cdot x \in CarOwners \Rightarrow x \in Polluters$$

It is worth noting that this statement does not exclude the possibility of the two sets containing exactly the same objects. This pattern is frequently required. Essentially it is saying that the first set *CarOwners* is completely contained in the second *Polluters*. We express this using the subset relation as *CarOwners* $\subseteq$ *Polluters*.

Definition 4.2 *We say that a set S is* a subset of *set T (written $S \subseteq T$) if every element of S is also in T:*

$$S \subseteq T \overset{\mathrm{df}}{=} \forall x : X \cdot x \in S \Rightarrow x \in T$$

Note that the objects contained in both sets S and T must be of a common type X, otherwise we say that they are incomparable. Note also that the definition does not exclude the possibility that S and T are equal. ∎

Proofs of a statement of the form $S \subseteq T$ always have the following form: from the hypothesis $x \in S$, and go on to prove that $x \in T$. Using implication and $\forall$-introduction, $X \subseteq Y$ would then follow.

Although we still have no way of defining concrete sets, we already have the mathematical tools to be able to prove properties and laws of set comparison.

Example We can prove that any set is a subset of itself: $S \subseteq S$. By definition of the subset operator, we have to prove that $\forall x : X \cdot x \in S \Rightarrow x \in S$.

$$x \in S$$
$$\Rightarrow \{\text{anything implies itself}\}$$
$$x \in S$$

Using $\forall$-introduction, we can conclude that $\forall x : X \cdot x \in S \Rightarrow x \in S$, or that $S \subseteq S$.

$\diamond$

This is a simple example of a rigorous proof. The single step in the proof is justified, and the proof makes no huge leaps of faith. However, there are a few lines hidden between the lines shown, hiding various applications of $\forall$-introduction and $\Rightarrow$-elimination.

Clearly, such a proof can be also be written as a formal one. However, as we add more and more operators and definitions, the size of the expressions grows, making reasoning formally about them very cumbersome. Rigorous proofs are detailed enough to enable a rather straightforward translation into a proper formal proof, but skip some details, which makes the proofs shorter and more understandable, for instance by allowing application of direct reasoning on subexpressions, which is forbidden in formal proofs. Let us look at another rigorous proof:

Example The subset relation is transitive:

$$R \subseteq S \wedge S \subseteq T \Rightarrow R \subseteq T$$

So we know that $R \subseteq S$ and that $S \subseteq T$. We would like to prove that $R \subseteq T$. As noted before, the proof proceeds by taking $x \in R$ and proving that $x \in T$:

$$x \in R$$
$$\Rightarrow \{(\forall x : X \cdot x \in R \Rightarrow x \in S) \text{ since } R \subseteq S\}$$
$$x \in S$$
$$\Rightarrow \{(\forall x : X \cdot x \in S \Rightarrow x \in T) \text{ since } S \subseteq T\}$$
$$x \in T$$

Hence, we can conclude that $\forall x : X \cdot x \in R \Rightarrow x \in T$ and hence $R \subseteq T$. $\diamond$

The application of rules to subexpressions in rigorous proofs requires special attention. If we know that two predicates are equivalent $\alpha \Leftrightarrow \beta$, then any formula $E(\alpha)$ containing α can be shown to be equivalent to the same formula with β instead, $E(\beta)$: $E(\alpha) \Leftrightarrow E(\beta)$. For example, since we know that $\neg P \vee \neg Q \Leftrightarrow \neg(P \wedge Q)$, in a rigorous proof, we can apply the bi-implication within the context of a negation (with $E(\alpha) = \neg \alpha$) to conclude that $\neg(\neg P \vee \neg Q) \Leftrightarrow \neg(\neg(P \wedge Q))$.

On the other hand, this approach does not always work with implications: from $\alpha \Rightarrow \beta$, it does not follow that $E(\alpha) \Rightarrow E(\beta)$. If, for example, we know that $P \Rightarrow Q$, and we try to extend this implication to the context of a negation ($E(\alpha) = \neg\alpha$), we will conclude that $\neg P \Rightarrow \neg Q$. However, as we have seen earlier, $\neg P \Rightarrow \neg Q$ does not necessarily follow from $P \Rightarrow Q$.

When can we use the rule that, if $\alpha \Rightarrow \beta$, then $E(\alpha) \Rightarrow E(\beta)$? A condition that guarantees correctness of the rule, is that α in $E(\alpha)$ appears only under conjunctions and disjunctions, and no negations. Thus, for example, if we know that $\alpha \Rightarrow \beta$, we can replace α by β in $(P \wedge \alpha) \vee \neg P$ to conclude that $((P \wedge \alpha) \vee \neg P) \Rightarrow ((P \wedge \beta) \vee \neg P)$, since negation does not appear above α. On the other hand, it is not safe to apply it to $\neg(P \wedge \alpha)$ or to $(\alpha \Rightarrow P) \wedge Q$.[1]

From this point onwards we will be using rigorous proofs.

4.3.2 Set Equality

What does it mean for two sets to be equal? Informally, we would like to say that $S \overset{\text{set}}{=} T$ when they contain the same elements—every element of S is in T and vice versa. We can give a definition similar to the one used for the subset relation:

$$\forall x : X \cdot x \in S \Leftrightarrow x \in T$$

However, we can also use the subset operator to define set equality.

Definition 4.3 *We say that a set S is equal to set T (written $S \overset{\text{set}}{=} T$) if S is contained in T and vice versa:*

$$S \overset{\text{set}}{=} T \overset{\text{df}}{=} S \subseteq T \wedge T \subseteq S$$

As before, the objects contained in both sets S and T must be of a common type X, otherwise we say that they are incomparable.

When it is clear from the context that we mean equality of sets, we will write $S = T$ instead of $S \overset{\text{set}}{=} T$. ∎

Example We will begin by showing that every set is equal to itself: $S = S$.

> {from the example in Sect. 4.3.1}
> $S \subseteq S$
> $\Rightarrow$ {∧ idempotent}
> $S \subseteq S \wedge S \subseteq S$
> $\Rightarrow$ {definition of set equality}
> $S = S$ ◇

---

[1] A stronger condition is to ensure that, if the formula is reduced to use only negation, conjunction and disjunction, then there is an even number of negations above α for it to be safe to replace it by β. For example, if $\alpha \Rightarrow \beta$, then $\alpha \wedge \neg(P \vee \neg\alpha)$ would imply $\beta \wedge \neg(P \vee \neg\beta)$, since all instances of α appear under an even number of negations (zero and two, respectively).

Example We will now show that set equality is symmetric—in other words, if $S = T$, then $T = S$.

$$S = T$$
$$\Rightarrow \{\text{definition of set equality}\}$$
$$S \subseteq T \wedge T \subseteq S$$
$$\Rightarrow \{\wedge \text{ commutative}\}$$
$$T \subseteq S \wedge S \subseteq T$$
$$\Rightarrow \{\text{definition of set equality}\}$$
$$T = S$$

$\diamond$

Exercises

4.1 Prove that set equality is transitive. If $R = S$ and $S = T$, then $R = T$.

4.4 Constructing Sets

The examples given in the previous section assumed that we were given sets to reason about. And yet, we have no means of building actual sets. This is the next important consideration. In general, we will use curly brackets—'{' and '}'—to construct sets.

4.4.1 Finite Sets

The simplest way to define a finite set is by listing all its elements, which must all be of the same type. For instance, the set of all whole numbers between 0 and 4 can be listed as $N_4 = \{0, 1, 2, 3, 4\}$ (all of type $\mathbb{N}$, the counting or natural numbers); the set of days of the week are $D = \{Monday, Tuesday, Wednesday, Thursday, Friday, Saturday, Sunday\}$ (all of type DAY). Obviously, to do so we would need to know all the elements of the set, which must be finite. Definitions with ellipsis (dot, dot, dot) to define infinite sets are not allowed as in, for example, the set of non-negative whole numbers $\{0, 1, 2, 3, \ldots\}$.

Now consider the predicate $3 \in \{0, 1, 2, 3, 4\}$. Clearly, this should be true. In general, when does the predicate $x \in \{0, 1, 2, 3, 4\}$ hold? To hold, x has to be equal to one of the finite number of values in the set—either $x = 0$ or $x = 1$ or $x = 2$ or $x = 3$ or $x = 4$. In general, we can define membership of an item in a finite listed set as follows.

Definition 4.4 *An object x of type X is said to be* an element of *a finite set of objects x_1 to x_n, all of type X, as defined in the following definition:*

$$x \in \{x_1, x_2, \ldots, x_n\} \stackrel{\text{df}}{=} x = x_1 \vee x = x_2 \vee \cdots \vee x = x_n$$

■

Example Let W be the set of weekend days: $W = \{Saturday, Sunday\}$. We will prove that $\forall x : \mathrm{DAY} \cdot x \in W \Rightarrow x \in D$.

$$x \in W$$
$$\Rightarrow \{\text{by definition of membership in listed set}\}$$
$$x = Saturday \vee x = Sunday$$
$$\Rightarrow \{\vee\text{-introduction and associativity}\}$$
$$x = Monday \vee x = Tuesday \vee x = Wednesday \vee$$
$$x = Thursday \vee x = Friday \vee x = Saturday \vee x = Sunday$$
$$\Rightarrow \{\text{by definition of membership in listed set}\}$$
$$x \in D \hspace{6cm} \diamond$$

Example Recall that in a set order is irrelevant, and sets may not contain more than one copy of an object. Therefore, the set $\{1, 1, 2\}$ should be provably equal to $\{2, 1\}$.

Recall that $\{1, 1, 2\} = \{2, 1\}$ is defined to be:

$$\{1, 1, 2\} \subseteq \{2, 1\} \wedge \{2, 1\} \subseteq \{1, 1, 2\}$$

We thus have to prove two properties, and then introduce the conjunction.

$$x \in \{1, 1, 2\}$$
$$\Rightarrow \{\text{definition of element of}\}$$
$$x = 1 \vee x = 1 \vee x = 2$$
$$\Rightarrow \{\vee \text{ idempotent}\}$$
$$x = 1 \vee x = 2$$
$$\Rightarrow \{\vee \text{ commutative}\}$$
$$x = 2 \vee x = 1$$
$$\Rightarrow \{\text{definition of element of}\}$$
$$x \in \{2, 1\}$$

Therefore, we can conclude that:

$$\forall x : \mathbb{N} \cdot x \in \{1, 1, 2\} \Rightarrow x \in \{2, 1\}$$

This is our definition of $\{1, 1, 2\} \subseteq \{2, 1\}$. The proof of $\{2, 1\} \subseteq \{1, 1, 2\}$ is practically identical, and is left as an exercise. From these two proofs, it follows that: $\{1, 1, 2\} = \{2, 1\}$. $\hspace{4cm} \diamond$

4.4.2 The Empty Set

One important set is the one which contains no elements: $\{\}$. We call this set the *empty set*, usually written as $\emptyset$. Since nothing is an element of the empty set, $x \in \emptyset$ is always false for whatever value of x: $x \in \emptyset \stackrel{\mathrm{df}}{=} false$.

Example An important property of the empty set is that it is a subset of any other set: $\emptyset \subseteq S$. By definition of set containment, we need to prove that $\forall x : X \cdot x \in \emptyset \Rightarrow x \in S$.

$$x \in \emptyset$$
$$\Rightarrow \{\text{definition of empty set}\}$$
$$false$$
$$\Rightarrow \{\text{false implies anything}\}$$
$$x \in S \qquad\qquad\qquad \diamond$$

This last example raises an important issue. From the proof, we can conclude that $\emptyset \subseteq \{1, 2\}$ and $\emptyset \subseteq \{Monday, Tuesday\}$. However, recall that we said that, for two sets to be comparable, they must be of the same type. From the first conclusion, we note that the (potential) elements of $\emptyset$ are of the same type as $\{1, 2\}$—natural numbers. From the second, that the elements of $\emptyset$ are of the same type as $\{Monday, Tuesday\}$—days of the week. Unless the natural numbers and the days of the week are the same type, which we know they are not, these two statements contradict each other. What has gone wrong? In actual fact, although we always write $\emptyset$, there is an empty set for each type, for example, $\emptyset_{\mathbb{N}}$ (the empty set containing no natural numbers) and $\emptyset_{\text{DAY}}$ (the empty set containing no days). Usually, we leave the type in the subscript out, since it is clear from the context.

Since our sets are typed, we do not have a single *universal set* which contains all possible objects. Instead, we have different universal sets, the types themselves. If a set S contains objects of a particular type X, we can always deduce that, given $x \in S$, then it follows that $x \in X$. Note that a type X is itself a set—the largest possible one with objects of that type.

4.4.3 Set Comprehensions

The use of set listing is rather limited. In general, we would like to define infinite sets, or define sets in terms of properties that their elements are expected to satisfy, such as, *the set of the mothers of computer scientists who are at least 50 years old*. The notation we will use to define such sets is called a *set comprehension*:

$$\{p : \text{PERSON} \mid p \in ComputerScientists \land age(p) \geq 50 \bullet motherOf(p)\}$$

Note that the symbols $\mid$ and $\bullet$ are used to separate different parts of the set comprehension. The first part of the set comprehension $p : \text{PERSON}$ declares the variables used in the definition. The second part $p \in ComputerScientists \land age(p) \geq 50$ is a predicate stating which values of the variables will be chosen, and the third part $motherOf(p)$ gives an expression representing the object we will put in the set. The set defined above thus corresponds to the set of all mothers of persons who are computer scientists and are 50 or older. In general, a set comprehension looks like this:

$$\{declarations \mid predicate \bullet term\}$$

Example We define C to be the set of all persons whose parents were both computer scientists:

$$C \stackrel{df}{=} \{p : \mathsf{PERSON} \mid motherOf(p) \in ComputerScientists \, \wedge \\ fatherOf(p) \in ComputerScientists \bullet p\}$$

Now some of these people may have married within this same set. We would like to study the set of age differences of such couples:

$$D \stackrel{df}{=} \{p : C \, , q : C \mid married(p, q) \wedge age(p) \geq age(q) \bullet age(p) - age(q)\}$$

We would now like the set N of computer scientists none of whose children are computer scientists:

$$N \stackrel{df}{=} \{p : ComputerScientists \mid \forall c : \mathsf{PERSON} \cdot \\ offspring(c, p) \Rightarrow c \notin ComputerScientists \bullet p\}$$

From the first example, we note that we can declare more than one variable in the $\diamond$ declaration part of the comprehension. Secondly, note that in these examples we are somewhat abusing the notion, in that in the declaration part of a set comprehension we are sometimes using sets rather than types (after the colon) to specify what kind of objects to use. This can be seen as simply a shorthand for using the type and adding the constraint of being an element of the set in the predicate. For example, the last example can be rewritten as:

$$N \stackrel{df}{=} \{p : \mathsf{PERSON} \mid p \in ComputerScientists \, \wedge \\ \forall c : \mathsf{PERSON} \cdot offspring(c, p) \Rightarrow c \notin ComputerScientists \bullet p\}$$

When the declaration uses only one variable, and that variable appears on its own in the term part of the comprehension, as in $\{n : \mathbb{N} \mid n \geq 17 \bullet n\}$, we may leave out the last part of the comprehension and write it simply as $\{n : \mathbb{N} \mid n \geq 17\}$.

Another recurring pattern is to have no constraints in the predicate part of the comprehension: $\{p : \mathsf{PERSON} \mid true \bullet age(p)\}$. Here, we may leave out the predicate part, to write $\{p : \mathsf{PERSON} \bullet age(p)\}$.

Exercises

4.2 Describe the following sets in English:

(i) $P \stackrel{df}{=} \{p, q : ComputerScientists \mid parent(p, q) \bullet p\}$

(ii) $Q \stackrel{df}{=} \{p : ComputerScientists \mid \exists q : ComputerScientists \cdot parent(p, q)\}$

(iii) $E \stackrel{df}{=} \{n : \mathbb{N} \mid \exists m : \mathbb{N} \cdot n = 2m\}$

(iv) $X \stackrel{df}{=} \{x : \mathbb{N} \bullet 2x\}$

(v) $Y \stackrel{df}{=} \{x : \mathbb{N} \mid false \bullet 2x\}$

4.3 From the definitions in the previous exercise, prove rigorously that $\forall x : ComputerScientists \cdot x \in P \Leftrightarrow x \in Q$. Rigorously prove that $\forall x : \mathbb{N} \cdot x \in E \Leftrightarrow x \in X$.

4.4 Define the following sets in terms of set comprehensions:

  (i) The set of all numbers which can be written as the sum of two square numbers.
 (ii) The set of all such numbers (as defined in (i)) which are themselves square.
(iii) The set of prime numbers—whole numbers larger than 1 and divisible only by 1 and themselves. Assume that we have a predicate available $divisor(n, m)$ saying that n is a divisor of m.

We can now define membership for sets defined in terms of set comprehensions. Consider the set comprehension $S = \{n : \mathbb{N} \mid n > 3 \bullet n^2\}$. Is 49 an element of this set?

$$49 \in \{n : \mathbb{N} \mid n > 3 \bullet n^2\}$$

One would argue that it is, since 49 can be written as 7^2, and $7 > 3$. In other words, we can choose an n which is greater than 3 such that $n^2 = 49$. We can express this in predicate calculus as:

$$49 \in S \text{ since } \exists n : \mathbb{N} \cdot n > 3 \wedge n^2 = 49$$

Similarly, one can argue that $38 \notin S$ since there is no natural number greater than 3 whose square is equal to 38:

$$38 \notin S \text{ since } \neg \exists n : \mathbb{N} \cdot n > 3 \wedge n^2 = 38$$

Combining these, we would expect x to be in S if and only if it satisfies the predicate:

$$x \in S \Leftrightarrow \exists n : \mathbb{N} \cdot n > 3 \wedge n^2 = x$$

We can generalise this to obtain a definition for what it means for an object to be an element of a set comprehension:

Definition 4.5 *An object x is an element of a set comprehension if we can find an instance of the declarations satisfying the predicate, and such that the term is equal to x:*

$$x \in \{decl \mid pred \bullet term\} \stackrel{\text{df}}{=} \exists decl \cdot pred \wedge x = term \qquad \blacksquare$$

Example Consider the set of square numbers: $Q \stackrel{\text{df}}{=} \{n : \mathbb{N} \bullet n^2\}$. Can we prove that $9 \in Q$?

  {basic arithmetic and propositional calculus}
  $true \wedge 9 = 3^2$
$\Rightarrow$ {$\exists$-introduction and $3 \in \mathbb{N}$}
  $\exists n : \mathbb{N} \cdot true \wedge 9 = n^2$
$\Rightarrow$ {definition of membership of set comprehension}
  $9 \in \{n : \mathbb{N} \mid true \bullet n^2\}$
$\Rightarrow$ {abbreviation of set comprehension}
  $9 \in \{n : \mathbb{N} \bullet n^2\}$
$\Rightarrow$ {definition of Q}
  $9 \in Q$

Although the way the proof reads in the forward direction may seem unusual, it would be constructed from the bottom (the conclusion) up. Reconstructing the proof bottom-up gives insight of how such proofs are written.   ◇

Example Now consider the set of squares of even numbers: $Q_E \overset{\mathrm{df}}{=} \{n : \mathbb{N} \bullet (2n)^2\}$. Can we prove that $9 \notin Q_E$?

$$\neg \exists n : \mathbb{N} \cdot 9 = 4n^2$$
$\Rightarrow$ {basic arithmetic}
$$\neg \exists n : \mathbb{N} \cdot 9 = (2n)^2$$
$\Leftrightarrow$ {basic propositional logic}
$$\neg \exists n : \mathbb{N} \cdot true \wedge 9 = (2n)^2$$
$\Rightarrow$ {definition of membership of set comprehension}
$$9 \notin \{n : \mathbb{N} \mid true \bullet (2n)^2\}$$
$\Rightarrow$ {abbreviation of set comprehension}
$$9 \notin \{n : \mathbb{N} \bullet (2n)^2\}$$
$\Leftrightarrow$ {definition of Q_E}
$$9 \notin Q_E$$
                                                                                  ◇

Example Consider the two sets $\{y : X \mid P \wedge Q\}$ and $\{y : X \mid P\}$. We would like to show that the first is a subset of the latter:

$$\{y : X \mid P \wedge Q\} \subseteq \{y : X \mid P\}$$

To prove set inclusion, we have to prove the following implication:

$$\forall x : X \cdot x \in \{y : X \mid P \wedge Q\} \Rightarrow x \in \{y : X \mid P\}$$

$$x \in \{y : X \mid P \wedge Q\}$$
$\Rightarrow$ {implicit y at end of comprehension}
$$x \in \{y : X \mid P \wedge Q \bullet y\}$$
$\Rightarrow$ {definition of element of}
$$\exists y : X \cdot P \wedge Q \wedge x = y$$
$\Rightarrow$ {law of existential quantification}
$$\exists y : X \cdot P \wedge x = y$$
$\Rightarrow$ {definition of element of}
$$x \in \{y : X \mid P \bullet y\}$$
$\Rightarrow$ {implicit y at end of comprehension}
$$x \in \{y : X \mid P\}$$
                                                                                  ◇

Sometimes, in a proof, when we would like to open up membership of a set comprehension, we have a potential name clash—consider, for example: $x \in \{x : \mathbb{N} \mid x \geq 17 \bullet x^2\}$. Opening this according to the definition of inclusion in a set specified as a comprehension, we get: $\exists x : \mathbb{N} \cdot x \geq 17 \wedge x = x^2$, which is clearly not what we want, since the two different variables named x are clashing together. To enable us to open up such an expression, we would need to rename the variables in the set comprehension. For instance, we would first rewrite the membership query as:

$x \in \{y : \mathbb{N} \mid y \geq 17 \bullet y^2\}$ (note that, not only do we rename the declaration part, but also, all free instances of x in the predicate and term are replaced by y). Are we sure that renaming variables leaves a set intact? Here is a proof that $\{x : \mathbb{N} \mid x \geq 17 \bullet x^2\}$ is equal to $\{y : \mathbb{N} \mid y \geq 17 \bullet y^2\}$:

$$n \in \{x : \mathbb{N} \mid x \geq 17 \bullet x^2\}$$
$\Leftrightarrow$ {definition of membership of set inclusion}
$$\exists x : \mathbb{N} \cdot x \geq 17 \wedge n = x^2$$
$\Leftrightarrow$ {by alpha-conversion of predicate logic}
$$\exists y : \mathbb{N} \cdot (x \geq 17 \wedge n = x^2)[x \leftarrow y]$$
$\Leftrightarrow$ {applying substitution}
$$\exists y : \mathbb{N} \cdot y \geq 17 \wedge n = y^2$$
$\Leftrightarrow$ {definition of membership of set inclusion}
$$n \in \{y : \mathbb{N} \mid y \geq 17 \bullet y^2\}$$

In general, we can define alpha-conversion of set comprehensions, in which we rename one variable in a responsible manner:

Definition 4.6 *Given a set comprehension*:

$$\{x_1 : X_1, \ x_2 : X_2, \ldots x : X, \ \ldots x_{n-1} : X_{n-1}, \ x_n : X_n \mid P \bullet t\}$$

and a variable y (different from all the variables x_i), such that $y \setminus P$ and $y \setminus t$, the alpha-conversion of the set comprehension from x to y is defined to be:

$$\{x_1 : X_1, \ x_2 : X_2, \ldots y : X, \ \ldots x_{n-1} : X_{n-1}, \ x_n : X_n \mid P[x \leftarrow y] \bullet t[x \leftarrow y]\} \quad \blacksquare$$

The proof we saw for a particular instance can be generalised to justify that alpha-conversion of sets leaves a set intact.

Exercises
4.5 Prove that, if $P \Rightarrow Q$, then $\{x : X \mid P \bullet t\} \subseteq \{x : X \mid Q \bullet t\}$ (t is an expression in terms of x).
4.6 Sometimes it is useful to say that a set S is contained in T without allowing for the possibility that they are equal (as $S \subseteq T$ does). We say that S is a *proper subset* of T, written as $S \subset T$ if S is contained in T, but not equal to it. Define $\subset$ in terms of $\subseteq$ and $=$.
4.7 Prove that $\{x : X \mid true \bullet 1\} = \{1\}$. Is it true that, for any P: $\{x : X \mid P \bullet 1\} = \{1\}$?

4.5 Set Operators

We have now got means of defining sets and means to compare them. However, defining every set using a set listing or a comprehension can be rather cumbersome, and having a number of operators to create new sets using ones already defined can be much more effective. We will be defining a number of operators in this section:

Set union: Given two sets S and T, their union is the set of all objects in at least one of the two. The union of two sets S and T is written as $S \cup T$.

Set intersection: The intersection of two sets S and T is the set of all objects present in both, and is written as $S \cap T$.

Set complement: The complement of a set S, written S^c, is the set of all objects not in S.

Set difference: The set difference of two sets S and T, is the set of objects in S but not in T. This is written as $S \setminus T$.

One important thing to note is that these operators are defined only when the sets are of the same type. For instance, the union of a set of numbers and a set of persons is not well defined. In particular, the complement of a set S will have the same type as S itself. For instance, the complement of the set of even numbers gives us the set of odd numbers and will not include *Peter* or *apple*.

Example Consider the set J of user accounts on a computer who use a Java compiler. The set H is another set of users, but who use a Haskell compiler. We can now define a number of other sets in terms of these:

Programmers: If we want to define the set of users who have access to at least one of the compilers, we may define it to be the union of H and J: $H \cup J$.

Polyglots: The users who can program in both languages and thus use both language compilers can be written using the intersection operator: $H \cap J$.

Haskell-challenged: Users who do not use the Haskell compiler, which can be expressed using the complement operator: H^c.

Language purists: Users who use Haskell but not Java can be described using set difference: $H \setminus J$. Similarly, Java users who do not use Haskell can be written as: $J \setminus H$. The language purists, who program in just one language, are thus the union of these two sets: $(H \setminus J) \cup (J \setminus H)$.

Non-programmers: Users who do not use either compiler can be expressed as the complement of the set of programmers: $(H \cup J)^c$.

Once we have defined these operators formally, we can formally express the predicate stating whether John is a Haskell purist as $John \in H \setminus J$. Similarly, if we are given the set S of users with student accounts, we can ask whether all students have used at least one of the compilers: $S \subseteq H \cup J$.   ◇

Exercises

4.8   A computer file system which has three classes of users: managers, employees and customers. The file system manages which users have access to which files. The sets R_m, R_e and R_c respectively contain files which the managers, employees and customers can read. Similarly, W_m, W_e and W_c are the sets of files to which the managers, employees and customers can write. All objects in these sets will be of a type FILE.

    (i) Formally express the following sets:

        (a) The set of files to which employees may write, and which customers may read.

        (b) Files which can be written to but not read by customers.

        (b) The set of files which all users may read.

        (b) The files which are readable only by managers.

        (b) The files which no one may read.

    (ii) Express the following statements as predicates:

        (a) The file *passwd* can only be written to by the managers.

        (b) Managers do not have less rights than employees: (i) if employees can read a file, then so can managers; (ii) if employees can write to a file, then so can the managers.

        (c) All files can be read by at least one of the classes of users.

        (d) No files may be written to but not read by a class of users.

        (e) Any file to which customers may both read and write must be readable by employees.

4.5.1 Set Union

The set operators will be defined using set comprehensions. The first operator, set union, is defined as follows:

Definition 4.7 *Given two sets S and T, both containing objects of type X, we define their union $S \cup T$ to be the set of all elements in either S or T (or both):*

$$S \cup T \stackrel{\mathrm{df}}{=} \{x : X \mid x \in S \vee x \in T\} \qquad \blacksquare$$

Proving properties directly using this definition results in reasoning about membership in set comprehensions. To avoid having to reason directly about existentially quantified formulae, we prove the following theorem, which enables us to bypass the use of the definition in our proofs:

Theorem 4.8 *An object is in the union of two sets if and only if it is an element of the first or of the second:*

$$\forall x : X \cdot x \in S \cup T \Leftrightarrow x \in S \vee x \in T$$

Proof The proof follows from the definition of set union and inclusion in a set comprehension:

$$y \in S \cup T$$
$\Leftrightarrow$ {by definition of set union}
$$y \in \{x : X \mid x \in S \vee x \in T\}$$
$\Leftrightarrow$ {implicit x at end of comprehension}
$$y \in \{x : X \mid x \in S \vee x \in T \bullet x\}$$
$\Leftrightarrow$ {by definition of membership of set comprehensions}
$$\exists x : X \cdot (x \in S \vee x \in T) \wedge x = y$$
$\Leftrightarrow$ {using the one-point rule}
$$y \in S \vee y \in T$$ ∎

Example We can now prove some interesting properties of set union in a rather straightforward manner. Consider commutativity of the union operator: $S \cup T = T \cup S$. We need to prove two things: $S \cup T \subseteq T \cup S$ and $T \cup S \subseteq S \cup T$. For the first, we need to prove that $\forall x : X \cdot x \in S \cup T \Rightarrow x \in T \cup S$, while for the second, we need to prove that $\forall x : X \cdot x \in T \cup S \Rightarrow x \in S \cup T$. As we write the proof, we will realise, however, that the two directions can be combined into one, by proving that:

$$\forall x : X \cdot x \in S \cup T \Leftrightarrow x \in T \cup S$$

The following proof thus suffices to prove the set equality. When writing such a proof, it is important to ensure that all the lines are really bi-implications, to ensure correctness.

$$x \in S \cup T$$
$\Leftrightarrow$ {theorem of union}
$$x \in S \vee x \in T$$
$\Leftrightarrow$ {disjunction is commutative}
$$x \in T \vee x \in S$$
$\Leftrightarrow$ {theorem of union}
$$x \in T \cup S$$ ◇

Exercises

4.9 Prove that the empty set acts as a zero for set union: $\emptyset \cup S = S \cup \emptyset = S$.
4.10 Prove that set union is idempotent: $S \cup S = S$.
4.11 Prove that set union is associative: $R \cup (S \cup T) = (R \cup S) \cup T$.

4.5.2 Set Intersection

Intersection can be defined in an almost identical fashion as union, except that conjunction is used rather than disjunction:

Definition 4.9 *Given two sets S and T, both containing objects of type X, we define their union $S \cap T$ to be the set of all elements in both S and T:*

$$S \cap T \overset{\text{df}}{=} \{x : X \mid x \in S \wedge x \in T\}$$ ∎

As before, we prove a theorem to enable us to reason about set intersection more easily.

Theorem 4.10 *An object is in the intersection of two sets if and only if it is an element of the first and of the second:*

$$\forall x : X \cdot x \in S \cap T \Leftrightarrow x \in S \wedge x \in T$$

The proof is almost identical to the one given for set union.

Example To start off with a simple example, let us prove that intersecting a set with another can only make it smaller: $S \cap T \subseteq S$.

$$
\begin{aligned}
& x \in S \cap T \\
\Leftrightarrow\ & \{\text{theorem of intersection}\} \\
& x \in S \wedge x \in T \\
\Rightarrow\ & \{\wedge\text{-elimination}\} \\
& x \in S
\end{aligned}
$$

Note that the last line in the proof is an implication, and not a bi-implication, which enables us to conclude that $x \in S \cap T \Rightarrow x \in S$, and thus that $S \cap T \subseteq S$. ◇

Example We now proceed to prove other properties of set intersection. Consider associativity of the intersection operator: $R \cap (S \cap T) = (R \cap S) \cap T$.

$$
\begin{aligned}
& x \in R \cap (S \cap T) \\
\Leftrightarrow\ & \{\text{theorem of intersection}\} \\
& x \in R \wedge x \in S \cap T \\
\Leftrightarrow\ & \{\text{theorem of intersection}\} \\
& x \in R \wedge (x \in S \wedge x \in T) \\
\Leftrightarrow\ & \{\text{conjunction is associative}\} \\
& (x \in R \wedge x \in S) \wedge x \in T \\
\Leftrightarrow\ & \{\text{theorem of intersection}\} \\
& x \in R \cap S \wedge x \in T \\
\Leftrightarrow\ & \{\text{theorem of intersection}\} \\
& x \in (R \cap S) \cap T
\end{aligned}
$$

◇

Example We can also prove properties relating union and intersection. For instance, consider the property that union distributes over intersection: $R \cup (S \cap T) = (R \cup S) \cap (R \cup T)$.

$$x \in R \cup (S \cap T)$$
$\Leftrightarrow$ {theorem of union}
$$x \in R \vee x \in S \cap T$$
$\Leftrightarrow$ {theorem of intersection}
$$x \in R \vee (x \in S \wedge x \in T)$$
$\Leftrightarrow$ {disjunction distributes over conjunction}
$$(x \in R \vee x \in S) \wedge (x \in R \vee x \in T)$$
$\Leftrightarrow$ {theorem of union applied twice}
$$x \in R \cup S \wedge x \in R \cup T$$
$\Leftrightarrow$ {theorem of intersection}
$$x \in (R \cup S) \cap (R \cup T)$$

$\diamond$

Exercises

4.12 Prove that set intersection is idempotent and commutative: $S \cap S = S$ and $S \cap T = T \cap S$.

4.13 Prove that $R \cap S \subseteq R \cup S$.

4.14 Prove that intersection distributes over union: $R \cap (S \cup T) = (R \cap S) \cup (R \cap T)$.

4.15 Prove Theorem 4.10.

4.5.3 Set Complement

Recall that the complement of a set is defined to the set of all objects of the original type, but which were not in the original set.

Definition 4.11 *Given a set S containing objects of type X, we define its complement S^c to be the set of all elements of type X which are not in S:*

$$S^c \stackrel{\mathrm{df}}{=} \{x : X \mid x \notin S\}$$

∎

The theorem to simplify reasoning about set complement relates it with negation:

Theorem 4.12 *An object is in the complement of a set if and only if it is not an element of the set:*

$$\forall x : X \cdot x \in S^c \Leftrightarrow \neg(x \in S)$$

Proof The proof is similar to the one for set union. As before, we will use variable y, instead of x, to avoid variable name clashes.

$\quad y \in S^c$
$\Leftrightarrow$ {by definition of set complement}
$\quad y \in \{x : X \mid x \notin S\}$
$\Leftrightarrow$ {implicit x at end of comprehension}
$\quad y \in \{x : X \mid x \notin S \bullet x\}$
$\Leftrightarrow$ {by definition of membership of set comprehensions}
$\quad \exists x : X \cdot x \notin S \wedge x = y$
$\Leftrightarrow$ {using the one-point rule}
$\quad y \notin S$
$\Leftrightarrow$ {definition of $\notin$}
$\quad \neg(y \in S)$ ∎

This theorem will be frequently used in rigorous proofs to reduce set comple-mention into negation of set membership. The following two examples will illustrate this.

Example To start off with a simple example, let us prove that complementing a set twice results in the original set: $(S^c)^c = S$.

$\quad x \in (S^c)^c$
$\Leftrightarrow$ {theorem of complement}
$\quad \neg(x \in S^c)$
$\Leftrightarrow$ {theorem of complement}
$\quad \neg(\neg(x \in S))$
$\Leftrightarrow$ {double negation}
$\quad x \in S$ ◇

Example As with disjunction and conjunction, union and intersection can be written in terms of each other using set complement: $S \cap T = (S^c \cup T^c)^c$.

$\quad x \in S \cap T$
$\Leftrightarrow$ {theorem of $\cap$}
$\quad x \in S \wedge x \in T$
$\Leftrightarrow$ {De Morgan's rule}
$\quad \neg(\neg(x \in S) \vee \neg(x \in T))$
$\Leftrightarrow$ {theorem of complement twice}
$\quad \neg(x \in S^c \vee x \in T^c)$
$\Leftrightarrow$ {theorem of $\cup$}
$\quad \neg(x \in S^c \cup T^c)$
$\Leftrightarrow$ {theorem of complement}
$\quad x \in (S^c \cup T^c)^c$

The simplest way to write such a proof is to open up the set operators using the theorems we have proved on the top and bottom lines, and then use propositional

logic rules to make the two ends meet, rather than start from the top and proceed
step by step to the bottom.   ◇

Exercises
4.16 Prove that $(S \cup T)^c \subseteq S^c$.
4.17 Prove that $S \cup T = (S^c \cap T^c)^c$.
4.18 Prove that $S \cap S^c = \emptyset$.
4.19 Prove that $S \cup S^c = \emptyset^c$.
4.20 If S contains objects of type X, prove that $S \cup S^c = X$. Recall that, if $x \in S$, it
      follows that $x \in X$.

4.5.4 Set Difference

Set difference can be defined in a very similar way to intersection and union using
set comprehension.

Definition 4.13 *The set difference of two sets S and T, written $S \setminus T$, is defined to
be the set of objects in set S, but not T. Formally, this can be defined as*:

$$S \setminus T \stackrel{\text{df}}{=} \{x : X \mid x \in S \wedge x \notin T\}$$ ∎

The corresponding theorem to simplify reasoning about set difference involves
reducing set difference to a combination of conjunction and negation.

Theorem 4.14 *An object is in the set difference of two sets if and only if it is an
element of the first but not the second*:

$$\forall x : X \cdot x \in S \setminus T \Leftrightarrow x \in S \wedge \neg(x \in T)$$

Exercises
4.21 Prove Theorem 4.14.
4.22 Prove the following:

   (i) $S \setminus S = \emptyset$.
   (ii) $S \setminus T = S \cap T^c$.
   (iii) $R \setminus (S \cup T) = (R \setminus S) \cup (R \setminus T)$.

4.23 The symmetric set difference of two sets S and T, written $S \otimes T$, is defined
      to be the set of objects in exactly one of the sets. Formally, this can be defined
      as:

$$S \otimes T \stackrel{\text{df}}{=} (S \setminus T) \cup (T \setminus S)$$

Prove that:

(i) $S \otimes S = \emptyset$.
(ii) Symmetric set difference is commutative.
(iii) Symmetric set difference is associative.

4.5.5 Other Properties of the Set Operators

This approach to reasoning about sets lends itself easily to proving properties of set operators and how they are related to each other. In this section we take a look at proving properties about the set operators given some information about the sets.

Example Let us start with a simple example. Given that $A \cup B \subseteq R \cap S$, let us prove that $A \subseteq R$.

From the given information, we know that $x \in A \cup B \Rightarrow x \in R \cap S$.

$$x \in A$$
$$\Rightarrow \{\vee\text{-introduction}\}$$
$$x \in A \vee x \in B$$
$$\Leftrightarrow \{\text{theorem of union}\}$$
$$x \in A \cup B$$
$$\Rightarrow \{\text{given information}\}$$
$$x \in R \cap S$$
$$\Leftrightarrow \{\text{theorem of intersection}\}$$
$$x \in R \wedge x \in S$$
$$\Rightarrow \{\wedge\text{-elimination}\}$$
$$x \in R \qquad\qquad\qquad \diamond$$

Example If S and T are two disjoint sets, that is $S \cap T = \emptyset$, let us prove that $S \subseteq T^c$.

From the known fact that $S \cap T = \emptyset$, we can conclude that $x \in S \cap T \Leftrightarrow x \in \emptyset$. We will now use this to obtain the desired result:

$$x \in S$$
$$\Rightarrow \{\ldots\}$$
$$\neg(x \in T)$$
$$\Leftrightarrow \{\text{theorem about complement}\}$$
$$x \in T^c$$

How can the proof be completed? If we were to argue informally about why we expect this missing part to be true, we would say something on the lines of "x cannot be in T since, if it were, then it would be in both S and T, but we know that $S \cap T = \emptyset$". We can transform this reasoning into a rigorous proof using the law of disjunction and negation to assert that $x \in T \vee \neg(x \in T)$. Here is the completed proof:

$$x \in S$$
$\Leftrightarrow$ {law of disjunction and negation}
$$x \in S \wedge (x \in T \vee \neg(x \in T))$$
$\Leftrightarrow$ {distributivity of conjunction over disjunction}
$$(x \in S \wedge x \in T) \vee (x \in S \wedge \neg(x \in T))$$
$\Leftrightarrow$ {theorem of intersection }
$$x \in S \cap T \vee (x \in S \wedge \neg(x \in T))$$
$\Leftrightarrow$ {given information}
$$x \in \emptyset \vee (x \in S \wedge \neg(x \in T))$$
$\Leftrightarrow$ {definition of empty set}
$$false \vee (x \in S \wedge \neg(x \in T))$$
$\Leftrightarrow$ {law of *false* and disjunction}
$$x \in S \wedge \neg(x \in T)$$
$\Rightarrow$ {$\wedge$-elimination}
$$\neg(x \in T)$$
$\Leftrightarrow$ {theorem about complement}
$$x \in T^c$$ $\diamond$

Exercises

4.24 Prove that, if $S \setminus T \subseteq T$, then $S \setminus T = \emptyset$.

4.25 Prove that, if $T \subseteq S \setminus T$, then $S \setminus T = S$.

4.26 Prove that $R \cap S = T$, $R \cap T = S$ and $S \cap T = R$ if and only if $R = S$ and $S = T$.

4.5.6 Generalised Set Operators

The set operators union and intersection both take two operands. In general, one sometimes needs to take the intersection or union of more than two sets. If the number of sets is finite, then the union or intersection can be written as repeated use of the binary operator, but this would not work in the case of an infinite number of sets. Even in the case of a finite number of sets, it is more convenient to have a notation to be able to write the generalised union or intersection more effectively.

Example For instance, let us write M_n to denote the set of all multiples of number n, and $P = \{p_1, p_2, \ldots\}$ to be the set of prime numbers—numbers larger than 1 not divisible by any number other than 1 and themselves. It is worth noting there is an infinite number of primes—we will prove this later on in the chapter about natural numbers. Now consider the set of all multiples of prime numbers:

$$N \stackrel{\text{df}}{=} M_{p_1} \cup M_{p_2} \cup \cdots$$

This definition of N is an informal one, since we have no way of writing an infinite number of unions yet. We will provide the notation to do so and write the infinite union as:

$$N \stackrel{\mathrm{df}}{=} \bigcup_{p \in P} M_p$$

This will correspond to the union of all the sets M_p, where $p \in P$. Note that repetition and order are lost when picking p from P, but luckily, we know that union is idempotent, commutative and associative, and therefore this does not matter much to us.   ◇

We can define generalised union and intersection using set comprehensions:

Definition 4.15 *Given an index set I and a set S_i for each $i \in I$, we define the generalised union and intersection as follows*:

$$\bigcup_{i \in I} S_i \stackrel{\mathrm{df}}{=} \{x : X \mid \exists i \in I \cdot x \in S_i\}$$

$$\bigcap_{i \in I} S_i \stackrel{\mathrm{df}}{=} \{x : X \mid \forall i \in I \cdot x \in S_i\}$$ ∎

It is worth noting that the definition quantifies over a set rather than a type. If the items in a set S are of type X, then $\exists x \in S \cdot P$ is considered to be shorthand for $\exists x : X \cdot x \in S \wedge P$. Similarly, $\forall x \in S \cdot P$ is shorthand for $\forall x : X \cdot x \in S \Rightarrow P$.

The notion of generalised union enables us to describe the notion of a partition of a set. A partition is a way of separating objects into different classes. Consider the following example:

Example Consider a multi-user operating system which has a notion of user groups. Each group consists of a number of users who may appear in different groups. Let U be the set of users registered on the system, and U_g be the set of users of a group g. These sets contain objects of type USERS.

Amongst the groups the system has, three important ones are the groups *Guest*, *Normal* and *Superuser*, which are used to classify users into one of three classes— guest users with minimal rights, normal users who may access various applications and access certain files, and superusers who can also access information about other users. These classes are said to partition the set of users since (i) all users must be in at least one of the classes, and (ii) users may not be in more than one of these classes.

If $G \stackrel{\mathrm{df}}{=} \{Guest, Normal, Superuser\}$, we can formally write the first property as follows:

$$U = \bigcup_{g \in G} U_g$$

The second property is equivalent to saying that no pair of these classes may have common elements. There are various ways of writing this. Given that we only have three classes, we can write it as:

$$(U_{Guest} \cap U_{Normal} = \emptyset) \wedge (U_{Superuser} \cap U_{Normal} = \emptyset) \wedge (U_{Superuser} \cap U_{Guest} = \emptyset)$$

Had we had many more classes, another way of writing this more compactly would have been:

$$\forall i, j : G \cdot i \neq j \Rightarrow U_i \cap U_j = \emptyset$$

There are also other groups defined on the system, such as *Database* (containing users who have access to a database), *External* (users allowed to connect to the system from elsewhere) and *Staff* (users who are members of staff), but they do not form a partition. ◇

Definition 4.16 *Given an index set I and set P_i for every $i \in I$, we say that the indexed sets partition a set S if (i) they give* complete *coverage of S, i.e. $\bigcup_{i \in I} P_i = S$; and (ii) they are pairwise disjoint—any two different partitions have no common elements*: $\forall i, j : I \cdot i \neq j \Rightarrow P_i \cap P_j = \emptyset$. ∎

Example One useful feature of computer memory is that, if we know the address where a piece of data is held, it takes a fixed amount of time to retrieve the information. Arrays use this feature to allow for fast access to data based on the index. One simple application of this is that, if we need to store information about employees to be accessed by their employee number which ranges from 1 to the number of employees, we can store this information in an array and fetch it quickly. A query of the form 'What is the monthly salary of employee 7?' can be answered quickly. On the other hand, if we want to ask questions, for example, based on the name of the employee, this approach does not help. To answer the question 'What is the monthly salary of Joe Smith?' we would have to search through the array to find which employee number pertains to Joe, and then access the salary information. As the number of employees increases, the more items we have to search through to answer the question, which is undesirable.

The solution typically used is that of storing the information in a *hash table*. In a hash table, the search term (in this case the name of the employee) is converted into a number where the data will be placed—the function used to convert the data into a number is called the *hash function*. For example, we may decide to take the first letter of the name and surname of the employee, convert them to numbers (with A being 1, and Z being 26) and add them together. Let us call this function h. Joe Smith would thus have the value $J + S = 10 + 19 = 29$, the location where his information will be stored and restored from. The solution works perfectly, until a person called Ken Roberts is also employed. Unfortunately, Ken's name also has value 29. In a hash table, the two are stored in the same location, meaning that every time we look for someone whose name has value 29, we need to go through a list of possibilities. A good hash function distributes the values well, so as to avoid many such *collisions*.

Let us write E_n to denote the set of employees with value n:

$$E_n \overset{\text{df}}{=} \{e : \text{EMPLOYEE} \mid h(e) = n\}$$

Note that each employee can have a value ranging from 2 (those with initials A.A.) to 52 (those with initials Z.Z.). The sets E_2 to E_{52} turn out to be a partition because: (i) all the employees appear in one of these sets, so their union covers all employees—$\text{EMPLOYEE} = \bigcup_{2 \le n \le 52} E_n$; and (ii) each employee e will only appear in $E_{h(e)}$, and thus any two such sets are disjoint—$\forall i, j : 2 \ldots 52 \cdot i \ne j \Rightarrow E_i \cap E_j = \emptyset$.

A hash table is therefore nothing but a partition of the data placed inside it; the better the hash function, the more balanced the sizes of the partitions are.   ◇

The notion of partitions has various other applications in mathematics and computer science, since it allows ways of classifying a large number of items into a smaller number of classes, allowing us to devise different ways of accessing and manipulating the items. One more example of the use of a partition is in chess-playing programs, where one approach is to start by classifying the game situation into being either in the (i) opening phase, (ii) middle game or (iii) endgame. Depending on the phase in which the current game is, one can then apply an appropriate strategy. It is important to note that this is a partition, since (i) any game position should fall under one of the classes, to ensure that the system knows how to deal with it; and (ii) a position cannot fall into more than one category, thus always making only one strategy applicable.

4.6 Sets as a Boolean Algebra

When we discussed propositional logic, we mentioned that it can be characterised using Boolean algebra (see Sect. 2.6.1). Recall that a Boolean algebra is (i) a collection of objects; with (ii) two binary operators $\odot$ and $\oplus$; (iii) a unary operator $\overline{\alpha}$; and (iv) a notion of equality, which satisfy the laws given on 49.

If we restrict our view to sets containing objects of just one type X, we can take the binary operators to be $\cup$ (for $\oplus$) and $\cap$ (for $\odot$), and complement as the unary operator. Using set equality, we can go through the laws of a Boolean algebra to check whether they are satisfied:

Commutativity: The binary operators have to be commutative, but this was proved for both $\cup$ and $\cap$.

Associativity: The binary operators also have to be associative, which has also been proved for set union and intersection.

Identities: There have to be special objects 1 and 0 which are, respectively, identities of $\odot$ and $\oplus$. Taking 1 to be X (the whole type) and 0 to be $\emptyset$, we note that they are in fact identities of intersection and union, respectively: $X \cap S = S$ and $\emptyset \cup S = S$.

Distributivity: The binary operators have to distribute over each other. It was shown that union distributes over intersection, and the proof of distributivity of intersection over union was left as an exercise.

Complement: The unary operator should interact with the binary operators in the following manner: $x \odot \bar{x} = 0$ and $x \oplus \bar{x} = 1$. These hold for the set operators: $S \cap S^c = \emptyset$, and $S \cup S^c = X$.

From this analysis, it follows that our notion of sets and their operators make them an instance of a Boolean algebra. In other words, we could have chosen to characterise sets using a Boolean algebra instead of the definitions we gave. Interestingly, since both the propositional operators and set operators obey the same laws, any equivalence deduced in one can automatically be carried over to the other. For instance, by proving that $P \vee (\neg P \wedge Q) \dashv\vdash P \vee Q$, it automatically follows that $P \cup (P^c \cap Q) = P \cup Q$. The converse is also true.

   Note that, in an algebra, everything hinges around the definition of equality. In our case, equality was defined in terms of the subset relation. Surprisingly, we could have done this the other way round. By first defining equality of sets, we could have then proceeded to define the subset relation in terms of equality. Consider the following theorem:

Theorem 4.17 *The subset relation between sets can be expressed in terms of set equality as follows:* $S \subseteq T$ *if and only if* $S \cap T = S$.

Proof Let us start with the forward direction. Assuming $S \subseteq T$, we will prove that $S \cap T = S$. The proof is split into two. We start by proving that $S \cap T \subseteq S$:

$$x \in S \cap T$$
$$\Leftrightarrow \{\text{theorem of intersection}\}$$
$$x \in S \wedge x \in T$$
$$\Rightarrow \{\wedge\text{-elimination}\}$$
$$x \in S$$

We can now prove that $S \subseteq S \cap T$:

$$x \in S$$
$$\Leftrightarrow \{\text{conjunction is idempotent}\}$$
$$x \in S \wedge x \in S$$
$$\Rightarrow \{\text{since } S \subseteq T, \text{ then } x \in S \Rightarrow x \in T\}$$
$$x \in S \wedge x \in T$$
$$\Leftrightarrow \{\text{theorem of intersection}\}$$
$$x \in S \cap T$$

From this it follows that, if $S \subseteq T$, then $S \cap T = S$.

We now prove the converse: if $S \cap T = S$, then $S \subseteq T$.

$$x \in S$$
$$\Rightarrow \{\text{since } S \cap T = S, \text{ then } S \subseteq S \cap T\}$$
$$x \in S \cap T$$
$$\Leftrightarrow \{\text{theorem of intersection}\}$$
$$x \in S \wedge x \in T$$
$$\Rightarrow \{\wedge\text{-elimination}\}$$
$$x \in T$$

It thus follows that $S \subseteq T \Leftrightarrow S \cap T = S$.   ∎

This result shows that we could have started with a definition of set equality (instead of the notion of subsets), based on which we could define the subset relation $S \subseteq T$ to be $S \cap T = S$.

4.7 More About Types

We have been using types since the chapter on predicate calculus. There are various questions one may ask at this point. Where do the types come from? Can we define new types? Can we do without them? In this section, we will address some of these questions.

4.7.1 Types and Sets

We started by saying that types are a collection of objects. But in that case, what is the difference between types and sets? The important distinction between a type and a set is that we assume our types to be mutually disjoint—no two types may contain objects in common. This gives a unique type to objects. Note that, due to the restrictions we placed on set operators, that operands must contain objects of the same type; union, intersection and set difference cannot be applied to sets of different types.

For the moment, we have assumed that we have started with a number of types available for our use. Later on, in Chap. 8, we will explore how new types can be defined. However, given a number of types, we can already create new types out of them.

4.7.1.1 Power Sets

Consider the set of even numbers less than 7: $E_7 = \{0, 2, 4, 6\}$, and the set of odd numbers $O = \{n : \mathbb{N} \bullet 2n + 1\}$. The sets both contain objects of the same type, namely the type of natural numbers, so the union, intersection and set difference of the sets are well defined. However, to go a step further, the two sets seem to be of the same type—both are sets of natural numbers. So we should be able to define a set containing both these sets: $\{E_7, O\}$. Note that this is different from the union of the two sets. For instance, this set contains only two elements, unlike the union of E_7 and O which contains an infinite number of elements. Furthermore, $O \in \{E_7, O\}$, while $O \subseteq E_7 \cup O$.

If $\mathbb{N}$ is the type of objects in set O, what is the type of objects contained in $\{E_7, O\}$?—in other words, what is the type of O itself? O is a set of objects in $\mathbb{N}$, and we say that O is an object in $\mathbb{PN}$, the so-called *power set* of the natural numbers. In general, given a type X, $\mathbb{P}X$ corresponds to the type of *sets containing objects in X*. Note that, with this notation, 1 is of type $\mathbb{N}$, while $\{1, 2, 4\}$ is of type $\mathbb{PN}$. Furthermore, $\{\{1, 2, 4\}, \{5, 2\}\}$ (which is a set of sets of natural numbers) is of type $\mathbb{PPN}$.

We can now say exactly what we mean by the statement that $x \in S$ is correctly typed: if x is of type X, then S has to be of type $\mathbb{P}X$.

This leads us back to the question which arose when discussing the empty set: What is the type of the empty set? Up till that point, we were still discussing sets with respect to the type of the objects they may contain. However, now we can give a type to a set. The set $\{1, 2\}$ is of type $\mathbb{PN}$, while $\{Monday, Tuesday\}$ is of type $\mathbb{P}DAY$. On one hand, we have proved that $\emptyset \subseteq \{1, 2\}$, and since we insisted that the subset relation works only on two sets of the same type, the empty set should be of the type $\mathbb{PN}$. On the other hand, we have also proved that $\emptyset \subseteq \{Monday, Tuesday\}$, and using the same argument, the empty set should be of type $\mathbb{P}DAY$. But we said that an object may only have a single type, which would lead us to the concluding that $\mathbb{PN}$ and $\mathbb{P}DAY$ are the same type, which is clearly nonsense. What is wrong with this argument? The problem is that we are assigning the empty set $\emptyset$ different types, which is not allowed in our model. How can we get around the problem? As we discussed earlier, we have, in fact, a different empty set for each type: $\emptyset_\mathbb{N}$ is the empty set of type $\mathbb{PN}$, while $\emptyset_{DAY}$ is the empty set of the type $\mathbb{P}DAY$. Using this notation, we can also define the empty set which contains no sets of numbers: $\emptyset_{\mathbb{PN}}$, which is of type $\mathbb{PPN}$. In practice, we usually leave the subscript out, as long as it is clear from the context.

The power set notation is sometimes also overloaded to be applicable to a set.

Definition 4.18 *Given a set S containing objects of type X, the power set of S, written as $\mathbb{P}S$, is defined to be the set of all subsets of S:*

$$\mathbb{P}S \stackrel{\text{df}}{=} \{T : \mathbb{P}X \mid T \subseteq S\} \qquad \blacksquare$$

For example, the power set of $\{1, 2\}$ is $\{\emptyset, \{1\}, \{2\}, \{1, 2\}\}$, consisting of four elements. Note that, if you are given a set S consisting of n distinct elements, its power set $\mathbb{P}S$ would contain 2^n elements.[2] In fact, another notation used for the power set of a set (or type) S is 2^S.

As we did with the other set operators, we can prove a useful theorem of power sets.

Theorem 4.19 *An set T is in the power set of S if and only if it is a subset of S:*

$$\forall T : \mathbb{P}X \cdot T \in \mathbb{P}S \Leftrightarrow T \subseteq S$$

The theorem can be shown to follow directly using the definition of power sets.

Exercises

4.27 Give the type of the following sets:

    (i) $\{1, 2\}$
    (ii) $\mathbb{P}\{1, 2\}$
    (iii) $\{\emptyset, \{1, 2\}\}$
    (iv) $\mathbb{P}\{\emptyset, \{1, 2\}\}$

4.28 List the contents of the following sets:

    (i) $\mathbb{P}\{1, 2\}$
    (ii) $\mathbb{P}\{1\}$
    (iii) $\mathbb{P}\emptyset$
    (iv) $\mathbb{P}\{\emptyset, \{1, 2\}\}$

4.7.1.2 Cartesian Product

A coordinate on a two-dimensional plane can be uniquely identified as a pair of numbers (x, y). What is the type of such a pair? If COORDINATES is the type of such two-dimensional coordinates, we need means of (i) combining a pair of numbers into an object of type COORDINATES, and (ii) breaking up an object of type COORDINATES into a pair of numbers. A general way of achieving this is to construct the type COORDINATES directly as pairs of natural numbers. The type COORDINATES is defined to be $\mathbb{N} \times \mathbb{N}$, pairs consisting of two numbers. This allows us to define sets of coordinates such as $\{x : \mathbb{N} \bullet (x, x^2)\}$.

In general, given two types X and Y, we will write $X \times Y$ to represent the type of pairs with the first item being of type X and the second being of type Y. This is called *the Cartesian product of types X and Y*.

In the case of COORDINATES, the two sets happened to be identical, but this is not necessarily the case. For instance, to represent the set of employees and their age we can use a set containing objects of type PERSON $\times \mathbb{N}$.

---

[2]Each of the n elements of the set may either be placed or not in a subset. This gives two possibilities for each element, and thus $2 \times 2 \times \cdots \times 2$ (one for each element) possibilities in total. Since we have n elements, we get 2^n.

Furthermore, sometimes pairs may not be sufficient. Consider coordinates in three-dimensional space, where we would need to write coordinates of the form (x, y, z). One way of encoding this is to use pairs within pairs, and write it as either $(x, (y, z))$ or $((x, y), z)$. However, this is not entirely satisfactory since we may want to distinguish between a pair with the first item being itself a pair $((x, y), z)$ from a triple (x, y, z). To handle such a situation, we allow taking the Cartesian product of more than two types. For example, the type of dates written as a triple of numbers (day, month, year) can be written as $\mathbb{N} \times \mathbb{N} \times \mathbb{N}$. If we were now to represent the set of employees and their date of birth we would use the type $\mathsf{PERSON} \times (\mathbb{N} \times \mathbb{N} \times \mathbb{N})$—a pair, with the second item being a triple.

As in the case of power sets, we can extend Cartesian product to be applicable to sets in the following way:

Definition 4.20 *Given a set* S *of type* $\mathbb{P}X$, *and set* T *of type* $\mathbb{P}Y$, *the Cartesian product (or simply* product*) of* S *and* T, *written as* $S \times T$, *is defined to be the set of all pairs with the first item being an element of* S, *and the second an element of* T:

$$S \times T \stackrel{\mathrm{df}}{=} \{x : X, \; y : Y \mid x \in S \wedge y \in T \bullet (x, y)\} \qquad \blacksquare$$

As we did with the other set operators, we can prove a useful theorem of the Cartesian product of two sets.

Theorem 4.21 *An object is in the Cartesian product of two sets if and only if it is a pair such that its first item is an element of the first set and its second item is an element of the second set:*

$$\forall x : X, \; y : Y \cdot (x, y) \in S \times T \Leftrightarrow x \in S \wedge y \in T$$

Proof The theorem can be proved as follows:

$$(x, y) \in S \times T$$
$\Leftrightarrow$ {definition of set product}
$$(x, y) \in \{x : X, \; y : Y \mid x \in S \wedge y \in T \bullet (x, y)\}$$
$\Leftrightarrow$ {alpha-conversion to avoid name clashes}
$$(x, y) \in \{a : X, \; b : Y \mid a \in S \wedge b \in T \bullet (a, b)\}$$
$\Leftrightarrow$ {definition of membership in set comprehension}
$$\exists a : X, \; b : Y \cdot a \in S \wedge b \in T \wedge (a, b) = (x, y)$$
$\Leftrightarrow$ {equality of pairs}
$$\exists a : X, \; b : Y \cdot a \in S \wedge b \in T \wedge a = x \wedge b = y$$
$\Leftrightarrow$ {using the one-point rule on a, then b}
$$x \in S \wedge y \in T \qquad \blacksquare$$

Example We can now prove that: $(R \cup S) \times T = (R \times T) \cup (S \times T)$. The proof is very similar to the ones we have already seen about union, intersection and the other set operators:

$$(x, y) \in (R \cup S) \times T$$
$$\Leftrightarrow \{\text{using the theorem of Cartesian products of sets}\}$$
$$x \in R \cup S \wedge y \in T$$
$$\Leftrightarrow \{\text{using the theorem of set union}\}$$
$$(x \in R \vee x \in S) \wedge y \in T$$
$$\Leftrightarrow \{\text{distributivity of disjunction over conjunction}\}$$
$$(x \in R \wedge y \in T) \vee (x \in S \wedge y \in T)$$
$$\Leftrightarrow \{\text{using the theorem of Cartesian products of sets}\}$$
$$(x, y) \in (R \times T) \vee (x, y) \in (S \times T)$$
$$\Leftrightarrow \{\text{using the theorem of set union}\}$$
$$(x, y) \in (R \times T) \cup (S \times T)$$

It is worth noting that we immediately started with a pair, rather than a single variable in the reasoning, since it could be deduced (by looking at the types of the sets) that objects in the set are pairs.   ◇

The use of the word *product* and the symbol $\times$, both usually used for multiplication, is intentionally. The reason is that, if sets S and T are both finite, the number of items in the set $S \times T$ is equal to the product of the number of items in S and the number of items in T.

Exercises

4.29 List the contents of the following sets:

    (i) $\{1, 2\} \times \{A, B\}$
    (ii) $\{1\} \times \{A, B\}$
    (iii) $\{1, 2\} \times \emptyset$

4.30 Prove that $(S^c \times T^c) \cup (S^c \times T) \cup (S \times T^c) = (S \times T)^c$.

4.7.2 On Why Types Are Desirable: Russell's Paradox

We have left this question till the end. Why bother with all the hassle of giving types to every single object and set? Cannot we simply leave types out and let the rest of the mathematics handle things for itself?

In 1903, as Gottlob Frege was about to publish a book on the theory of sets, Bertrand Russell discovered a paradox which allowed one to write contradictory statements in Frege's logic, making it inconsistent. Russell's paradox can be resolved in various ways, one of which is through the use of types.

Let us explore the paradox. Let the set of chapters in this book be C, the set of letters written on this page be L and the set of definitions in this chapter be D. Now, let $\mathcal{T}$ consist of all the sets which contain at least three elements. You can easily check for yourself that $C \in \mathcal{T}$, $L \in \mathcal{T}$ and $D \in \mathcal{T}$. This means that $\mathcal{T}$ is a set which contains at least three objects and thus $\mathcal{T} \in \mathcal{T}$.

On the other hand, clearly the set of letters on this page L is itself not a letter, and thus not a letter on this page: $L \notin L$.

This shows that sets sometimes, though not always, are elements of themselves. Sets not elements of themselves will be called normal, and those which are elements of themselves will be called abnormal. L is normal, while $\mathcal{T}$ is abnormal.

Let us now define the set of normal sets (sets which are not elements of themselves): $\mathcal{N} \overset{\mathrm{df}}{=} \{S \mid S \notin S\}$. From the previous discussion, $L \in \mathcal{N}$ since $L \notin L$, while $\mathcal{T} \notin \mathcal{N}$ since $\mathcal{T} \in \mathcal{T}$.

Now, is $\mathcal{N}$ normal or abnormal? To answer this question, we need to ask whether $\mathcal{N}$ is an element of itself? Using the law of the excluded middle, we know that $\mathcal{N}$ is either abnormal ($\mathcal{N} \in \mathcal{N}$) or abnormal ($\neg(\mathcal{N} \in \mathcal{N})$). Let us consider the two statements in turn:

$\mathcal{N}$ **is abnormal** ($\mathcal{N} \in \mathcal{N}$): Since, by definition, $\mathcal{N}$ is the set of normal sets, all its elements are normal sets. In particular, since $\mathcal{N} \in \mathcal{N}$, it follows that $\mathcal{N}$ is normal. But we started by saying that $\mathcal{N}$ is abnormal, which contradicts this conclusion.
   More formally, the argument goes as follows. Since $\mathcal{N} \in \mathcal{N}$, by the definition of $\mathcal{N}$, we know that $\mathcal{N} \in \{S \mid S \notin S\}$. In other words, $\mathcal{N}$ must satisfy the condition in the set comprehension, implying that $\mathcal{N} \notin \mathcal{N}$, which contradicts the original statement that $\mathcal{N} \in \mathcal{N}$.

$\mathcal{N}$ **is normal** ($\mathcal{N} \notin \mathcal{N}$): Since $\mathcal{N}$ is normal, it must be in the set of normal sets. But the set of normal sets is $\mathcal{N}$ itself and therefore $\mathcal{N} \in \mathcal{N}$, making it abnormal, and contradicting the statement we started off with.
   More formally, Since $\mathcal{N} \notin \mathcal{N}$, then $\mathcal{N} \in \{S \mid S \notin S\}$. But we defined the set comprehension to be $\mathcal{N}$, and therefore $\mathcal{N} \in \mathcal{N}$, contradicting the original statement that $\mathcal{N} \notin \mathcal{N}$.

In other words, we have just defined a set $\mathcal{N}$ which is neither an element of itself nor not an element of itself. The statement $\mathcal{N} \in \mathcal{N}$ is neither true nor false, contradicting the result that $\vdash P \vee \neg P$. What went wrong?

The problem is in the innocuous looking statement that we started off with, discussing whether or not a set S is an element of itself. Recall that, for a statement $x \in S$ to be type correct, we require S to be of type $\mathbb{P}X$, where X is the type of x. Therefore, if S is of type X, to write $S \in S$, we would require the type $\mathbb{P}X$ to be the same as X, which it clearly is not. In typed set theory, this statement is thus not possible to formulate, avoiding the paradox altogether.

Exercises

4.31 Sometimes, Russell's paradox is explained in a less formal way through the story of a male barber in a town in which all males shave themselves if and only if they are not shaved by the barber. Does the barber shave himself? Follow the reasoning in the answer to this question to find the paradox, and relate this version to the formal version given in this section. What corresponds to set $\mathcal{N}$ in this story?

4.32 Another popularisation of the paradox is formulated in terms of the adjectives *autological* and *heterological*. An adjective is said to be autological if it refers to itself. For instance, polysyllabic is an autological adjective since it is itself polysyllabic. On the other hand, heterological adjectives are ones which do not refer to themselves. For instance the adjective long is not a long word, and is thus heterological. Now, is the adjective *heterological* autological or heterological? As before, follow the reasoning to find the paradox, and relate to Russell's paradox.

4.8 Summary

In this chapter we have formalised typed set theory, not by axiomatising it as we did with predicate calculus, but by reducing everything to predicates using definitions. In a certain sense, our axioms are all in the form of definitions. Whereas before it was our responsibility to ensure that the axioms really corresponded to basic truths in what we were modelling, when using definitions, it is still important to ensure that the definitions also correspond to the objects being modelled.

As we shall see in the coming chapters, sets will be used to define further mathematical notions such as functions, sequences and graphs. In the rest of the book, these other notions will be expressed in terms of sets, which allows us to formulate things, in turn, back in predicate logic. The beauty of this approach is that proofs can be reduced simply to unfolding or folding of definitions, and then to the application of the few axioms of propositional and predicate calculus given in the previous chapters.

Chapter 5
Relations

Through the use of sets, we now have a way of describing collections of objects in a formal way. Beyond collections, another frequently encountered notion is that of relating objects to each other. For instance, one may want to model file access control in an operating system, in which one relates a user with the files he or she has access to, or the notion of friendship between persons (whether in a real-life setting, or in a social network one).

Sets give us a way of describing a collection of objects all related to one another. The set of all employees of a particular company are related in that they all have the same employer. However, without additional structure in the sets, this is a very limited way of relating objects to each other. In this chapter we will explore how relations can be expressed in terms of structured sets. In this manner, we still inherit the mathematical results we have proved for sets, but describe more complex mathematical structures.

5.1 An Informal View of Relations

Consider an access-control module in an operating system, which stores information regarding which users can access which files. In this setting, we would like to relate users of a system with the files they have access to.

Concretely, one would implement such a relation as a database storing user-filename pairs to keep track of who has access to what. Conceptually, we can visualise this relation as a collection of lines connecting users with files. A line from a user u to a file f is used to indicate that user u has access to file f. Let us call the relation *hasAccessTo*.

Graphically, this relation can be depicted in the following manner:

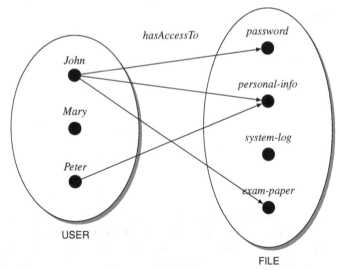

The relation *hasAccessTo* corresponds, not only to the first collection of users, and neither solely to the second collection of files, but rather to the arrows in between the two collections. Note that users may be related to any number of objects, not necessarily just one—for instance, *John* has access to the *password*, *personal-info* and *exam-paper* files, while *Mary* has access to none of the files. The same observation may be made about the objects in the second collection. For instance the file *system-log* is not accessible by anyone, while *personal-info* is accessible by both *John* and *Peter*.

From our discussion on sets, we can also question whether we need ordering and repetition in describing relations. Clearly, the order in which the objects are related in this example is important. It does not make sense to say that the file *password* has access to the user *John*. On the other hand, the related pairs (the arrows in the diagram) are not in any particular order. Similarly, repetition of arrows between objects makes no sense in this case—having 27 arrows from *John* to *personal-info* is not any different from having just 1. In certain circumstances, repetition may be important. For instance, a relation between users and files *accessed* may use an arrow between a user and a file for each time that user has accessed that file. However, when discussing basic relations between objects, we will not be dealing with cases which require repetition—in this chapter we will only be considering relations in which neither order nor repetition of the arrows is important.

5.2 Formalising Relations

We can now define this notion of relations in terms of mathematical structures we have already formalised.

5.2.1 The Type of a Relation

The first question we have to address is how to encode a relation. Let us start with encoding the pairs of related objects (via an arrow in the diagram). Using sets to relate the source and destination of an arrow does not work—for instance *John* has access to the file *password*, but we cannot write {*John, password*} for two reasons. Firstly, the two objects are of different types, and thus cannot be put into the same set. Secondly, we lose the order of the pair: is it *John* who has access to the *password*, or *password* which has access to *John*? The construction we have encountered which enables us to talk about order between two objects, which may be of different types, is the Cartesian product of two types. The arrow between *John* and *password* can be encoded as the pair (*John, password*).

As to the whole relation, recall that order and repetition of arrows is not important for to us. This indicates that using sets of arrows (pairs) can be sufficient to describe the relation.

$$hasAccessTo = \{(John,\ password),\ (John,\ personal\text{-}info),$$
$$(John,\ exam\text{-}paper),\ (Peter,\ personal\text{-}info)\}$$

The type of the relation *hasAccessTo* is thus a set of pairs of users and files:

$$hasAccessTo \subseteq \mathsf{USER} \times \mathsf{FILE}$$

Recall that, if a set S contains objects of type X—in other words S is a subset of a type X, then the type of S is $\mathbb{P}X$.

$$hasAccessTo : \mathbb{P}(\mathsf{USER} \times \mathsf{FILE})$$

Since it is cumbersome to write, and understand each and every time, that a power set of a Cartesian product corresponds to a relation, we will be using a shorthand notation:

$$hasAccessTo : \mathsf{USER} \leftrightarrow \mathsf{FILE}$$

In general, we define the type of relations between types X and Y as follows:

Definition 5.1 *The type of relations between type X and type Y, written $X \leftrightarrow Y$, is defined to be the type of all subsets of the Cartesian product of X and Y:*

$$X \leftrightarrow Y \stackrel{\mathrm{df}}{=} \mathbb{P}(X \times Y) \qquad \blacksquare$$

Using this notation, we can thus define the meaning of a relation between any two arbitrary types X and Y. This does not rule out the possibility that X and Y happen to be the same type. Consider the relation $<$, which relates two numbers if the first is smaller than the second. The type of $<$ is $\mathbb{N} \leftrightarrow \mathbb{N}$, relating objects of the same type. Similarly, the relation *friendOf*, which relates two persons if and only if they are friends on a social networking site, has type $\mathsf{PERSON} \leftrightarrow \mathsf{PERSON}$.

We call relations between objects of the same type, such as $<$ and *friendOf*, *homogeneous*. Relations between different types, such as *hasAccessTo*, are called *heterogeneous*.

5.2.2 Some Basic Relations

The use of sets to encode relations gives us the opportunity to use set comprehensions and set notation to describe relations directly.

Consider the most trivial relation, in which nothing is related to anything. For instance, consider the relation *hasAccessTo* in a scenario where no one is allowed to access any file. This would simply correspond to the empty set.

Another simple relation is the one which relates items in a particular set only with themselves. This can be defined using a set comprehension:

Definition 5.2 *The identity relation for a set S (containing items of type X), written as id_S, is a homogeneous relation from X to X, and relates all items in S with themselves*:

$$id_S : X \leftrightarrow X$$

$$id_S \stackrel{\mathrm{df}}{=} \{x : X \mid x \in S \bullet (x, x)\} \qquad \blacksquare$$

As when we were defining sets, we usually would like to avoid reasoning with set comprehensions directly, and propositions making reasoning easier about these newly defined objects will be proved.

Proposition 5.3 *Two objects are related to each other by id_S if and only if they are equal to each other, and elements of set S.*

$$\forall x, y : X \cdot (x, y) \in id_S \Leftrightarrow x = y \wedge x \in S$$

Proof The proof follows in a rather straightforward manner from the definitions:

$$(x, y) \in id_S$$
$$\Leftrightarrow \{\text{definition of } id_S\}$$
$$(x, y) \in \{z : X \mid z \in S \bullet (z, z)\}$$
$$\Leftrightarrow \{\text{membership of set comprehension}\}$$
$$\exists z : X \cdot z \in S \wedge (z, z) = (x, y)$$
$$\Leftrightarrow \{\text{definition of equality of pairs}\}$$
$$\exists z : X \cdot z \in S \wedge z = x \wedge z = y$$
$$\Leftrightarrow \{\text{using the one-point rule on } z = x\}$$
$$x \in S \wedge x = y$$
$$\Leftrightarrow \{\text{commutativity of conjunction}\}$$
$$x = y \wedge x \in S \qquad \blacksquare$$

5.2.3 Relational Equality

Whenever we model a new type of object, one question to ask is what equality means. In the case of sets, we defined equality in terms of subsets. What about relations? When is a relation $r : X \leftrightarrow Y$ equal to a relation $s : X \leftrightarrow Y$?

The most natural way of expressing such equality between relations is to say that any pair of objects related by r should also be related by s, and vice versa:

$$\forall x : X, \; y : Y \cdot (x, y) \in r \Leftrightarrow (x, y) \in s$$

This is, however, the same as checking for set equality between r and s.

Definition 5.4 *Relations r and s, both of type $X \leftrightarrow Y$, are said to be equal (written $r \stackrel{rel}{=} s$), if and only if they are the same set:*

$$r \stackrel{rel}{=} s \stackrel{df}{=} r \stackrel{set}{=} s$$

As in the case of sets, if the relations are not of the same type, then we say that they are incomparable. Also, when it is clear from the context that we are referring to relational equality, we will write $r = s$ instead of $r \stackrel{rel}{=} s$. ■

5.2.4 Combining Relations as Sets

Another advantage of using sets to represent relations is that we can automatically use the set operators union, intersection, complementation and set difference to combine relations of the same type.

Example Relations *canRead* and *canWrite*, both of type USER $\leftrightarrow$ FILE, give information as to which users can, respectively, read from and write to files. Using these relations, we can express a number of new relations in terms of these two, and the set operators:

- A user is related to a file by relation *canAccess* if that user can either read or write to the file. This can be defined as:

$$canAccess \stackrel{df}{=} canRead \cup canWrite$$

- The relation *canFullyAccess* relates users with files to which they can both read and write.

$$canFullyAccess \stackrel{df}{=} canRead \cap canWrite$$

- The relation which gives the information as to which user may not read a particular file can be defined as follows:

$$cannotRead \stackrel{df}{=} canRead^c$$

- The relation *canReadButNotWrite* relates a user to file if and only if the user may read the file, but not write to it:

$$canReadButNotWrite \overset{\text{df}}{=} canRead \setminus canWrite$$

- The relation between users and files which they cannot in any way access can be defined as:

$$neitherReadNorWrite \overset{\text{df}}{=} (canRead \cup canWrite)^c \qquad \diamond$$

Using sets to represent relations not only enables us to *use* the set operators, but also inherit the laws of the set operators, such as commutativity of union and intersection, and that complementing a relation twice gives us back the original relation.

The empty and identity relations can be proved to satisfy a number of useful properties.

Example Consider the identity relation over the union of two sets. This should intuitively be the union of the identity relation over the first and second set, respectively: $id_{S \cup T} = id_S \cup id_T$.

In this proof we have to show relational equality, which is just the same as set equality—however, since we know that the sets contain pairs of objects, we consider the object to be a pair.

$$(x, y) \in id_{S \cup T}$$
$\Leftrightarrow$ {proposition about the identity relation}
$$x = y \wedge x \in S \cup T$$
$\Leftrightarrow$ {theorem of set union}
$$x = y \wedge (x \in S \vee x \in T)$$
$\Leftrightarrow$ {distributivity of conjunction over disjunction}
$$(x = y \wedge x \in S) \vee (x = y \wedge x \in T)$$
$\Leftrightarrow$ {proposition about the identity relation twice}
$$(x, y) \in id_S \vee (x, y) \in id_T$$
$\Leftrightarrow$ {theorem of set union}
$$(x, y) \in id_S \cup id_T \qquad \diamond$$

Example As a second example, we show that the identity over the complement of a set is contained in the complement of the identity over the set: $id_{S^c} \subseteq (id_S)^c$.

The proof is similar to the previous one, except that we only require to show implication in one direction. As before, we take pairs in the first relation, to show that they have to be in the second.

$$(x, y) \in id_{S^c}$$
$\Leftrightarrow$ {proposition about the identity relation}
$$x = y \land x \in S^c$$
$\Leftrightarrow$ {theorem of set complement}
$$x = y \land \neg(x \in S)$$
$\Rightarrow$ {elimination of conjunction}
$$\neg(x \in S)$$
$\Rightarrow$ {introduction of disjunction}
$$\neg(x = y) \lor \neg(x \in S)$$
$\Leftrightarrow$ {De Morgan's law}
$$\neg(x = y \land x \in S)$$
$\Leftrightarrow$ {proposition about the identity relation}
$$\neg((x, y) \in id_S)$$
$\Leftrightarrow$ {theorem of set complement}
$$(x, y) \in (id_S)^c \qquad\qquad \diamond$$

Exercises

5.1 Given the relation *supervises*, which relates persons with those they supervise, and the relation *sharesOffice* which relates two persons if they work in the same office, use set operators to:

(a) Define the relation which relates a person with their supervisors who work in the same room as they do.
(b) Define the relation which relates two persons if they do not share a room.
(c) Hence or otherwise define the relation with relates a person with their supervisors who work in a different office.

Informally explain what the following predicates mean:

(a) $id_{\text{PERSON}} \subseteq sharesOffice$
(b) $supervises \setminus sharesOffice = \emptyset$
(c) $supervises \subseteq (id_{\text{PERSON}})^c$

5.2 Given the relations on numbers $<$ and $>$, such that $n < m$ if n is less than m, and similarly, $n > m$ if n is greater than m,[1] define the following relations between numbers using only these two relations and set operators:

(a) The relation $\leq$ over numbers, such that $n \leq m$ if n is not more than m.
(b) Similarly, the relation $\geq$, such that $n \geq m$ if n is not less than m.
(c) The equality relation eq over numbers.

5.3 Prove that $id_{S \setminus T} = id_S \setminus id_T$.

5.4 Prove that, if $S \subseteq T$, then $id_S \subseteq id_T$.

5.5 Prove that $id_\emptyset = \emptyset$.

---

[1] Note that we are writing $n < m$ as shorthand for $(n, m) \in <$. Similarly for $>$.

5.3 Domain and Range

As we have seen, we can define relations and combine them using set operators. We will now move on to define relation-specific operators. To start off, we will define two operators to obtain information about a relation $r : X \leftrightarrow Y$—the domain of r, corresponding to the objects in X related to at least one object in Y via relation r, and the range of r, conversely corresponding to objects in Y to which some object in X is related through relation r. For example, the domain of relation $canWrite :$ USER $\leftrightarrow$ FILE is the set of users who can write to at least one file. Similarly, the range of the relation is the set of files to which at least one user may write. This will allow us to write certain properties in a concise and precise way. For instance, the statement that 'all files may be written to by some users' may be expressed as saying that the range of $canWrite$ is the whole type FILE.

Definition 5.5 *The domain of a relation* $r : X \leftrightarrow Y$ *(written $dom(r)$) is the set of all items in X related to some object in Y via relation r. Similarly, the range of r (written $ran(r)$) is the set of all items in Y to which some object in X is related via r:*

$$dom(r) \stackrel{\mathrm{df}}{=} \{x : X \mid \exists y : Y \cdot (x, y) \in r\}$$

$$ran(r) \stackrel{\mathrm{df}}{=} \{y : Y \mid \exists x : X \cdot (x, y) \in r\} \qquad \blacksquare$$

Example Recall the relations $canRead$ and $canWrite$, both of type USER $\leftrightarrow$ FILE, relating users with the files they can read from and write to, respectively. We can now express statements about these relations in a mathematical way:

(i) All files can be written to by at least one user: $ran(canWrite) =$ FILE.
(ii) All users may read at least one file: $dom(canRead) =$ USER.
(iii) Every file which may be written to by at least one person must also be readable by at least one person: $ran(canWrite) \subseteq ran(canRead)$. $\diamond$

On the basis of the definition we have just given, we will prove a proposition about domain and range, which enables us to prove things without resorting back to set comprehensions:

Proposition 5.6 *An object x is in the domain of a relation r if and only if it is related to an object in Y via r, and similarly, y is in the range of r if and only if an object in X is related to it via relation r:*

$$x \in dom(r) \leftrightarrow \exists y : Y \cdot (x, y) \in r$$

$$y \in ran(r) \leftrightarrow \exists x : X \cdot (x, y) \in r$$

Proof The proofs of these statements follow directly from the definitions and set comprehension. Consider the case of the domain of a relation:

$$x \in dom(r)$$
$\Leftrightarrow$ {definition of domain}
$$x \in \{z : X \mid \exists y : Y \cdot (z, y) \in r\}$$
$\Leftrightarrow$ {definition of set comprehension membership}
$$\exists z : X \cdot \exists y : Y \cdot (z, y) \in r \wedge z = x\}$$
$\Leftrightarrow$ {using the one-point rule}
$$\exists y : Y \cdot (x, y) \in r$$

The proof for the range is similar and is left as an exercise. ∎

Using this proposition, further properties of domain and range can be proved:

Example The range of the identity relation id_S is set S: $ran(id_S) = S$. The proof requires us to show equality of two sets, so the proof proceeds similarly to the ones we saw in the previous chapter:

$$z \in ran(id_S)$$
$\Leftrightarrow$ {proposition about the range of a relation}
$$\exists x : X \cdot (x, z) \in id_S$$
$\Leftrightarrow$ {proposition about the identity relation}
$$\exists x : X \cdot z = x \wedge z \in S$$
$\Leftrightarrow$ {using the one-point rule}
$$z \in S$$ ◇

Exercises

5.6 Prove the second part of Proposition 5.6.
5.7 Prove that $dom(id_S) = S$.
5.8 Prove that $r \cap id_S = id_{dom(r) \cap S}$.

5.4 Building New Relations from Old Ones

The only ways to construct relations we have till now are set comprehensions and the set operators. In this section, we will look at other operators to construct new relations based on ones we have already defined.

5.4.1 Composition of Relations

Let us look at an operating system which has groups of users set up for security purposes. A group, such as *student* or *staff*, would be related to each user who belongs to that group—*groupMember* : GROUP ↔ USER. This relation is shown in the figure below, alongside the relation *hasAccessTo* : USER ↔ FILE relating users of a system with the files they have access to.

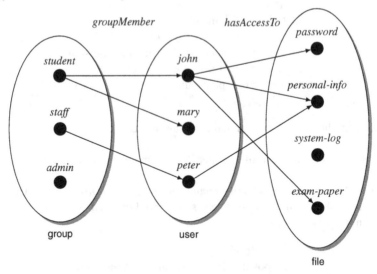

Looking at the relations above, we can deduce that some students have access to the exam paper, which is clearly not desirable. In this case, it is John who is the culprit, being both a student and having access to the exam paper. We are effectively combining the two relations to obtain the following new one:

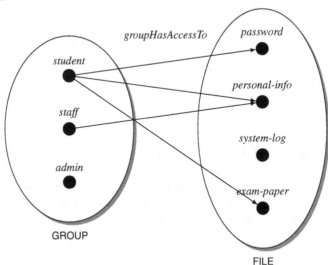

In general, the observation that some student has access to the exam paper can be formulated mathematically as follows:

$$\exists u : \text{USER} \cdot (\textit{student}, u) \in \textit{groupMember} \land (u, \textit{exam-paper}) \in \textit{hasAccessTo}$$

Rather than having to quantify explicitly over the objects lying in between, we would like compose the two relations to obtain a new one, leaving out the intermediate type altogether. How can we define relation *groupHasAccessTo*? We observe that a group g is related to a file f if there is at least one user u such that $(g, u) \in groupMember$ and $(u, f) \in hasAccessTo$. The composition of the two relations can be written using set comprehension:

$$groupHasAccessTo \stackrel{df}{=}$$
$$\{(g, f) : \mathsf{GROUP} \times \mathsf{FILE} \mid$$
$$\exists u : \mathsf{USER} \cdot (g, u) \in groupMember \land (u, f) \in hasAccessTo\}$$

We can generalise this pattern and define the composition of two relations as follows:

Definition 5.7 *Given relations* $r : X \leftrightarrow Y$ *and* $s : Y \leftrightarrow Z$, *the relational composition of* r *and* s, *written as* $r \mathbin{\raise0.2ex\hbox{$\circ$}\kern-0.4em\raise-0.2ex\hbox{$\circ$}} s$, *is a relation from* X *to* Z *and defined as*:

$$r \mathbin{\raise0.2ex\hbox{$\circ$}\kern-0.4em\raise-0.2ex\hbox{$\circ$}} s : X \leftrightarrow Z$$

$$r \mathbin{\raise0.2ex\hbox{$\circ$}\kern-0.4em\raise-0.2ex\hbox{$\circ$}} s \stackrel{df}{=} \{(x, z) : X \times Z \mid \exists y : Y \cdot (x, y) \in r \land (y, z) \in s\} \qquad \blacksquare$$

To avoid having to reason in terms of set comprehensions we prove the following proposition:

Proposition 5.8 *Given two relations* $r : X \leftrightarrow Y$ *and* $s : Y \leftrightarrow Z$, *the pair* (x, z) *is in the relational composition of* r *and* s *if and only if, for some element* y *of* Y, r *relates* x *with* y *and* s *relates* y *with* z:

$$(x, z) \in r \mathbin{\raise0.2ex\hbox{$\circ$}\kern-0.4em\raise-0.2ex\hbox{$\circ$}} s \Leftrightarrow \exists y : Y \cdot (x, y) \in r \land (y, z) \in s$$

Proof The proof follows from the definitions of relational composition and set comprehensions.

$$(x, z) \in r \mathbin{\raise0.2ex\hbox{$\circ$}\kern-0.4em\raise-0.2ex\hbox{$\circ$}} s$$
$\Leftrightarrow$ {definition of relational composition and variable renaming}
$$(x, z) \in \{(x', z') : X \times Z \mid \exists y : Y \cdot (x', y) \in r \land (y, z') \in s\}$$
$\Leftrightarrow$ {definition of membership of a set comprehension}
$$\exists (x', z') : X \times Z \cdot \exists y : Y \cdot$$
$$(x', y) \in r \land (y, z') \in s \land (x, z) = (x', z')$$
$\Leftrightarrow$ {Cartesian product equality}
$$\exists x' : X \cdot \exists z' : Z \cdot \exists y : Y \cdot$$
$$(x', y) \in r \land (y, z') \in s \land x = x' \land z = z'$$
$\Leftrightarrow$ {using the one-point rule twice, on x' and z'}
$$\exists y : Y \cdot (x, y) \in r \land (y, z) \in s \qquad \blacksquare$$

Using this proposition, various laws of relational composition can be proved.

Theorem 5.9 *Relational composition is associative. Given three relations $r : W \leftrightarrow X$, $s : X \leftrightarrow Y$ and $t : Y \leftrightarrow Z$, then $r \mathbin{\stackrel{\circ}{,}} (s \mathbin{\stackrel{\circ}{,}} t) = (r \mathbin{\stackrel{\circ}{,}} s) \mathbin{\stackrel{\circ}{,}} t$.*

Proof Proposition 5.8 allows us to prove this property in a direct manner:

$$(w, z) \in r \mathbin{\stackrel{\circ}{,}} (s \mathbin{\stackrel{\circ}{,}} t)$$
$\Leftrightarrow$ {proposition about composition}
$$\exists x : X \cdot (w, x) \in r \land (x, z) \in s \mathbin{\stackrel{\circ}{,}} t$$
$\Leftrightarrow$ {proposition about composition}
$$\exists x : X \cdot (w, x) \in r \land \exists y : Y \cdot (x, y) \in s \land (y, z) \in t$$
$\Leftrightarrow$ {y is not free in the first conjunct}
$$\exists x : X \cdot \exists y : Y \cdot (w, x) \in r \land (x, y) \in s \land (y, z) \in t$$
$\Leftrightarrow$ {existential quantifiers commute}
$$\exists y : Y \cdot \exists x : X \cdot (w, x) \in r \land (x, y) \in s \land (y, z) \in t$$
$\Leftrightarrow$ {x is not free in the last conjunct}
$$\exists y : Y \cdot (\exists x : X \cdot (w, x) \in r \land (x, y) \in s) \land (y, z) \in t$$
$\Leftrightarrow$ {proposition about composition}
$$\exists y : Y \cdot (w, y) \in r \mathbin{\stackrel{\circ}{,}} s \land (y, z) \in t$$
$\Leftrightarrow$ {proposition about composition}
$$(w, z) \in (r \mathbin{\stackrel{\circ}{,}} s) \mathbin{\stackrel{\circ}{,}} t \qquad \blacksquare$$

In mathematics, whenever we introduce a new operator, we try to identify the object which acts as an identity for that operator. For instance, zero is the identity of addition, since $0 + x = x + 0 = x$. Similarly, 1 is the identity of multiplication, since $1 \times x = x \times 1 = x$. Now consider the division operator ($\div$). Can we find a number α such that $x \div \alpha = x$ for any value of x? Yes, 1 is once again the identity element. However, now try to identify a fixed α such that $\alpha \div x = x$. No matter how hard you try, no single value fits the bill. In this case, we say that 1 is the right identity of division, but division has no left identity. In the cases of addition and multiplication, the operators had identical left and right identities. Does relational composition have left and right identity elements? For a particular relational type, we can identify left and right identity elements which are different:

Proposition 5.10 *Given a relation $r : X \leftrightarrow Y$, $id_X \mathbin{\stackrel{\circ}{,}} r = r \mathbin{\stackrel{\circ}{,}} id_Y = r$.*

Proof We will prove that $id_X \mathbin{\stackrel{\circ}{,}} r = r$:

$$(x, z) \in id_X \mathbin{\stackrel{\circ}{,}} r$$
$\Leftrightarrow$ {proposition about composition}
$$\exists y : Y \cdot (x, y) \in id_X \land (y, z) \in r$$
$\Leftrightarrow$ {proposition about the identity relation}
$$\exists y : Y \cdot x = y \land x \in X \land (y, z) \in r$$
$\Leftrightarrow$ {$x \in X$ follows from the type}
$$\exists y : Y \cdot x = y \land (y, z) \in r$$
$\Leftrightarrow$ {one-point rule}
$$(x, z) \in r$$

The second part of the proof, that $r \mathbin{\mathring{,}} id_Y = r$, follows similarly and is left as an exercise.   ∎

Exercises

5.9   Prove the second part of Proposition 5.10: $r \mathbin{\mathring{,}} id_Y = r$.

5.10 Prove that, for relations $r : X \leftrightarrow Y$ and $s : Y \leftrightarrow Z$, the following statement holds:

   (i) $dom(r \mathbin{\mathring{,}} s) \subseteq dom(r)$

   (ii) $ran(r \mathbin{\mathring{,}} s) \subseteq ran(s)$

5.11 Prove that the set operators distribute over relational composition:

   (i) $(r \cup s) \mathbin{\mathring{,}} t = (r \mathbin{\mathring{,}} t) \cup (s \mathbin{\mathring{,}} t)$

   (ii) $(r \cap s) \mathbin{\mathring{,}} t = (r \mathbin{\mathring{,}} t) \cap (s \mathbin{\mathring{,}} t)$

5.12 The left and right identities of a relation $r : X \leftrightarrow Y$ are not unique, and can be refined further than as given in Proposition 5.10, by using the domain and range of the relation. Prove that:

$$id_{dom(r)} \mathbin{\mathring{,}} r = r \mathbin{\mathring{,}} id_{ran(r)} = r$$

5.4.2 Relational Inverse

The intuition behind a relation is that it gives information as to which pairs of objects are related to each other. Consider the relation *isParentOf* which relates person p with another c whenever p is the parent of c. So, if $(John, Jane) \in isParentOf$, it means that *John* is a parent of *Jane*. From our experience of how these things work, *Jane* is certainly not a parent of *John* in this case—so the order of the pairs in a relation is important. In relations where the types are different the order is even more important, since the type of the pair would not even match.

Since order is important, inverting the order of all the pairs gives us a new relation. For instance, inverting the pairs in the relation *isParentOf*, we obtain a new relation which contains a pair of persons (c, p) if and only if c is a child of p. We write the inverted relation as $isParentOf^{-1}$.

Note that the inverse of a relation is not the same as the complement of the relation. Consider the relation $isParentOf^c$. If *John* is related to *Jane* by $isParentOf^c$ it just means that *John* is not the parent of *Jane*. Contrast this with the case when *John* is related to *Jane* via $isParentOf^{-1}$, which would mean that *Jane* is the parent of *John*. This difference becomes even more pronounced with heterogeneous relations. Given a relation $r : X \leftrightarrow Y$, its complement r^c would still be of type $X \leftrightarrow Y$, while its inverse r^{-1} would be of type $Y \leftrightarrow X$.

Example Recall the relation *hasAccessTo* : USER $\leftrightarrow$ FILE which gives the information of which users have access to which files, and *groupMember* : GROUP $\leftrightarrow$ USER which identifies the members of the user groups.

To which groups does a user belong: The relation *memberOf*, which relates a user to the groups he or she is a member of, is exactly the inverse of the relation *groupMember*, and can thus be defined as:

$$memberOf \stackrel{\mathrm{df}}{=} groupMember^{-1}$$

Members of which groups may access a file: If we need to define a relation *mayBeAccessedByGroup* : FILE $\leftrightarrow$ GROUP, this corresponds to the inverse of the relation *groupHasAccessTo* : GROUP $\leftrightarrow$ FILE which was defined earlier as *groupMember* $\mathring{,}$ *hasAccessTo*:

$$mayBeAccessedByGroup \stackrel{\mathrm{df}}{=} (groupMember \mathring{,} hasAccessTo)^{-1}$$

Relating users in the same group: We now require a relation *shareGroup* which relates two users if and only if there is at least one group to which both belong. A first attempt to write this could be using predicate logic, to say that two users u_1 and u_2 are related if

$$\exists g : \text{GROUP} \cdot (g, u_1) \in groupMember \wedge (g, u_2) \in groupMember$$

However, with some manipulation, we may realise that this can be rewritten as

$$\exists g : \text{GROUP} \cdot (u_1, g) \in memberOf \wedge (g, u_2) \in groupMember$$

This matches the definition of relational composition, allowing us to write the relation as *memberOf* $\mathring{,}$ *groupMember*, or (to show the relational inverse):

$$shareGroup \stackrel{\mathrm{df}}{=} groupMember^{-1} \mathring{,} groupMember$$

Note that the composition of the inverses of two relations is not the same as the inverse of their composition. $\diamond$

We can now proceed to formally define relational inverse, to enable us to prove laws of the operator, and its use.

Definition 5.11 *Given a relation* $r : X \leftrightarrow Y$, *the relational inverse of* r, *written as* r^{-1}, *is a relation from* Y *to* X *and defined as*:

$$r^{-1} : Y \leftrightarrow X$$
$$r^{-1} \stackrel{\mathrm{df}}{=} \{(y, x) : Y \times X \mid (x, y) \in r\} \qquad \blacksquare$$

As before, we will prove a property of the operator to enable simpler proofs.

Proposition 5.12 *Given a relation* $r : X \leftrightarrow Y$, *the pair* (y, x) *is in* r^{-1} *if and only if* (x, y) *is in* r:

$$(y, x) \in r^{-1} \Leftrightarrow (x, y) \in r$$

Proof The proof follows from the definitions of relational inverse and set comprehensions.

$$(y, x) \in r^{-1}$$
$\Leftrightarrow$ {definition of relational inverse and renaming variables}
$$(y, x) \in \{(y', x') : Y \times X \mid (x', y') \in r\}$$
$\Leftrightarrow$ {definition of membership of a set comprehension}
$$\exists (y', x') : Y \times X \cdot (x', y') \in r \wedge (y, x) = (y', x')$$
$\Leftrightarrow$ {Cartesian product equality}
$$\exists y' : Y \cdot \exists x' : X \cdot (x', y') \in r \wedge y = y' \wedge x = x'$$
$\Leftrightarrow$ {using the one-point rule twice, on x' and y'}
$$(x, y) \in r \qquad \blacksquare$$

We can now proceed to prove laws of relational inverse. The first says that taking the inverse of the inverse of a relation gives, as would be expected, the original relation:

Theorem 5.13 *Since relational inverse inverts the direction of the arrows, applying it twice should result in the original relation: Given a relation $r : X \leftrightarrow Y$, $(r^{-1})^{-1} = r$.*

Proof The proof follows simply by applying the proposition about relational inverse twice:

$$(x, y) \in (r^{-1})^{-1}$$
$\Leftrightarrow$ {by proposition about relational inverse}
$$(y, x) \in r^{-1}$$
$\Leftrightarrow$ {by proposition about relational inverse}
$$(x, y) \in r \qquad \blacksquare$$

Relational inverse does not distribute into relational composition directly. In fact, given relations $r : X \leftrightarrow Y$ and $s : Y \leftrightarrow Z$, the statement $(r \,{}^{\circ}_{9}\, s)^{-1} = r^{-1} \,{}^{\circ}_{9}\, s^{-1}$ does not even type check, since r^{-1} is of type $Y \leftrightarrow X$ while s^{-1} is of type $Z \leftrightarrow Y$, making the relational composition inapplicable. However, if we switch the order of r^{-1} and s^{-1}, the equivalence holds:

Theorem 5.14 *The relational inverse of the composition of two relations is the same as the composition of the inverted relations in the opposite order: Given relations $r : X \leftrightarrow Y$ and $s : Y \leftrightarrow Z$, $(r \,{}^{\circ}_{9}\, s)^{-1} = s^{-1} \,{}^{\circ}_{9}\, r^{-1}$.*

Proof The proof uses the propositions about relational composition and inverse:

$$(x, z) \in (r \,{}^{\circ}_{\circ}\, s)^{-1}$$
$\Leftrightarrow$ {by proposition about relational inverse}
$$(z, x) \in r \,{}^{\circ}_{\circ}\, s$$
$\Leftrightarrow$ {by proposition about relational composition}
$$\exists y : Y \cdot (z, y) \in r \land (y, x) \in s$$
$\Leftrightarrow$ {by proposition about relational inverse twice}
$$\exists y : Y \cdot (y, z) \in r^{-1} \land (x, y) \in s^{-1}$$
$\Leftrightarrow$ {commutativity of conjunction}
$$\exists y : Y \cdot (x, y) \in s^{-1} \land (y, z) \in r^{-1}$$
$\Leftrightarrow$ {by proposition about relational composition}
$$(x, z) \in s^{-1} \,{}^{\circ}_{\circ}\, r^{-1}$$ ∎

Example For a more complex example, we may consider what happens when we follow a relation $r : X \leftrightarrow Y$, and then its inverse to obtain: $r \,{}^{\circ}_{\circ}\, r^{-1}$. Any outgoing arrow from the domain of r can be followed back in r^{-1} giving back the original object—ensuring that $r \,{}^{\circ}_{\circ}\, r^{-1}$ includes all the identity arrows over the domain of r: $id_{dom(r)} \subseteq r \,{}^{\circ}_{\circ}\, r^{-1}$. Let us prove it:

$$(x, y) \in id_{dom(r)}$$
$\Leftrightarrow$ {proposition about the identity relation}
$$x = y \land x \in dom(r)$$
$\Leftrightarrow$ {proposition about domain}
$$x = y \land \exists z : Y \cdot (x, z) \in r$$
$\Rightarrow$ {since $x = y$}
$$\exists z : Y \cdot (x, z) \in r \land (y, z) \in r$$
$\Leftrightarrow$ {proposition about relational inverse}
$$\exists z : Y \cdot (x, z) \in r \land (z, y) \in r^{-1}$$
$\Leftrightarrow$ {proposition about relational composition}
$$(x, y) \in r \,{}^{\circ}_{\circ}\, r^{-1}$$ ◇

Exercises

5.13   Prove that $dom(r) = ran(r^{-1})$. Hence prove that $dom(r^{-1}) = ran(r)$.

5.14   Consider the relation *isParentOf* : PERSON $\leftrightarrow$ PERSON, which relates two persons whenever the first is the parent of the second. Explain informally the meaning of the following mathematical statements:

(i)   *isParentOf*$^{-1}$
(ii) *isParentOf* $\,{}^{\circ}_{\circ}\,$ *isParentOf*
(iii) *isParentOf*$^{-1}$ $\,{}^{\circ}_{\circ}\,$ *isParentOf*$^{-1}$
(iv) *isParentOf* $\,{}^{\circ}_{\circ}\,$ *isParentOf*$^{-1}$
(v)   *isParentOf*$^{-1}$ $\,{}^{\circ}_{\circ}\,$ *isParentOf*

5.15   Consider the relations *canRead* and *canWrite*, both of type USER $\leftrightarrow$ FILE, relating users to files they can read from and write to, respectively. Let us assume that information can flow from a file to another if and only if there is some user who may read the first file and also write to the second. Express the relation *flow* : FILE $\leftrightarrow$ FILE, which relates two files if and only if information can flow from the first to the second. Try to express it using only the relational operators and the two given relations.

5.16 Consider the situation where we are just given the relation $<$ over natural numbers, relating a number with another if the first is strictly smaller than the second.

    (i) Define the relation $>$, which relates two numbers if the first is bigger than the second using relational inverse.

   (ii) Define the relation $\neq_{\mathbb{N}}$ over natural numbers, which relates two numbers if they are not equal to each other. Hint: Two numbers are not equal to each other if either the first is bigger than the second, or it is smaller than the second.

  (iii) Hence define the relation $=_{\mathbb{N}}$, natural number equality, using the relation $\neq_{\mathbb{N}}$.

5.17 Prove that relational inverse distributes over union:

$$(r \cup s)^{-1} = r^{-1} \cup s^{-1}$$

Prove similar laws saying that relational inverse also distributes over intersection and set difference.

5.18 Prove that $id_S^{-1} = id_S$.

5.19 Prove that $id_{ran(r)} \subseteq r^{-1} \mathbin{\overset{\circ}{,}} r$, where $r : X \leftrightarrow Y$.

5.4.3 Repeated Relations

Consider the relation *linked* : WEBPAGE $\leftrightarrow$ WEBPAGE, which relates two webpages if and only if the first page contains a direct link to the second—making it reachable in one step. Sometimes, we would like to know which webpages may be reached in two steps—the first page has a link to a page which has a link to the second page: *linked$_2$* : WEBPAGE $\leftrightarrow$ WEBPAGE. One initial attempt to write this may be by formally writing that p_1 is linked to p_2 in two steps if there is some webpage m to which p_1 is linked directly, and which is itself linked to p_2:

$$(p_1, p_2) \in linked_2 \leftrightarrow \exists m : \text{WEBPAGE} \cdot (p_1, m) \in linked \wedge (m, p_2) \in linked$$

The pattern used matches that of relational composition, indicating that we can define *linked$_2$* as follows:

$$linked_2 \overset{\mathrm{df}}{=} linked \mathbin{\overset{\circ}{,}} linked$$

This also indicates a pattern we can use to define links via three, four or any number of pages:

$$linked_n \stackrel{\text{df}}{=} \underbrace{linked \mathbin{\substack{\circ\\\circ}} linked \mathbin{\substack{\circ\\\circ}} \cdots \mathbin{\substack{\circ\\\circ}} linked}_{n \text{ copies}}$$

This process can be generalised to any homogeneous relation $r : X \leftrightarrow X$, and we will write r^n to denote the composition of n copies of r. Defining the term using ellipsis is not sufficiently formal to enable us to prove things formally, so we will be using a recursive definition—r repeated n times, is the same as repeating r for $(n-1)$ times, and adding an additional r composed to the relation:

$$r^n \stackrel{\text{df}}{=} r \mathbin{\substack{\circ\\\circ}} r^{n-1}$$

The problem is when to stop this recursive process. We can opt to define r^1 to be equal to r. Alternatively, we can choose to define r^0, so as to give a more complete definition. What is a reasonable definition for r^0? We would like r^1 to be equal to r, but if we choose 0 to be the base case of the recursive definition, r^1 is defined in terms of r^0:

$$r^1 = r \mathbin{\substack{\circ\\\circ}} r^0$$

What choice of r^0 guarantees that $r \mathbin{\substack{\circ\\\circ}} r^0$ is the same as r? In Proposition 5.10, we proved that id_X acts as the identity of relational composition. This thus proves to be a good choice for r^0.

Definition 5.15 *Given a homogeneous relation $r : X \leftrightarrow X$ and a natural number n, the repetition or iteration of r for n times, written r^n, is defined as follows:*

$$r^n : X \leftrightarrow X$$
$$r^0 \stackrel{\text{df}}{=} id_X$$
$$r^{n+1} \stackrel{\text{df}}{=} r \mathbin{\substack{\circ\\\circ}} r^n \qquad \blacksquare$$

Note that the definition of r^n is given in two separate lines. Clearly, to ensure that the definition is consistent, in other words that different lines do not give different results, we must ensure that they are mutually exclusive. In this case, it is clear that the first line only gives us a solution to r^n when $n = 0$, while the second line defines r^{n+1} for any natural number n, thus effectively defining it for n being at least 1.

The definition is also recursive—in that it depends on itself. Not all such definitions are meaningful. If we were to (wrongly) define r^{n+1} as $r \mathbin{\substack{\circ\\\circ}} r^{n+1}$, we would not have been giving any definition, for example, to r^7. As we will see later on in the book, a recursive definition is said to be well founded if all the instances of defined operator on the right take a parameter which becomes progressively smaller than the one appearing on the left-hand side of the definition. Since a natural number cannot become smaller infinitely often, termination is guaranteed. Such definitions are called well-founded.

Example It is important to use the definition only in the way it is stated. For instance, it may seem obvious that, for any relation $r : X \leftrightarrow X$, it is true that $r^{n+1} = r^n \mathbin{\mathring{\,}} r$. But obvious statements are not always true, and thus need to be proved. Since we have a statement which should be true for all natural numbers n, a proof technique we may want to use is numeric induction. We will be going into more detail about numeric induction in a later chapter, but for the moment we will be simply using it as a tool to prove this statement.

To prove that $r^{n+1} = r^n \mathbin{\mathring{\,}} r$, we will use numeric induction on n.

Base case: The base case is when $n = 0$. We thus need to prove that $r^{0+1} = r^0 \mathbin{\mathring{\,}} r$:

$$
\begin{aligned}
&r^{0+1} \\
=\ & \{\text{using the definition of } r^{n+1}\} \\
&r \mathbin{\mathring{\,}} r^0 \\
=\ & \{\text{using the definition of } r^0\} \\
&r \mathbin{\mathring{\,}} id_X \\
=\ & \{id_X \text{ is the right identity of relational composition }\} \\
&r \\
=\ & \{id_X \text{ is the left identity of relational composition}\} \\
&id_X \mathbin{\mathring{\,}} r \\
=\ & \{\text{using the definition of } r^0\} \\
&r^0 \mathbin{\mathring{\,}} r
\end{aligned}
$$

Inductive hypothesis: We now assume that the statement holds for a particular value k: $r^{k+1} = r^k \mathbin{\mathring{\,}} r$.

Inductive case: We finally try to prove the statement for $n = k + 1$. We require to prove that $r^{(k+1)+1} = r^{k+1} \mathbin{\mathring{\,}} r$.

$$
\begin{aligned}
&r^{(k+1)+1} \\
=\ & \{\text{using the definition of } r^{n+1}\} \\
&r \mathbin{\mathring{\,}} r^{k+1} \\
=\ & \{\text{by the inductive hypothesis}\} \\
&r \mathbin{\mathring{\,}} (r^k \mathbin{\mathring{\,}} r) \\
=\ & \{\text{by Theorem 5.9 relational composition is associative}\} \\
&(r \mathbin{\mathring{\,}} r^k) \mathbin{\mathring{\,}} r \\
=\ & \{\text{using the definition of } r^{n+1}\} \\
&r^{k+1} \mathbin{\mathring{\,}} r
\end{aligned}
$$

This completes the proof for any natural number n. The obvious statement turned out to be true.   ◇

Going back to the example about links between webpages, we would expect that composing the relation which relates pages n links away from each other, and another which relates pages m links away, we should get back the relation between pages $n + m$ links apart. In fact, this holds for any relation r—$r^n \mathbin{\mathring{\,}} r^m$ is the same relation as r^{n+m}, as we prove in the following theorem.

Theorem 5.16 *For any relation $r : X \leftrightarrow X$ and natural numbers n and m, the following equation holds*:

$$r^{n+m} = r^n \,\mathring{,}\, r^m$$

Proof The theorem will be proved by induction on n.

Base case: Taking $n = 0$, we need to prove that $r^{0+m} = r^0 \,\mathring{,}\, r^m$:

$$r^{0+m}$$
$= \{\text{since } 0 + m = m\}$
$$r^m$$
$= \{id_X \text{ is the left identity of relational composition}\}$
$$id_X \,\mathring{,}\, r^m$$
$= \{\text{using the definition of } r^0\}$
$$r^0 \,\mathring{,}\, r^m$$

Inductive hypothesis: We now assume that the statement holds for a particular value k: $r^{k+m} = r^k \,\mathring{,}\, r^m$.

Inductive case: We finally try to prove the statement for $n = k + 1$. We require to prove that $r^{(k+1)+m} = r^{k+1} \,\mathring{,}\, r^m$.

$$r^{(k+1)+m}$$
$= \{\text{using the fact that addition is associative and commutative}\}$
$$r^{((k+m)+1}$$
$= \{\text{using the definition of } r^{n+1}\}$
$$r \,\mathring{,}\, r^{k+m}$$
$= \{\text{by the inductive hypothesis}\}$
$$r \,\mathring{,}\, (r^k \,\mathring{,}\, r^m)$$
$= \{\text{by Theorem 5.9 relational composition is associative}\}$
$$(r \,\mathring{,}\, r^k) \,\mathring{,}\, r^m$$
$= \{\text{using the definition of } r^{n+1}\}$
$$r^{k+1} \,\mathring{,}\, r^m$$

This completes the proof of the theorem.   ∎

Exercises

5.20 Prove, once again, that $r^{n+1} = r^n \,\mathring{,}\, r$, without using induction, but using Theorem 5.16.

5.21 Prove that for any set S and natural number n, $id_S^n = id_S$.

5.22 Using induction on m, prove that $(r^n)^m = r^{n \times m}$.

5.23 Given a homogeneous relation $r : X \leftrightarrow X$ we call r^2 the double of r. Given relation r and natural number n, we define the operator $r^{[n]}$, which doubles the relation for n times as follows:

$$r^{[k]} : X \leftrightarrow X$$

$$r^{[0]} \overset{\text{df}}{=} r$$

$$r^{[n+1]} \overset{\text{df}}{=} r^{[n]} \mathbin{\overset{\circ}{\underset{9}{}}} r^{[n]}$$

Prove that repetition and this new operator are related in the following way: $r^{[n]} = r^{(2^n)}$.

5.4.4 Closure

Recall the *linked* relation between webpages. If we want to define the relation *reachable*, which relates two webpages p and q if, starting on page p, one can reach page q through any number of intermediate pages, this can be defined as: $(r \cup r^2 \cup r^3 \cup \cdots)$. If, we also want to relate webpages to themselves (reasoning that a page is reachable from itself by following exactly zero links), it is simply a matter of adding id_{WEBPAGE}, equal to r^0, to the relation: $(r^0 \cup r^1 \cup r^2 \cup r^3 \cup \cdots)$. This notion of repeating a homogeneous relation arbitrarily often is encapsulated in the transitive closures of a relation.

Definition 5.17 *Given a homogeneous relation* $r : X \leftrightarrow X$, *we define the reflexive transitive closure of the relation, written as* r^*, *as follows:*

$$r^* \overset{\text{df}}{=} \bigcup_{n \in \mathbb{N}} r^n$$

The transitive closure of r, *written as* r^+, *is similarly defined as follows:*

$$r^+ \overset{\text{df}}{=} \bigcup_{n \in \mathbb{N} \setminus \{0\}} r^n$$

■

Using the definition of generalised set union from the previous chapter, we can prove the following proposition:

Proposition 5.18 *Transitive closures of a relation* $r : X \leftrightarrow X$ *satisfy the following equivalences:*

$$(x, y) \in r^* \Leftrightarrow \exists n : \mathbb{N} \cdot (x, y) \in r^n$$

$$(x, y) \in r^+ \Leftrightarrow \exists n : \mathbb{N} \cdot (x, y) \in r^n \wedge n > 0$$

Theorem 5.19 *The reflexive transitive closure of a relation* $r : X \leftrightarrow X$ *can be expressed in terms of the irreflexive transitive closure:* $r^* = id_X \cup r^+$.

Proof The proof follows using the previous proposition and reasoning about the natural numbers:

$$(x, y) \in r^*$$
$\Leftrightarrow$ {by the Proposition 5.18 about r^*}
$$\exists n : \mathbb{N} \cdot (x, y) \in r^n$$
$\Leftrightarrow$ {n can be either zero or larger than zero}
$$(x, y) \in r^0 \vee \exists n : \mathbb{N} \cdot (x, y) \in r^n \wedge n > 0$$
$\Leftrightarrow$ {by definition of r^0 and Proposition 5.18 about r^+}
$$(x, y) \in id_X \cup r^+ \qquad \blacksquare$$

Theorem 5.20 *Similarly, the irreflexive transitive closure of a relation $r : X \leftrightarrow X$ can be expressed in terms of the reflexive transitive closure: $r^+ = r \mathbin{\mathring{,}} r^*$.*

Proof The proof uses the definition of relational composition, repetition of relations and transitive closure:

$$(x, y) \in r^+$$
$\Leftrightarrow$ {by the Proposition 5.18 about r^+}
$$\exists n : \mathbb{N} \cdot (x, y) \in r^n \wedge n > 0$$
$\Leftrightarrow$ {basic arithmetic}
$$\exists n : \mathbb{N} \cdot (x, y) \in r^{n+1}$$
$\Leftrightarrow$ {definition of r^{n+1}}
$$\exists n : \mathbb{N} \cdot (x, y) \in r \mathbin{\mathring{,}} r^n$$
$\Leftrightarrow$ {by proposition about relational composition}
$$\exists n : \mathbb{N} \cdot \exists z : X \cdot (x, z) \in r \wedge (z, y) \in r^n$$
$\Leftrightarrow$ {pushing in existential quantification}
$$\exists z : X \cdot (x, z) \in r \wedge \exists n : \mathbb{N} \cdot (z, y) \in r^n$$
$\Leftrightarrow$ {by Proposition 5.18 about r^*}
$$\exists z : X \cdot (x, z) \in r \wedge (z, y) \in r^*$$
$\Leftrightarrow$ {by proposition about relational composition}
$$(x, y) \in r \mathbin{\mathring{,}} r^* \qquad \blacksquare$$

The following proposition shows how a result can be proved by first proving other intermediate results within the proof:

Proposition 5.21 *The reflexive transitive closure of a relation $r : X \leftrightarrow X$ is the same as taking the reflexive transitive closure twice: $(r^*)^* = r^*$.*

Proof We start by proving that $r^* \mathbin{\overset{\circ}{\scriptscriptstyle 9}} r^* = r^*$.

$$(x, y) \in r^* \mathbin{\overset{\circ}{\scriptscriptstyle 9}} r^*$$
$\Leftrightarrow$ {by proposition about relational composition}
$$\exists z : X \cdot (x, z) \in r^* \wedge (z, y) \in r^*$$
$\Leftrightarrow$ {by Proposition 5.18 about r^* applied twice}
$$\exists z : X \cdot (\exists n : \mathbb{N} \cdot (x, z) \in r^n) \wedge (\exists m : \mathbb{N} \cdot (z, y) \in r^m)$$
$\Leftrightarrow$ {moving existential quantifiers outside}
$$\exists n, m : \mathbb{N} \cdot \exists z : X \cdot (x, z) \in r^n \wedge (z, y) \in r^m$$
$\Leftrightarrow$ {by proposition about relational composition}
$$\exists n, m : \mathbb{N} \cdot (x, y) \in r^n \mathbin{\overset{\circ}{\scriptscriptstyle 9}} r^m$$
$\Leftrightarrow$ {by theorem about relational composition and repetition}
$$\exists n, m : \mathbb{N} \cdot (x, y) \in r^{n+m}$$
$\Leftrightarrow$ {every number can be written as the sum of two numbers and
         the sum of every two numbers is itself a number}
$$\exists p : \mathbb{N} \cdot (x, y) \in r^p$$
$\Leftrightarrow$ {by Proposition 5.18 about r^*}
$$(x, y) \in r^*$$

We can now prove that this is true for any number of repetitions larger than zero: $(r^*)^{n+1} = r^*$. The proof is by induction on n.

Base case: We require to prove the result for $n = 0$: $(r^*)^{0+1} = r^*$.

$$(r^*)^{0+1}$$
$= $ {definition of r^{n+1}}
$$r^* \mathbin{\overset{\circ}{\scriptscriptstyle 9}} r^0$$
$= $ {definition of r^0}
$$r^* \mathbin{\overset{\circ}{\scriptscriptstyle 9}} id_X$$
$= $ {id_X is the identity of relational composition}
$$r^*$$

Inductive hypothesis: We assume that the statement holds for $n = k$: $(r^*)^{k+1} = r^*$.
Inductive case: It remains to prove that the statement holds for $n = k + 1$: $(r^*)^{(k+1)+1} = r^*$.

$$(r^*)^{(k+1)+1}$$
$= $ {definition of r^{n+1}}
$$r^* \mathbin{\overset{\circ}{\scriptscriptstyle 9}} r^{k+1}$$
$= $ {by the inductive hypothesis}
$$r^* \mathbin{\overset{\circ}{\scriptscriptstyle 9}} r^*$$
$= $ {by the first result}
$$r^*$$

If thus follows that, for any natural number n, $(r^*)^{n+1} = r^*$. Finally, we can prove the result sought after:

$$(x, y) \in (r^*)^*$$
$$\Leftrightarrow \{\text{by Proposition 5.18 about } r^*\}$$
$$\exists n : \mathbb{N} \cdot (x, y) \in (r^*)^n$$
$$\Leftrightarrow \{\text{by the second result}\}$$
$$\exists n : \mathbb{N} \cdot (x, y) \in r^*$$
$$\Leftrightarrow \{\text{by Proposition 5.18 about } r^*\}$$
$$(x, y) \in r^* \qquad \blacksquare$$

Exercises

5.24 Prove that $(r^+)^+ = r^+$.

5.25 Prove that $(r^+)^* = (r^*)^+ = r^*$.

5.26 Prove that $r^* \mathbin{\substack{\circ \\ 9}} r^+ = r^+ \mathbin{\substack{\circ \\ 9}} r^* = r^+$.

5.27 Prove that if r is reflexive, then $r^+ = r^*$.

5.5 Other Relational Operators

Various other relational operators can be useful when specifying relations. Consider, for instance, relational application. Given a relation between users telling us whether two users are connected as 'friends' on a social networking site, one would want to apply the friendship relation to extract the set of all 'friends' of a particular user. Similarly, given a relation which relates users to the files they have access to, one would want to request which files a particular user can access. A relation $r : X \leftrightarrow Y$ can be applied to an item $x : X$ to return a set of objects in Y. In general, we can extend this notion to take a set of objects in $S : \mathbb{P}X$, and return the set of objects in Y to which at least one of the objects in S is related. For instance, applying the friends relation to the set $\{John, Peter\}$ would return all those who are friends of either John or Peter (or both).

Definition 5.22 *Given a relation $r : X \leftrightarrow Y$ and a set $S : \mathbb{P}X$, the relational application of r to S, written $r \langle\!\langle S \rangle\!\rangle$, is the set of all objects in Y related to some object in S:*

$$r \langle\!\langle S \rangle\!\rangle \stackrel{\mathrm{df}}{=} \{y : Y \mid \exists x : X \cdot x \in S \wedge (x, y) \in r\} \qquad \blacksquare$$

As before, to avoid opening set comprehensions in proofs, we start by proving a simple proposition encapsulating the whole meaning of relational application. The proof is straightforward, and left as an exercise to the reader.

Proposition 5.23 *An object is in the relational application of $r : X \leftrightarrow Y$ to set S if and only if it is related to some object in S:*

$$y \in r \langle\!\langle S \rangle\!\rangle \Leftrightarrow \exists x : X \cdot x \in S \wedge (x, y) \in r$$

Relational application satisfies various laws which we can now prove using this formal definition. We will only prove one such law here. Proofs of others laws are given as exercises.

Theorem 5.24 *Applying a relation $r : X \leftrightarrow Y$ to the whole of the domain of r returns the whole range of r: $r \langle\langle dom(r) \rangle\rangle = ran(r)$.*

Proof We will show that objects in the relational application of the domain of r must be in the range of r and vice versa:

$$y \in r \langle\langle dom(r) \rangle\rangle$$
$\Leftrightarrow$ {proposition about relational application}
$$\exists x : X \cdot x \in dom(r) \wedge (x, y) \in r$$
$\Leftrightarrow$ {proposition about domain of relations}
$$\exists x : X \cdot (\exists y : Y \cdot (x, y) \in r) \wedge (x, y) \in r$$
$\Leftrightarrow$ {propositional logic reasoning}
$$\exists x : X \cdot (x, y) \in r$$
$\Leftrightarrow$ {proposition about range of relations}
$$y \in ran(r) \qquad \blacksquare$$

Consider the example of the relation *canRead*, which says which users may read which files. One may want to reason about a constrained version of this relation. For instance, we may want to say things such as 'if we consider only student users, the relation must ensure that no two users may read the same file' or 'if we just consider system files, no user may read more than five such files'. In both cases, we want to talk about a relation—*canRead*, but constrained to just student users, and *canRead*, but constrained to just system files. Note that, for instance in the second example, it is acceptable for a user to be able to read more than five files, as long as no more than five of them are system files. To enable these forms of looking at a smaller part of the relation, we will introduce two notions of relational restriction—a *domain restriction* says which elements of the domain we are interested in, thus discarding all others, while a *range restriction* constrains the relation to the elements of the range we are interested in. If we are only interested in student users in the relation *canRead* : USER $\leftrightarrow$ FILE, we would use a domain restriction, while if we want to talk about system files, we would use range restriction.

Definition 5.25 *Given a relation $r : X \leftrightarrow Y$ and a set $S : \mathbb{P}X$, the domain restriction of r to S, written as $S \lhd r$, is the subset of relation r such that the first objects of the pairs come from set S. Similarly, given a set $T : \mathbb{P}Y$, the range restriction of r to T, written as $r \rhd T$, is the subset of relation r such that the second objects of the pairs come from set T:*

$$S \lhd r \stackrel{df}{=} \{(x, y) : X \times Y \mid x \in S \wedge (x, y) \in r\}$$

$$r \rhd T \stackrel{df}{=} \{(x, y) : X \times Y \mid y \in T \wedge (x, y) \in r\} \qquad \blacksquare$$

Domain and range restriction satisfy the following proposition:

Proposition 5.26 *A pair of elements x and y are related by the domain restriction of a relation* $r : X \leftrightarrow Y$ *to a set S if and only if* (i) *they are related in r, and* (ii) *x lies in set S. Similarly,* (x, y) *lies in the range restriction of r to a set T if and only if the pair lies in r, and y in T:*

$$(x, y) \in S \lhd r \Leftrightarrow x \in S \land (x, y) \in r$$

$$(x, y) \in r \rhd T \Leftrightarrow y \in T \land (x, y) \in r$$

The proof of this proposition involves just applying the definition of set comprehension, and is left as an exercise. As in the case of relational application, relation restriction satisfies various laws—below we show a typical proof of one such law:

Theorem 5.27 *Domain restricting a relation* $r : X \leftrightarrow Y$ *to its whole domain gives back the original relation. Similarly, range restricting r to its whole range gives back the whole original relation:*

$$dom(r) \lhd r = r$$

$$r \rhd ran(r) = r$$

Proof We will prove the first of these propositions. The proof follows from the propositions about domain restriction and the domain of a relation:

$$(x, y) \in dom(r) \lhd r$$
$$\Leftrightarrow \{\text{proposition about domain restriction}\}$$
$$x \in dom(r) \land (x, y) \in r$$
$$\Leftrightarrow \{\text{proposition about domain}\}$$
$$(\exists x : Y \cdot (x, y) \in r) \land (x, y) \in r$$
$$\Leftrightarrow \{\text{predication logic reasoning}\}$$
$$(x, y) \in r$$

The second follows similarly and is left as an exercise.   ∎

Relation restriction is not only useful when stating properties which may hold only on a subset of a relation, but can be useful in practice when, for performance or security reasons, we may want to implement or have access to only a subset of the relation.

Example A multi-national company stores the information about their customers only at the location where the customer resides. However, a central database is maintained, giving information as to which customer records each of their employees should have access to. So as to avoid having to access the remote central database every time an employee wants to access information, it has been decided that a local copy of the database is to be made on a regular basis. The programmers realise that the system can be made more efficient by reducing the size of the database—discarding information about users who do not log in from that particular location. This measure also addresses privacy concerns, since a user in a particular location will not be able to check what rights other users elsewhere have. However, does the restricted database really give the same results as the original one?

Let us call the relation giving access rights to users $canRead$: USER $\leftrightarrow$ FILE, and the set of local users $LocalUsers$. The programmers are proposing to reduce the relation to a smaller one: $LocalUsers \vartriangleleft canRead$. All requests made locally will always be about local users, so any application of the relation will on a subset U of the local users: $U \subseteq LocalUsers$. Is it true that applying the global relation $canRead \langle\!\langle U \rangle\!\rangle$ is the same as applying the reduced relation $(LocalUsers \vartriangleleft canRead)\langle\!\langle U \rangle\!\rangle$?

We will prove this in general. Given a relation $r : X \leftrightarrow Y$, and two sets S and T, both subsets of X, and $T \subseteq S$, applying to set T the relation r is the same as applying the restriction of r to S:

$$(S \vartriangleleft r)\langle\!\langle T \rangle\!\rangle = r \langle\!\langle T \rangle\!\rangle$$

The proof of the statement follows from the propositions we have already proved and set theory:

$$
\begin{aligned}
& y \in (S \vartriangleleft r)\langle\!\langle T \rangle\!\rangle \\
\Leftrightarrow\ & \{\text{proposition about relational application}\} \\
& \exists x : X \cdot x \in T \wedge (x, y) \in S \vartriangleleft r \\
\Leftrightarrow\ & \{\text{proposition about domain restriction}\} \\
& \exists x : X \cdot x \in T \wedge x \in S \wedge (x, y) \in r \\
\Leftrightarrow\ & \{\text{proposition about intersection}\} \\
& \exists x : X \cdot x \in T \cap S \wedge (x, y) \in r \\
\Leftrightarrow\ & \{\text{since } T \subseteq S, \text{ it follows that } T \cap S = T\} \\
& \exists x : X \cdot x \in T \wedge (x, y) \in r \\
\Leftrightarrow\ & \{\text{proposition about relational application}\} \\
& y \in r \langle\!\langle T \rangle\!\rangle
\end{aligned}
$$

So, yes, the programmers are justified in keeping the smaller local database. ◇

Exercises

5.28 Prove that set union distributes over relational application: $r\langle\!\langle S \cup T \rangle\!\rangle = r\langle\!\langle S \rangle\!\rangle \cup r \langle\!\langle T \rangle\!\rangle$.

5.29 Prove that $r\langle\!\langle S \rangle\!\rangle \subseteq ran(r)$.

5.30 Prove that $dom(S \vartriangleleft r) = S$ and $ran(r \vartriangleright T) = T$.

5.31 Range restriction can, in fact, be expressed using domain restriction and relational inverse. Prove that, for a relation $r : X \leftrightarrow Y$, $r \vartriangleright S = (S \vartriangleleft r^{-1})^{-1}$.

5.32 Relational application can also be expressed in terms of domain restriction. Given a relation $r : X \leftrightarrow Y$, prove that $r\langle\!\langle S \rangle\!\rangle = ran(S \vartriangleleft r)$.

5.33 The reverse application of a relation $r : X \leftrightarrow Y$ to a set $S : \mathbb{P}Y$, written as $\langle\!\langle S \rangle\!\rangle r$, returns the set of objects in X related to some object in S.

(i) Define reverse relational application using a set comprehension.

(ii) Prove that $x \in \langle\langle S \rangle\rangle r \Leftrightarrow \exists y : Y \cdot y \in S \wedge (x, y) \in r$.

(iii) Prove that normal and reverse relational application are related in the following manner:

$$r \langle\langle S \rangle\rangle = \langle\langle S \rangle\rangle r^{-1}$$

(iv) Hence or otherwise prove that $\langle\langle ran(r) \rangle\rangle r = dom(r)$.

5.34 Relational application was defined to return the set of all objects related to *at least one object* in the given set. Thus $canRead \langle\langle \{John, Peter\} \rangle\rangle$ gives the files which either John or Peter can read.

A different form of relational application is to return the objects which *all items* in the given set are related to e.g. $canRead \langle\langle \{John, Peter\} \rangle\rangle_\forall$ gives the files which both John and Peter can read.

The formal definition of universal relational application can be given as follows:

$$r \langle\langle S \rangle\rangle_\forall \stackrel{\mathrm{df}}{=} \{ y : Y \mid \forall x : X \cdot x \in S \Rightarrow (x, y) \in r \}$$

(i) Prove that $x \in r \langle\langle S \rangle\rangle_\forall \Leftrightarrow \forall x : X \cdot x \in S \Rightarrow (x, y) \in r$.

(ii) Prove that $r \langle\langle S \rangle\rangle_\forall \subseteq r \langle\langle S \rangle\rangle$.

(iii) Prove that normal and universal relational application give the same result when applied to a set with one element:

$$r \langle\langle \{x\} \rangle\rangle = r \langle\langle \{x\} \rangle\rangle_\forall$$

5.6 Beyond Binary Relations

Sometimes we need to relate more than two objects to each other. For instance, rather than using different relations *canRead*, *canWrite*, *canAppend* and others to relate users with files, one may opt to have a single relation relating users, files and access rights. There are multiple options as to how this can be expressed as a binary relation:

USER $\leftrightarrow$ FILE $\times$ RIGHTUSER $\times$ FILE $\leftrightarrow$ RIGHTFILE $\leftrightarrow$ USER $\times$ RIGHT

Each of these options can be found useful in different scenarios. For instance, if we are mainly interested in a user-centric view, the first encoding may be appropriate.

However, as with Cartesian products, encoding triples and other n-tuples using pairs can be inconvenient and confusing since we are grouping together objects that may not necessarily be more closely related together. For instance, are users and files somehow more closely related than files and rights in the second example of the above relation types?

To avoid this issue, relations can be extended to work over more than two objects using n-tuples instead of pairs. A relation using n-tuples is called an n-ary relation. For instance, a relation between users, files and access rights can be expressed as:

$$\{(John, user.txt, read), (John, user.txt, write),$$
$$(Jane, user.txt, delete), (Jane, accounts.txt, read), \ldots\}$$

The type of the relation *accessRights* is thus $\mathbb{P}(\text{FILE} \times \text{USER} \times \text{RIGHT})$. This can be generalised to enable the description of relations between any number of objects. A relation between types X_1, X_2 to X_n will thus have type $\mathbb{P}(X_1 \times X_2 \times \cdots X_n)$. We write this as

$$rel(X_1, X_2, \ldots, X_n) \overset{\text{df}}{=} \mathbb{P}(X_1 \times X_2 \times \cdots X_n)$$

Such a relation corresponds to a table in a database, with n fields with types X_1 to X_n.

With this approach we have to redefine, and in some cases rethink afresh, the relational operators we have seen in this chapter. For instance, what is the domain of the relation *accessRights* : $rel(\text{FILE, USER, RIGHT})$? One may decide to choose it to be the set of files related to some user and right, with the range being a set of rights assigned to some file and user. However, in this manner we can no longer extract the set of users related to some file with some right. Similarly, one may ask what the composition of two relations over more than two types is.

The solution to this problem which is adopted in databases is to give unique names to the fields. This enables generalisation of the notions of domain and range of a relation by requesting the values that appear in a particular named field in the relation. Similarly, to compose two generalised relations, we would have to specify which field in the first relation is to be matched to which field in the second relation. For instance, one may want to compose the relation: *accessRights* : $rel(\text{FILE, USER, RIGHT})$ and a relation between users, their status and supervisor: *userInformation* : $rel(\text{USER, STATUS, USER})$. To do so, one would have to specify that we will associate the second field of the first relation to the first field in the second, to obtain a new relation of type $rel(\text{FILE, RIGHT, USER, STATUS, USER})$. This is usually called a *join* over two tables. In fact, the position of the user field has been (arbitrarily) placed in this example. Using the approach used in databases with naming of fields, there is no implicit ordering of the fields since the only way to refer to them is through the use of their names. The problem of which ordering to give to the fields of the result of the join is thus circumvented.

5.7 Summary

In this chapter we have looked into how we can encode a relationship between different objects through the use of sets and Cartesian products. Through the use of relational operators, relations can be constructed from simpler ones, and thanks to the formal definitions, properties of these operators and their practical applications can be mathematically proved.

The definition of relations we have given is a very liberal one to cover most cases one may encounter in practice. In the next chapter we will be looking at particular classes of relations. In this manner, we will be able to show interesting and useful laws which a particular class of relations would satisfy. For instance, one particular class of relations corresponds to the notion of mathematical functions. By focussing only on this class of relations, we will be able to prove results about functions, not all of which may hold for an arbitrary relation.

Chapter 6
Classifying Relations

The notion of relations is extremely general, and allows for various concepts to be represented. Despite the desirability of having a very abstract notion which is applicable for different concrete systems, more specific notions have the advantage that more properties can be proved about them. In this section we look at different classes of relations, and prove properties which hold for these classes.

6.1 Classes of Relations

Relations are classified on the basis of regularities in the arrows between the source and destination types. We identify four classes of relations—two for relations which cover the whole source or destination type, and two for relations which have no more than one arrow connected to each item. The first guarantee completeness of coverage—given an item, we are guaranteed to find another item to which it is related. The latter guarantee uniqueness—given an item, there is at most a single item to which it is related. We will see various applications of these classes of relations in the rest of the book.

6.1.1 Totality

A relation does not have to relate every object to another. Relations which do relate every object in the source type are called *total* relations. For instance, consider the relation *isChildOf* which relates every person to his or her parents. Since every person has parents, the relation is total. On the other hand, the relation *isParentOf*, which relates a person to his or her children, is not total, since there are persons who have no children.

Definition 6.1 *A relation $r : X \leftrightarrow Y$ is said to be total if each element of X is related to at least one element of Y: $\forall x : X \cdot \exists y : Y \cdot (x, y) \in r$.* ∎

G.J. Pace, *Mathematics of Discrete Structures for Computer Science*,
DOI 10.1007/978-3-642-29840-0_6, © Springer-Verlag Berlin Heidelberg 2012

The definition is a direct formalisation of the description of totality. However, another, more concise way of characterising totality is to say that a relation is total if its domain is equal to the whole source type. We can prove that the two definitions are identical.

Proposition 6.2 *A relation* $r : X \leftrightarrow Y$ *is total if and only if* $dom(r) = X$.

Proof Let us first consider the proof in the forward direction. Assuming that r is total, we will prove that $dom(r) = X$. From the type of the domain or relation r, we know that $dom(r) \subseteq X$. We thus have to prove that $X \subseteq dom(r)$:

$$z \in X$$
$$\Rightarrow \{r \text{ is total}\}$$
$$z \in X \wedge \forall x : X \cdot \exists y : Y \cdot (x, y) \in r$$
$$\Rightarrow \{\text{specialising universal quantification to } z\}$$
$$\exists y : Y \cdot (z, y) \in r$$
$$\Leftrightarrow \{\text{proposition about domain of a relation}\}$$
$$z \in dom(r)$$

In the other direction, assuming that $dom(r) = X$, we will prove that r is total. Since $dom(r) = X$, we know that $\forall x : X \cdot x \in dom(r)$.

$$\forall x : X \cdot x \in dom(r)$$
$$\Leftrightarrow \{\text{proposition about domain of a relation}\}$$
$$\forall x : X \cdot \exists y : Y \cdot (x, y) \in r$$
$$\Leftrightarrow \{\text{definition of totality}\}$$
$$r \text{ is total} \qquad \blacksquare$$

Example We can now prove that, if a relation $r : X \leftrightarrow Y$ is total, then so is $r \, \mathring{,} \, r^{-1}$. To prove this we need to prove that $dom(r \, \mathring{,} \, r^{-1}) = X$. From the type, it follows that $dom(r \, \mathring{,} \, r^{-1}) \subseteq X$, so we only need to prove set inclusion in the opposite direction:

$$x \in X$$
$$\Rightarrow \{\text{since } r \text{ is total}\}$$
$$\exists y : Y \cdot (x, y) \in r$$
$$\Leftrightarrow \{\text{proposition about relational inverse}\}$$
$$\exists y : Y \cdot (x, y) \in r \wedge (y, x) \in r^{-1}$$
$$\Rightarrow \{\text{proposition about relational composition}\}$$
$$(x, x) \in r \, \mathring{,} \, r^{-1}$$
$$\Rightarrow \{\text{taking } x' \text{ to be } x\}$$
$$\exists x' : X \cdot (x, x') \in r \, \mathring{,} \, r^{-1}$$
$$\Rightarrow \{\text{definition of the domain of a relation}\}$$
$$x \in dom(r \, \mathring{,} \, r^{-1})$$

As an application of this result, consider a memory management algorithm—given a set of users and memory pages, the algorithm manages which users have read and write access to which memory pages. Let us call this relation *access* $\in$ USER $\leftrightarrow$

MEMPAGE. If the relation is total, it means that every user must have access to at least one memory page.

Now consider the relation $comm \in$ USER $\leftrightarrow$ USER, defined as $comm = access \, {}^{\circ}_{9} \, access^{-1}$. This relates together two users if they share at least one page that both can read from and write to—meaning that both can write messages on that memory page for the other to read, thus communicating with one another. Using the result we have just proved, it follows that $comm$ is a total relation—everyone can communicate via memory with someone, even if it is possibly themselves. ◇

Exercises

6.1 Which of the relations $\leq, \geq, <$ and $>$ on natural numbers are total? What about the same relations on real numbers?

6.2 The two relations $isChildOf$ and $isParentOf$ are inverses of each other—one is total, while the other is not. Identify a relation which is total, and so is its inverse. Identify another relation which is not, and neither is its inverse.

6.3 Prove that id_X is a total relation.

6.4 Prove that, if both $r : X \leftrightarrow Y$ and $s : Y \leftrightarrow Z$ are total, then so is $r \, {}^{\circ}_{9} \, s$.

6.1.2 Surjectivity

While totality ensures that everything in the source type is related, surjectivity is the dual notion focussing on the right-hand side of the relation. A relation is said to be surjective if every object in the destination type has an object which it relates to. For instance, the relation $isChildOf$ is not surjective since some individuals never appear on the right-hand side of the relation (because they are not parents), while the relation $isParentOf$ is surjective because, for every person, there is someone who is their parent.

Definition 6.3 *A relation $r : X \leftrightarrow Y$ is said to be surjective (or onto) if all objects of type Y are mapped to via the relation r:* $\forall y : Y \cdot \exists x : X \cdot (x, y) \in r$. ∎

As with totality, there is a more concise but equivalent way of expressing surjectivity:

Proposition 6.4 *A relation $r : X \leftrightarrow Y$ is surjective if and only if $ran(r) = Y$.*

The proof is very similar to the one of totality, and is left as an exercise. When switching the direction of a relation (by taking the relational inverse) it is easy to see that a total relation becomes a surjective one, while a surjective relation becomes total.

Theorem 6.5 *A relation $r : X \leftrightarrow Y$ is surjective if and only if r^{-1} is total. Similarly, $r : X \leftrightarrow Y$ is total if and only if r^{-1} is surjective.*

Proof The proof of the first statement follows from the propositions about surjectivity and totality, and a result from the previous chapter: $ran(r) = dom(r^{-1})$:

> r surjective
> $\Leftrightarrow$ {proposition about surjectivity of relations}
> $ran(r) = Y$
> $\Leftrightarrow$ {since $ran(r) = dom(r^{-1})$}
> $dom(r^{-1}) = Y$
> $\Leftrightarrow$ {proposition about totality of relations}
> r^{-1} total

The second can either be proved in a similar way, or using the result we have just proved and the fact that by taking the inverse of a relation twice we end up with the original relation. We take the latter approach here:

> r total
> $\Leftrightarrow$ {since $r = (r^{-1})^{-1}$}
> $(r^{-1})^{-1}$ total
> $\Leftrightarrow$ {the first part of the proof}
> r^{-1} surjective   ∎

Example Returning back to the memory management algorithm using relation $access \in$ USER $\leftrightarrow$ MEMPAGE, if the relation is surjective, it means that every memory page can be accessed by some user. If, for instance, we have a super user *superuser* who can access all memory $\forall p :$ MEMPAGE $\cdot$ (*superuser*, p) $\in access$, the surjectivity of the relation is easy to prove:

> {the given information about the super user}
> $\Leftrightarrow \forall p :$ MEMPAGE $\cdot$ (*superuser*, p) $\in access$
> $\Rightarrow$ {by taking u to be *superuser*}
> $\forall p :$ MEMPAGE $\cdot \exists u :$ USER $\cdot (u, p) \in access$
> $\Leftrightarrow$ {definition of surjectivity}
>   *access* is surjective   ◇

Exercises
  6.5 Prove Proposition 6.4.
  6.6 Which of the relations $\leq, \geq, <$ and $>$ on natural numbers are surjective? What
       about the same relations on real numbers?
  6.7 Prove that, if both $r : X \leftrightarrow Y$ and $s : Y \leftrightarrow Z$ are surjective, then so is $r \,\mathring{,}\, s$.

6.1.3 Injectivity

With totality and surjectivity we can characterise relations for which each element in the source or destination type is connected by at least one arrow. We now look at another pair of classes of relations, for which each element in the source or destination type is connected by *at most* one arrow.

Let us start by looking at the class of injective relations in which, items in the destination type cannot have more than one incoming arrow. Consider the relation var : VARNAME $\leftrightarrow$ MEMORY, which relates a variable name with the memory locations where the variable is stored. Although a single variable may use more than one memory location to store its value (for instance if we want to store an array of information), we cannot have more than one variable stored in the same location, since updating one would change the other, which is undesirable. The relation should thus be injective, since no memory location should be associated with more than one variable. On the other hand, consider the relation $size$: VARNAME $\leftrightarrow$ $\mathbb{N}$, which relates each variable with the amount of memory locations required to store it. Since there can be more than one variable with the same size, this relation is not necessarily injective.

The way we formalise injectivity is to ensure that, if two objects are related to a single object in the destination type, then those two objects must be identical:

Definition 6.6 *A relation $r : X \leftrightarrow Y$ is said to be injective if no two distinct objects in the domain map to the same object in the range*:

$$\forall x_1, x_2 : X \cdot \forall y : Y \cdot ((x_1, y) \in r \wedge (x_2, y) \in r) \Rightarrow x_1 = x_2$$

Relations which are both injective and surjective are called bijective relations. ∎

If we combine two injective relations by relational composition, we end up with an injective relation:

Theorem 6.7 *Given relations $r : X \leftrightarrow Y$ and $s : Y \leftrightarrow Z$, both of which are injective, $r \,\mathring{,}\, s$ is also injective.*

Proof We are required to prove that $r \,\mathring{,}\, s$ is injective:

$$\forall x_1, x_2 : X \cdot \forall z : Z \cdot ((x_1, z) \in r \,\mathring{,}\, s \wedge (x_2, z) \in r \,\mathring{,}\, s) \Rightarrow x_1 = x_2$$

The proof follows by taking the left-hand side of the implication and proves the right-hand side—we consider two values x_1 and x_2 related to z and prove that they have to be identical.

$(x_1, z) \in r \,\mathring{,}\, s \wedge (x_2, z) \in r \,\mathring{,}\, s$
$\Rightarrow$ {proposition about relational composition applied twice}
$\quad (\exists y_1 : Y \cdot (x_1, y_1) \in r \wedge (y_1, z) \in s) \wedge$
$\quad (\exists y_2 : Y \cdot (x_2, y_2) \in r \wedge (y_2, z) \in s)$
$\Rightarrow$ {moving quantifiers to the top level and
$\qquad\qquad$ commutativity of conjunction}
$\quad \exists y_1, y_2 : Y \cdot (x_1, y_1) \in r \wedge (x_2, y_2) \in r \wedge (y_1, z) \in s \wedge (y_2, z) \in s$
$\Rightarrow$ {using the injectivity of s}
$\quad \exists y_1, y_2 : Y \cdot (x_1, y_1) \in r \wedge (x_2, y_2) \in r \wedge y_1 = y_2$
$\Rightarrow$ {applying the one-point rule on y_2}
$\quad \exists y_1 : Y \cdot (x_1, y_1) \in r \wedge (x_2, y_1) \in r$
$\Rightarrow$ {using the injectivity of r}
$\quad x_1 = x_2$
$\hfill$ ∎

Exercises

6.8 Prove that, for any set S, $id_S : X \leftrightarrow X$ is an injective relation.

6.9 Prove that, if $r \subseteq s$ and $s : X \leftrightarrow Y$ is an injective relation, then so is r.

6.10 Is the intersection of two injective relations always an injective relation? What about the union? In both cases, either prove it is true, or give a counterexample to show it is false.

6.1.4 Functionality

The dual concept of injectivity is to ensure that no item in the source type has more than one outgoing arrow. Such relations are called functional, or simply functions.

Returning to the relation var : VARNAME $\leftrightarrow$ MEMORY, in which a variable name may require more than one memory location in which to be stored. This means that var is not necessarily a function. On the other hand, any variable name should have a single fixed size. The relation $size$: VARNAME $\leftrightarrow \mathbb{N}$ is thus functional.

The definition of functionality is similar to that of injectivity:

Definition 6.8 *A relation* $r : X \leftrightarrow Y$ *is said to be* functional *if no object in the domain of* r *is mapped onto two distinct objects in the range*:

$$\forall x : X \cdot \forall y_1, y_2 : Y \cdot ((x, y_1) \in r \wedge (x, y_2) \in r) \Rightarrow y_1 = y_2$$

We write $r : X \to Y$ *to express that* $r : X \leftrightarrow Y$ *and that* r *is functional.*

An injective function is also called a one-to-one function. ∎

Since a function f relates every object in the domain to, at most, one object in the range, we know that, using relation application: $f\langle\!\langle\{x\}\rangle\!\rangle$ is equal to either $\{y\}$ or $\emptyset$. When dealing with functions, we will be using the more familiar notation $f(x)$ (where x is an element of the domain of f) to represent the unique value y to which x is related. Note that, if f were not functional due to some value x in the domain having more than one outgoing arrow, writing $f(x)$ would be ambiguous and is thus not allowed.

Both injectivity and functionality constrain the number of arrows to be at most one—the former constrains the arrows on the range, while the latter on the domain. If we were to switch the direction of the arrows, we would thus go from an injective relation to a functional one, and vice versa:

Theorem 6.9 *A relation* $r : X \leftrightarrow Y$ *is injective if and only if* r^{-1} *is functional. Conversely,* $r : X \leftrightarrow Y$ *is functional if and only if* r^{-1} *is injective.*

Proof The proof of the first part follows from the fact that functionality and injectivity are duals of each other:

> r is injective
> $\Leftrightarrow$ {definition of injectivity}
> $\quad \forall x_1, x_2 : X \cdot \forall y : Y \cdot ((x_1, y) \in r \wedge (x_2, y) \in r) \Rightarrow x_1 = x_2$
> $\Leftrightarrow$ {proposition about relational inverse}
> $\quad \forall x_1, x_2 : X \cdot \forall y : Y \cdot ((y, x_1) \in r^{-1} \wedge (y, x_2) \in r^{-1}) \Rightarrow x_1 = x_2$
> $\Leftrightarrow$ {universal quantifiers commute}
> $\quad \forall y : Y \cdot \forall x_1, x_2 : X \cdot ((y, x_1) \in r^{-1} \wedge (y, x_2) \in r^{-1}) \Rightarrow x_1 = x_2$
> $\Leftrightarrow$ {definition of functionality}
> $\quad r^{-1}$ is functional

The proof of the second part follows from the first using the additional fact that inverting a relation twice yields back the original relation:

> r is functional
> $\Leftrightarrow$ {the inverse of the inverse of a relation is the original relation}
> $\quad (r^{-1})^{-1}$ is functional
> $\Leftrightarrow$ {using the previous proof}
> $\quad r^{-1}$ is injective   ∎

A notation frequently used to define functions is lambda abstraction, where we will write $\lambda x : X \cdot e$ to denote the function which relates x to expression e. For example, $\lambda n : \mathbb{N} \cdot n^2$ is the function which returns the square of a given number. Using the variable storage example, we can write $\lambda v : \mathsf{VARNAME} \cdot (min(var\langle\!\langle \{v\} \rangle\!\rangle), max(var\langle\!\langle \{v\} \rangle\!\rangle))$ to define the function which identifies the lowest and highest memory address where variable v is located. Since functions are special forms of relations, which are themselves sets of pairs, lambda notation is simply shorthand for a set comprehension:

$$\lambda x : X \cdot e \stackrel{\mathrm{df}}{=} \{x : X \bullet (x, e)\}$$

It is worth noting that a lambda abstraction defines a function, since any value x will only be related to a single result obtained from the expression e.

Sometimes, it is convenient to use conditions in a lambda abstraction. Consider a machine on which the least amount of memory which can be allocated is a word of 16 bits. We can define $realSize \in \mathsf{VARNAME} \to \mathbb{N}$ in terms of the function $size$ to return the actual amount of memory taken by a variable:

$$realSize \stackrel{\mathrm{df}}{=} \lambda v : \mathsf{VARNAME} \cdot \begin{cases} 16 & \text{if } size(v) \leq 16 \\ size(v) & \text{otherwise} \end{cases}$$

The *otherwise* clause is simply shorthand for the negation of the disjunction of all the previous conditions. It is important that no two conditions can hold for a single value of the input—they are pairwise mutually exclusive. The conditions can be formalised in the set comprehension using a conditional predicate:

$$realSize \stackrel{\mathrm{df}}{=} \{v : \mathsf{VARNAME} \mid size(v) \leq 16 \bullet 16\} \cup$$
$$\{v : \mathsf{VARNAME} \mid \neg(size(v) \leq 16) \bullet size(v)\}$$

Frequently, we would like to update the output of a function for some of the inputs, leaving the rest unchanged. For instance, consider the function giving the value of three numeric variables: $val = \{(x, 4), (y, 3), (z, 9)\}$. We would like to update the value of the variables after executing two assignments which set the value of x to 7 and that of y to 1: $update = \{(x, 7), (y, 1)\}$. We will use function overriding to enable us to express the result as: $val \oplus update$—which represents the function which, given a variable, returns the value according to val unless it was updated in $update$, in which case it returns the new value.

Definition 6.10 *Given two relations $r, s : X \leftrightarrow Y$, both of which are functional, we define the overriding of r by s, written $r \oplus s$, to be the function which acts like s when defined, and like r otherwise:*

$$r \oplus s \stackrel{\text{df}}{=} \lambda x : X \cdot \begin{cases} s(x) & \text{if } x \in dom(s) \\ r(x) & \text{otherwise} \end{cases} \qquad \blacksquare$$

Since function overriding is defined in terms of a lambda abstraction with mutually disjoint conditions, it immediately follows that overriding a function with another returns a function.

Example Consider once again the function which gives the amount of memory each variable consumes: $size$, and let us consider different scenarios to update the memory requirements:

- To update a particular variable v to consume 10 memory units: $size \oplus \{(v, 10)\}$.
- To update variable v to consume 10 more units then it consumed before is similar, except that the new amount will have to be $size(v) + 10$, making the resulting memory requirements: $size \oplus \{(v, size(v) + 10)\}$.
- Finally, to ensure that the memory consumption function is total, mapping the size to zero if not used can be expressed using function overriding and lambda abstraction: $(\lambda x : \text{VARNAME} \cdot 0) \oplus size$.   ◇

Exercises

6.11 Prove that the identity relation, for any set S, $id_S : X \leftrightarrow X$ is functional.

6.12 Given functions $r : X \to Y$ and $s : Y \to Z$, prove that $r \,\mathring{,}\, s$ is also functional.

6.13 Given two functions $r, s : X \to Y$, with disjoint domains ($dom(r) \cap dom(s) = \emptyset$), prove that taking their union is equivalent to overriding one by the other: $r \cup s = r \oplus s = s \oplus r$.

6.1.5 Combining the Results

Totality and surjectivity ensure that there is at least one arrow leaving and entering each object in the source and destination types. On the other hand, functionality and injectivity ensure that there is not more than one. Combining them together, a relation which is a total bijective function has exactly one arrow leaving or entering each item in the source and destination types. Since this property is symmetric between the two types, switching the direction of the relation does not break it.

Theorem 6.11 *The inverse of a total bijective function is itself a total bijective function.*

Proof Let f be a (i) total, (ii) surjective, (iii) injective, (iv) function. We have to prove that f^{-1} has the same properties. By using Theorem 6.5, from (i) and (ii) we know that f^{-1} is surjective and total. Similarly, Theorem 6.9 allows us to conclude that f^{-1} is functional and injective from (iii) and (iv), respectively. ∎

6.2 Relating a Type to Itself

We have already focussed on homogeneous relations in a previous chapter. Such relations allow us to perform operators such as repeated application and inverting them retaining the same type. This permits us to identify further classes of homogeneous relations over and above the ones we already identified for general relations.

6.2.1 Properties of These Relations

The first subclass of homogeneous relations we identify is that of reflexive relations, which includes relations which relate every object to itself. For instance, the relation $\leq$ is reflexive, since every number n satisfies $n \leq n$, while the relation $<$ is not. Consider a relation which relates two numbers n and m if and only if $n \times m > 0$. Although every positive and negative number is related to itself, we note that zero is not, implying that the relation is not reflexive.

Definition 6.12 *A homogeneous relation $r : X \leftrightarrow X$ is said to be reflexive, if all objects in X are related to themselves via r: $\forall x : X \cdot (x, x) \in r$.* ∎

An equivalent way of characterising reflexive functions is to check whether the identity relation is included in the relation. The proof of the following proposition is left as an exercise.

Proposition 6.13 *A homogeneous relation $r : X \leftrightarrow X$ is reflexive if and only if $id_X \subseteq r$.*

Another important subclass of homogeneous relations is that of symmetric relations, in which the order of the pair in the relation does not matter. For example, while equality is symmetric (since $x = y$ is equivalent to $y = x$), the relation $\leq$ is not (since, for example, $3 \leq 7$ while it is not the case that $7 \leq 3$).

Definition 6.14 *A homogeneous relation* $r : X \leftrightarrow X$ *is said to be symmetric if the inverse of any pair of objects related in r is also in r:*

$$\forall x, y : X \cdot (x, y) \in r \Rightarrow (y, x) \in r \qquad\qquad \blacksquare$$

There are also other ways of characterising symmetric relations.

Proposition 6.15 *The following three statements about a homogeneous relation* $r :$ $X \leftrightarrow X$ *are equivalent:* (i) r *is symmetric,* (ii) $r^{-1} \subseteq r$, (iii) $r^{-1} = r$.

Proof We start by proving that the first condition implies the second. Given that r is symmetric, we want to prove that $r^{-1} \subseteq r$. To prove set inclusion:

$(x, y) \in r^{-1}$
$\Leftrightarrow$ {proposition about relational inverse}
$(y, x) \in r$
$\Rightarrow$ {since r is symmetric}
$(x, y) \in r$

We can now proceed to prove that the second condition implies the first. Assuming that $r^{-1} \subseteq r$, we need to prove that r is symmetric, or that:

$$\forall x, y : X \cdot (x, y) \in r \Rightarrow (y, x) \in r$$

The proof thus starts with $(x, y) \in r$, from which we conclude $(y, x) \in r$:

$(x, y) \in r$
$\Leftrightarrow$ {proposition about relational inverse}
$(y, x) \in r^{-1}$
$\Rightarrow$ {since $r^{-1} \subseteq r$}
$(y, x) \in r$

It thus follows that $(x, y) \in r \Rightarrow (y, x) \in r$, from which it follows that r is symmetric.

The first and second condition are thus equivalent.

Clearly, the third condition implies the second. If we manage to show that the second implies the third, we would thus be able to conclude that the second and third are equivalent. Since the first and the second are equivalent, all three would thus be equivalent.

There remains to prove that $r^{-1} \subseteq r$ implies that $r^{-1} = r$. Since $r^{-1} = r$ is, by definition, $r^{-1} \subseteq r \land r \subseteq r^{-1}$, we need to prove that $r^{-1} \subseteq r$ implies $r \subseteq r^{-1}$:

$(x, y) \in r$
$\Leftrightarrow$ {proposition about relational inverse}
$(y, x) \in r^{-1}$
$\Rightarrow$ {since $r^{-1} \subseteq r$}
$(y, x) \in r$
$\Leftrightarrow$ {proposition about relational inverse}
$(x, y) \in r^{-1}$

This completes the proof of equivalence of the three statements. $\blacksquare$

A related subclass is that of anti-symmetric relations. A relation r is said to be anti-symmetric if, when both (x, y) and (y, x) are related by r, then x must be equal to y. For instance, the relation $\leq$ is anti-symmetric, since from $x \leq y$ and $y \leq x$, we can conclude that $x = y$. On the other hand, the relation $=_{\pm 1}$, which relates two numbers if their difference is exactly one, is not. A counterexample showing that it is not anti-symmetric is that, despite the fact that $1 =_{\pm 1} 2$ and $2 =_{\pm 1} 1$, it is not the case that $1 = 2$. Anti-symmetry is not to be confused with asymmetric relations—a relation is said to be asymmetric if it is not symmetric.

The last subclass of homogeneous relations we will identify is that of transitive relations. Such relations satisfy that, if x is related to y and y is related to z, then x must also be related to z. The relation $<$ is transitive, since from $l < m$ and $m < n$ we can conclude that $l < n$. On the other hand, the relation $=_{\pm 1}$ is not, since $1 =_{\pm 1} 2$ and $2 =_{\pm 1} 3$, while $1 =_{\pm 1} 3$ does not hold.

Definition 6.16 *A homogeneous relation $r : X \leftrightarrow X$ is said to be transitive if any pair of objects related by following r twice are already included in r itself:*

$$\forall x, y, z : X \cdot ((x, y) \in r \wedge (y, z) \in r) \Rightarrow (x, z) \in r \qquad \blacksquare$$

Another way of characterising transitive relations is to use relational composition to compose r with itself:

Proposition 6.17 *A homogeneous relation $r : X \leftrightarrow X$ is transitive if and only if $r \,\mathring{,}\, r \subseteq r$.*

Proof We start by proving the result in the forward direction—proving that $r \,\mathring{,}\, r \subseteq r$ if we know that r is transitive:

$$(x, z) \in r \,\mathring{,}\, r$$
$\Leftrightarrow$ {proposition about relational composition}
$$\exists y : X \cdot (x, y) \in r \wedge (y, z) \in r$$
$\Rightarrow$ {since r is transitive}
$$\exists y : X \cdot (x, z) \in r$$
$\Leftrightarrow$ {since y is not free in the body of the quantifier}
$$(x, z) \in r$$

In the other direction, we assume that $r \,\mathring{,}\, r \subseteq r$, and prove that r is transitive:

$$\forall x, y, z : X \cdot ((x, y) \in r \wedge (y, z) \in r) \Rightarrow (x, z) \in r$$

This is proved as follows:

$$(x, y) \in r \wedge (y, z) \in r$$
$\Rightarrow$ {$\exists$-introduction}
$$\exists y : X \cdot (x, y) \in r \wedge (y, z) \in r$$
$\Leftrightarrow$ {proposition about relational composition}
$$(x, z) \in r \,\mathring{,}\, r$$
$\Rightarrow$ {since $r \,\mathring{,}\, r \subseteq r$}
$$(x, z) \in r$$

It thus follows that $((x, y) \in r \land (y, z) \in r) \Rightarrow (x, z) \in r$, from which we can conclude that r is transitive.   ∎

In this section we identified four important classes of homogeneous relations—reflexive, symmetric, anti-symmetric and transitive relations. We will be see how relations from these classes case be used to compare items in a set.

Exercises

6.14 Prove Proposition 6.13.

6.15 Consider the subset-or-equal-to relation $\subseteq$. Is it (i) reflexive, (ii) symmetric, (iii) anti-symmetric, (iv) transitive?

6.16 Identify a relation which is both symmetric and anti-symmetric.

6.17 Unlike symmetry, transitivity of a relation r is equivalent to $r \, \r{\circ}\, r \subseteq r$, but not to $r \, \r{\circ}\, r = r$. Give an example of a relation r which is transitive but does not satisfy $r \, \r{\circ}\, r = r$.

6.18 Consider the three classes reflexivity, symmetry and transitivity. A relation may fall under any of the categories. This gives eight different types of relations. Identify a relation for each of these eight classes.

6.2.2 Order-Inducing Relations

Relations which are reflexive, anti-symmetric and transitive are called *partial orders*. Two common examples of a partial order are the not-larger-than relation ($\leq$) on numbers and the subset-or-equal-to relation on sets ($\subseteq$).

Direct descendant: Consider a relation *descendant* $\in$ PERSON $\leftrightarrow$ PERSON, which relates a person with another if the first one is a direct descendant of the second. For the purposes of the relation, a person is considered a direct descendant of themselves. This relation is (i) reflexive since everyone is a direct descendant of themselves; (ii) anti-symmetric, since if a is a direct descendant of b and b is a direct descendant of a, then clearly a must be the same person as b; and (iii) transitive, since if a is the direct descendant of b who is a direct descendant of c, then a must also be a direct descendant of c. Therefore, *descendant* is a partial order.

Natural number divisors: Now consider the relation *divides* $\in \mathbb{N} \leftrightarrow \mathbb{N}$, which relates two natural numbers n and m if n is an exact divisor of m. Clearly, every natural number is an exact divisor of itself, meaning that the relation is reflexive. Also, since a divisor of a number n can never be larger than n itself, if two numbers are exact divisors of each other, it follows that the numbers are equal. The relation is thus anti-symmetric. Finally, if l divides m which, in turn, divides n, we can conclude that l divides n, which means that the relation is also transitive. Hence it is a partial order.

Semantic entailment between propositions: If we consider semantic entailment $\models$ as a relation between propositions, with equality of propositions being semantic equivalence, it is easy to verify that $\models$ is a reflexive, anti-symmetric and transitive relation over well-formed propositional formulae. It is thus, a partial order.

Such a relation is called an *order* because transitivity ensures a notion of ordering of the objects which is preserved. Anti-symmetry, together with transitivity, is included to disallow cycles of objects which are not equal to each other. Consider a short cycle in an anti-symmetric, transitive relation r: (a, b), (b, c), (c, a). By transitivity, (b, a), (c, b) and (a, c) are also in r, implying, using anti-symmetry, that $a = b = c$.

These orders are called *partial* to distinguish them from *total orders*, which add the constraint that the relation ensures that any pair of values are comparable: for any two values a and b, it must either be the case that $(a, b) \in r$ or that $(b, a) \in r$. Note that this is not the same notion of relational totality we defined earlier in this chapter. While not-larger-than and subset relations are total orders, the three other examples are not; for instance, 3 is not a divisor of 5, and neither is 5 a divisor of 3.

Exercises

6.19 Prove that a relation which ensures comparability of any pair of values (the third condition in a total order) must be reflexive.

6.2.3 Equivalence Relations

We will now look at the type of relations which provide a notion of equivalence, which enables us to classify the related items into partitions. For instance, the relation *sameFiletype* : FILE $\leftrightarrow$ FILE, which relates two files if they are of the same type, is an equivalence because it can be used to separate all the files into partitions depending on their filetype, and with each partition containing files which are all related to each other.

These three properties are necessary for the notion of partitioning the type since (i) an item x is related to itself (since it must be in the same partition as itself); (ii) if x is in the same partition as y, and hence x is related to y, then y must be related to x since y is in the same partition as x; and finally (iii) if x is in the same partition as y, and y in the same partition as z, then x has to be in the same partition as z. An equivalence relation is thus a relation which is reflexive, symmetric and transitive. The notion of equality over numbers is an equivalence relation, since it satisfies all three constraints. Similarly, so is the relation *sameFiletype* we have just seen.

We can show that an equivalence relation $r : X \leftrightarrow X$ automatically gives us a means of partitioning items in X. By applying the relation to each object in the type, we get classes of objects related to each other. We will proceed to prove that these classes induce a partition of X.

Theorem 6.18 *An equivalence relation over X induces a partition of X: given an equivalence relation $r : X \leftrightarrow X$, the sets $P_x = r \langle\!\langle \{x\} \rangle\!\rangle$ partition X.*

Proof Recall the definition of a partition—the sets P_x are a partition of X if (i) they cover the whole of X, i.e. $\bigcup P_x = X$; and (ii) any two non-equal partitions are disjoint.

Complete coverage: We need to prove that the union of the sets in P_x is the whole type X. Since every object in the sets P_x is of type X, it immediately follows that $\bigcup P_x \subseteq X$. The proof of the opposite direction is as follows:

$$x \in X$$
$$\Rightarrow \{\text{since } r \text{ is a reflexive relation}\}$$
$$(x, x) \in r$$
$$\Rightarrow \{\text{by definition of relational application}\}$$
$$x \in r \langle\!\langle \{x\} \rangle\!\rangle$$
$$\Rightarrow \{\text{by definition of } P_x\}$$
$$x \in P_x$$
$$\Rightarrow \{\text{by definition of generalised union}\}$$
$$x \in \bigcup_x P_x$$

Hence, it follows that $X = \bigcup P_x$.

Disjoint partitions: The other property which must be satisfied by a partition is that each pair of non-equal partitions are mutually disjoint: $P_x \neq P_y \Rightarrow P_x \cap P_y = \emptyset$. We will prove the contrapositive of this statement, which we know is equivalent: $P_x \cap P_y \neq \emptyset \Rightarrow P_x = P_y$.

Let us assume that $P_x \cap P_y \neq \emptyset$, from which we will prove that $P_x = P_y$. Let z be an element of $P_x \cap P_y$ (which we assumed is non-empty). Since $z \in P_x$, it follows that $(x, z) \in r$. Similarly, since $z \in P_y$, we know that $(y, z) \in r$. Since r is symmetric we also know that $(z, x) \in r$ and $(z, y) \in r$.

We will use these facts in the following proof:

$$a \in P_x$$
$$\Rightarrow \{\text{by definition of } P_x\}$$
$$(x, a) \in r$$
$$\Rightarrow \{\text{since } (z, x) \in r \text{ and } r \text{ is transitive}\}$$
$$(z, a) \in r$$
$$\Rightarrow \{\text{since } (y, z) \in r \text{ and } r \text{ is transitive}\}$$
$$(y, a) \in r$$
$$\Rightarrow \{\text{by definition of } P_y\}$$
$$a \in P_y$$

This proof allows us to conclude that $P_x \subseteq P_y$. The proof in the other direction is similar.   ∎

6.3 Summary

In this chapter we have seen how relations fall under different classes, about which we can prove various properties. Classes of relations such as orders and equivalences frequently appear in mathematics and computer science, making it desirable to have laws which we know they satisfy.

In computer science, for instance, one can identify different equivalence and order relations over computer programs. For example, a useful equivalence relation is one which relates two programs which, given the same input, produce the same output. An important partial order on programs is one which relates a program with another if the first is not more efficient than the second. Using these two relations, one could see the process of optimisation as a task which starts with a program and attempts to produce another which (i) is in the same equivalence class as the original program according to the equivalence relation, but (ii) is more efficient (larger) than the original program according to the partial order.

Chapter 7
More Discrete Structures

As we have seen, the notion of sets and tuples allows the encoding of, and reasoning about, notions such as relations and functions without having to axiomatise them directly. In this chapter, we will see three other mathematical structures which can be encoded using the notions of sets, relations and functions.

7.1 Multisets

We characterised sets as collections of objects in which order and repetition do not matter. For instance, we wanted to ensure that the set $\{1, 2\}$ is equivalent to $\{2, 1, 1\}$. Although this is useful in various settings, sometimes we need to recognise repetition or order. Consider keeping track of the items a user is buying from an online store. The order in which the items were selected is not important, but recognising that a user is buying multiples of a single item is. In this case, we need a mathematical structure to keep track of repetition but not order. In this section we will formalise the notion of bags, or multisets, to be able to describe such structures. For example, if we use a notation similar to that used for finite sets but with sharp-edged parentheses, to represent multisets, the multiset $\lbrace 1, 2, 1 \rbrace$ would be expected to be equal to $\lbrace 2, 1, 1 \rbrace$, but not to $\lbrace 2, 2, 1 \rbrace$.

Definition 7.1 *A multiset, or bag, over type X is a collection of items from type X such that it is possible to query the number of instances of an item in the multiset, but not their order. We write $M \in \mathbb{M}\, X$ to denote that M is a multiset over type X.*

A multiset over type X can be encoded as a total function from X to the natural numbers, such that applying the function to an item $x \in X$ will give the number of instances of x in the multiset: $\mathbb{M}\, X \stackrel{\mathrm{df}}{=} X \to \mathbb{N}$. ∎

For example, the empty multiset $\varnothing$ contains zero copies of any item x of type X. This can be defined as $\varnothing \stackrel{\mathrm{df}}{=} \lambda x : X \cdot 0$. This representation allows us to query how many copies of a single object appear in a multiset using just function application. The number of copies of an object $x : X$ appearing in a multiset $M \in \mathbb{M}\, X$ is $M(x)$.

G.J. Pace, *Mathematics of Discrete Structures for Computer Science*,   157
DOI 10.1007/978-3-642-29840-0_7, © Springer-Verlag Berlin Heidelberg 2012

Based on such a representation, we can now define notions such as membership in a multiset.

Definition 7.2 *An object x of type X is said to be an* element *of a multiset M (over type X), written $x \in M$, if M contains at least one copy of x:*

$$x \in S \stackrel{\text{df}}{=} S(x) > 0$$

Given two multisets M_1 and M_2, both over type X, we say that M_1 is a submultiset of M_2, written $M_1 \sqsubseteq M_2$, if any item occurs in M_2 at least as many times as it occurs in M_1:

$$M_1 \sqsubseteq M_2 \stackrel{\text{df}}{=} \forall x : X \cdot M_1(x) \leq M_2(x)$$

Multiset equality is defined in terms of the submultiset relation:

$$M_1 \stackrel{\text{mset}}{=} M_2 \stackrel{\text{df}}{=} M_1 \sqsubseteq M_2 \wedge M_2 \sqsubseteq M_1$$

As usual, we will write $=$, when it is clear from the context that multisets are being compared for equality. ∎

It is worth noting that this definition of multiset equality is equivalent to having $M_1(x) \leq M_2(x)$ and $M_2(x) \leq M_1(x)$, and thus $M_1(x) = M_2(x)$ (for all values of x).
We can use these definitions to prove basic properties of multisets.

Example Just as with sets, we can prove that the empty multiset is a submultiset of any multiset: $\emptyset \sqsubseteq M$. By the definition of submultiset, we need to prove that $\forall x : X \cdot \emptyset(x) \leq M(x)$.

$$\emptyset(x)$$
$$= \{\text{by definition of } \emptyset\}$$
$$(\lambda x : X \cdot 0)(x)$$
$$= \{\text{function application}\}$$
$$0$$
$$\leq M(x) \qquad\qquad\qquad\qquad\qquad\qquad\qquad\qquad \diamond$$

Just as we did with sets, we can define operators to combine multisets together. The union of two multisets is rather straightforward: combine all items together into one set. For example taking the union of $\{1, 2, 2\}$ and $\{2, 3\}$ would give $\{1, 2, 2, 2, 3\}$—any item will occur in the union of two multisets as many times as it occurred in the first, in addition to as many times as it occurred in the second.

Definition 7.3 *The union of two multisets M_1 and M_2 over type X, written $M_1 \sqcup M_2$, is defined to be $\lambda x : X \cdot M_1(x) + M_2(x)$.* ∎

Formally defining the union of two multisets allows us to prove laws of the operator, such as that it is commutative.

Theorem 7.4 *Multiset union is commutative:* $M_1 \sqcup M_2 = M_2 \sqcup M_1$.

Proof To prove the set equality, we show that any object x occurs as many times in $M_1 \sqcup M_2$ as it does in $M_2 \sqcup M_1$:

$$(M_1 \sqcup M_2)(x)$$
$$= \{\text{definition of multiset union}\}$$
$$(\lambda x : X \cdot M_1(x) + M_2(x))(x)$$
$$= \{\text{function application}\}$$
$$M_1(x) + M_2(x)$$
$$= \{\text{addition is commutative}\}$$
$$M_2(x) + M_1(x)$$
$$= \{\text{function application}\}$$
$$(\lambda x : X \cdot M_2(x) + M_1(x))(x)$$
$$= \{\text{definition of multiset union}\} \qquad \blacksquare$$
$$(M_2 \sqcup M_1)(x)$$

The intersection of two multisets is slightly more involved. For example, the intersection of $\{1, 2, 2, 2\}$ and $\{2, 2, 3\}$ would give $\{2, 2\}$. Neither 1 nor 3 appear in the intersection, since they do not appear in one of the multisets. The number 2 occurs twice, since it occurs three times in the first and twice in the second multiset. No other item appears in the intersection since it does not appear in either multiset. In general, if x occurs n times in the first multiset and m times in the second, the number of copies occurring in the intersection is the smaller value of n and m.

Definition 7.5 *The intersection of two multisets M_1 and M_2 over type X, written $M_1 \sqcap M_2$, is defined to be* $\lambda x : X \cdot min(M_1(x), \ M_2(x))$. $\qquad \blacksquare$

As before, we can use this definition to prove properties of intersection.

Theorem 7.6 *The intersection of two multisets is included in their union:* $M \sqcap N \sqsubseteq M \sqcup N$.

Proof We will show that, for any object x, the number of instances of x in $M \sqcap N$ does not exceed the number of instances in $M \sqcup N$.

$$(M \sqcap N)(x)$$
$$= \{\text{by definition of multiset intersection}\}$$
$$(\lambda x : X \cdot min(M(x), \ N(x)))(x)$$
$$= \{\text{function application}\}$$
$$min(M(x), \ N(x))$$
$$\leq \{\text{basic arithmetic}\}$$
$$M(x) + N(x)$$
$$= \{\text{function application}\}$$
$$(\lambda x : X \cdot M(x) + N(x))(x)$$
$$= \{\text{by definition of multiset union}\}$$
$$(M \sqcup N)(x) \qquad \blacksquare$$

We have thus seen how multisets can also be encoded using sets and relations, allowing us to prove formal results about them using the mathematical tools we have identified in the previous chapters.

Exercises

7.1 Prove that the submultiset relation is reflexive and transitive.

7.2 Prove that the empty multiset acts as the zero of intersection: $\emptyset \sqcap S = S \sqcap \emptyset = \emptyset$, and the identity element of multiset union: $\emptyset \sqcup S = S \sqcup \emptyset = S$.

7.3 Prove that multiset intersection is idempotent, commutative and associative.

7.4 Prove that multiset union is idempotent and associative.

7.5 The difference of two multisets, written $M_1 - M_2$, is the operation which, for each instance of an item in the second operand, removes an instance from the first operand. For example $\{1,\ 2,\ 2,\ 2,\ 3\} - \{1,\ 1,\ 2,\ 2,\ 4\}$ would result in $\{2,\ 3\}$—two copies of 1 and 2 and one copy of 4 are removed from the first multiset. Formally define multiset difference, and prove that $M - M = \emptyset$ and that $M - \emptyset = M$.

7.2 Sequences

Multisets allow us to describe collections which may admit repetition, but are not adequate if the order is important. In computer science, structures in which order is crucial are also very frequent—print queues, arrays, prioritised search results and DNA sequences in bioinformatics are typical examples in which content order is important. In this section, we will formalise sequences, sometimes also called lists—collections in which order is important and repetition is possible. Unlike sets and multisets, the sequence $\langle 1,\ 2,\ 1 \rangle$ is distinct from $\langle 1,\ 2 \rangle$, $\langle 2,\ 2,\ 1 \rangle$ and $\langle 1,\ 1,\ 2 \rangle$. Note that we use angle brackets to mark the beginning and end of a sequence.

To formalise finite sequences, we use functions from the natural numbers to the type of items that will appear in the sequence. Applying the function to a number n will return the item at that position. For example, the sequence $\langle a,\ b,\ a \rangle$ can be encoded as the function $xs = \{(0, a),\ (1, b),\ (2, a)\}$. Note that we start counting from 0, and thus, to obtain the first item we write $xs(0)$. Also note that the function is partial, in that it is not defined for position 3 or higher, and its domain contains all numbers less than some value N (in this case, 3). If we define $upto(N)$ to be the natural numbers less than N: $\{n : \mathbb{N} \mid n < N\}$—we can formalise finite sequences as follows:

Definition 7.7 *A finite sequence or list xs over type X is a collection of items from type X in which both order and repetition is important. We will write $xs \in seq\ X$ to denote that xs is a finite sequence over type X. A finite sequence over type X can be encoded as a function from the natural numbers to X, such that applying the* *function to a number n will return the item at position n. Furthermore, the domain must also be the natural numbers up to some value N:*

$$seq\ X \stackrel{\mathrm{df}}{=} \{xs : \mathbb{N} \to X \mid \exists N : \mathbb{N} \cdot dom(xs) = upto(N)\}$$

The empty sequence is defined to be the empty function: $\langle\rangle \stackrel{\mathrm{df}}{=} \emptyset.$ ■

Based on this, we can define operations on finite sequences.

Definition 7.8 *The head of a sequence xs, written head(xs), is a function returning the first item in the sequence:* $head(xs) \stackrel{\mathrm{df}}{=} xs(0)$. *Similarly, the tail of a sequence xs, written tail(xs), is also a function returning the same list, but excluding the first item:* $tail(xs) \stackrel{\mathrm{df}}{=} \lambda n \cdot xs(n+1)$. *Neither head nor tail are defined for the empty sequence, and are thus not total functions.*

Combining a single item x to the front of a list xs, written cons(x, xs), is defined as follows:

$$cons(x, xs) \stackrel{\mathrm{df}}{=} \lambda n \cdot \begin{cases} x & \textit{if } n = 0 \\ xs(n-1) & \textit{otherwise} \end{cases}$$

The length of a sequence xs, written length(xs), is defined to be zero if the sequence is empty, and one more than the largest index in its domain otherwise:

$$length(xs) \stackrel{\mathrm{df}}{=} \begin{cases} 0 & \textit{if } xs = \langle\rangle \\ 1 + max(dom(xs)) & \textit{otherwise} \end{cases}$$

Using the length of a sequence, we can define a function to obtain the last item in a non-empty sequence: $last(xs) \stackrel{\mathrm{df}}{=} xs(length(xs) - 1)$. ■

Of these functions, *tail* and *cons* return a sequence. Since sequences have a well-formedness condition—that their domain is the set of all natural numbers less than some particular value—we must show that the functions are well defined, that they really return a sequence. Without going into the details of the proof, we outline the argument one would use to show that *tail* is well defined—we note that for a non-empty *xs*, if the domain of *xs* is *upto(N)*, then the domain of *tail(xs)* is (by definition of *tail*) the domain of $\lambda n \cdot xs(n+1)$, which can be shown to be *upto(N − 1)*.

Theorem 7.9 *The length of a non-empty sequence xs is one more than the length of its tail:* $length(xs) = length(tail(xs)) + 1$.

Proof Consider a non-empty sequence *xs*, with domain *upto(N +1)*. The proof will take into account two cases: (i) when $N = 0$ and thus $tail(xs) = \langle\rangle$, and (ii) when $N > 0$. The proof for $N = 0$ is straightforward, and is left as an exercise.

The case when $N > 0$ is given below. Note that, since the domain is $upto(N + 1)$ and $N > 0$, we know that $tail(xs) \neq \langle\rangle$.

$$
\begin{aligned}
&length(xs) \\
&= \{\text{definition of } length \text{ and } xs \neq \langle\rangle\} \\
&\quad max(dom(xs)) + 1 \\
&= \{\text{domain of } xs \text{ is } upto(N + 1)\} \\
&\quad max(upto(N + 1)) + 1 \\
&= \{\text{definition of } upto \text{ and basic arithmetic}\} \\
&\quad N + 1 \\
&= \{\text{definition of } upto \text{ and basic arithmetic}\} \\
&\quad max(upto(N)) + 2 \\
&= \{\text{the domain of the following function is } upto(N)\} \\
&\quad (max(dom(\lambda n \cdot xs(n + 1))) + 1) + 1 \\
&= \{\text{definition of } length \text{ and } tail(xs) \neq \langle\rangle\} \\
&\quad length(\lambda n \cdot xs(n + 1)) + 1 \\
&= \{\text{definition of } tail\} \\
&\quad length(tail(xs)) + 1 \qquad\qquad\qquad\qquad\qquad\qquad\qquad\blacksquare
\end{aligned}
$$

When defining functions over sequences, it is frequently convenient to define them recursively. For instance, an alternative definition of the length of a sequence xs can be expressed as: (i) the length is 0 if xs is the empty sequence; (ii) the length is $1 + length(tail(xs))$ otherwise. Formally, this can be written as:

$$
length(ns) \overset{df}{=} \begin{cases} 0 & \text{if } ns = \langle\rangle \\ 1 + length(tail(ns)) & \text{otherwise} \end{cases}
$$

An important issue to consider is that the recursive definition is applied to progressively smaller sequences. In this case, as in most others, to the *tail* of the original sequence. A similar definition can be used to sum a list of natural numbers:

$$
sum(ns) \overset{df}{=} \begin{cases} 0 & \text{if } ns = \langle\rangle \\ head(ns) + sum(tail(ns)) & \text{otherwise} \end{cases}
$$

This approach allows us to give different definitions to useful operators. For instance, the catenation of two sequences, written $xs \mathbin{+\mkern-8mu+} ys$, returns a sequence composed of the items in xs (in order) followed by the items in ys (also in order). For example $\langle a, b \rangle \mathbin{+\mkern-8mu+} \langle c, b, a \rangle$ should give $\langle a, b, c, b, a \rangle$. One way of defining catenation is by defining the function directly:

$$
xs \mathbin{+\mkern-8mu+} ys \overset{df}{=} \lambda n \cdot \begin{cases} xs(n) & \text{if } n \in dom(xs) \\ ys(n - length(xs)) & \text{otherwise} \end{cases}
$$

Another way is to define the catenation function recursively on its first operand:

$$
xs \mathbin{+\mkern-8mu+} ys \overset{df}{=} \begin{cases} ys & \text{if } xs = \langle\rangle \\ cons(head(xs), tail(xs) \mathbin{+\mkern-8mu+} ys) & \text{otherwise} \end{cases}
$$

This second definition removes elements from the first operand, catenating the rest, then adding them to the result. Seeing it work on an example illustrates how:

$$\langle a,\ b \rangle \mathbin{+\!\!\!+} \langle c,\ b,\ a \rangle$$
$= \{\text{definition of } \mathbin{+\!\!\!+}\}$
$\quad cons(head(\langle a,\ b \rangle), tail(\langle a,\ b \rangle)) \mathbin{+\!\!\!+} \langle c,\ b,\ a \rangle)$
$= \{\text{definition of } head \text{ and } tail\}$
$\quad cons(a, \langle b \rangle \mathbin{+\!\!\!+} \langle c,\ b,\ a \rangle)$
$= \{\text{definition of } \mathbin{+\!\!\!+}\}$
$\quad cons(a, cons(head(\langle b \rangle)), tail(\langle b \rangle)) \mathbin{+\!\!\!+} \langle c,\ b,\ a \rangle)$
$= \{\text{definition of } head \text{ and } tail\}$
$\quad cons(a, cons(b, \langle \rangle \mathbin{+\!\!\!+} \langle c,\ b,\ a \rangle)$
$= \{\text{definition of } \mathbin{+\!\!\!+}\}$
$\quad cons(a, cons(b, \langle c,\ b,\ a \rangle)$
$= \{\text{definition of } cons\}$
$\quad cons(a, \langle b,\ c,\ b,\ a \rangle)$
$= \{\text{definition of } cons\}$
$\quad \langle a,\ b,\ c,\ b,\ a \rangle$

One advantage of defining two versions of the same function is that we can prove that the two are equivalent—confirming that we defined the operator in a correct manner. However, the proof of equivalence of the two definitions of catenation requires the use of techniques which we will explore later on in Chap. 8.

When discussing sets and multisets, we defined the notion of what a subset and a submultiset are. In sequences we have a number of different definitions we can adopt, depending on the application. For example, consider the sequence $qs = \langle a,\ b,\ c,\ d \rangle$. There are different notions of subsequences we can adopt:

Prefix: A sequence xs is a prefix of ys, written $xs \preceq_p ys$, if xs is the first part of ys. For example, $\langle a,\ b,\ c \rangle$ is a prefix of qs. Formally, we can define the prefix relation as: $xs \preceq_p ys \stackrel{\text{df}}{=} \exists zs : seq\ X \cdot xs \mathbin{+\!\!\!+} zs = ys$.

Suffix: Similarly, one can talk about a suffix of a sequence (written $xs \preceq_s ys$), relating a sequence to another if it appears at the end. For example, $\langle c,\ d \rangle$ is a suffix of qs. Formally, we can define the suffix relation as: $xs \preceq_s ys \stackrel{\text{df}}{=} \exists zs : seq\ X \cdot zs \mathbin{+\!\!\!+} xs = ys$.

Exact subsequence: An exact subsequence must occur somewhere within the main sequence. For example, the sequence $\langle b,\ c \rangle$ appears within qs, and is thus an exact subsequence of qs. We define the notion that xs is an exact subsequence of ys, written $xs \preceq ys$, as: $xs \preceq ys \stackrel{\text{df}}{=} \exists zs_1, zs_2 : seq\ X \cdot zs_1 \mathbin{+\!\!\!+} xs \mathbin{+\!\!\!+} zs_2 = ys$.

Which containment operator to use largely depends on the application. Furthermore, one can define more complex subsequence relations. For instance, xs is an interleaved subsequence ys if the items in xs appear in ys in the right order but possibly with additional items interleaved. For example, one finds $\langle b,\ d \rangle$ interleaved in sequence qs—with b and d appearing in the right order but with additional items inserted in between.

In this section we have seen a way of formalising and reasoning about finite sequences. In Chap. 8, we will look at a different way of formalising sequences, allowing us to reason about them in a more effective way. However, at this stage, it is important to see that collections in which both order and repetition matter can be reasoned about mathematically with the tools we have at hand.

It is worth noting that we did not look at collections in which order, but not repetition matters. To formalise these, we can use sequences, but with an additional constraint that no item appears twice in a sequence—in other words, that the function from the natural numbers to X is injective.

Another constraint is that our sequences are of finite length. If we were to define infinite sequences over a type X, we would look at *total* functions from $\mathbb{N}$ to X, which allow us to reason about infinite ordered collections in a formal manner: $seq_\infty X \stackrel{df}{=} \{s : \mathbb{N} \to X \mid s \text{ is total}\}$.

Exercises

7.6 Prove the case when $N = 0$ of Theorem 7.9.

7.7 Prove that, for a sequence of length 1, the head and the last item of the sequence are the same: $length(xs) = 1 \Rightarrow head(xs) = last(xs)$.

7.8 Define the operator $x \lessdot xs$, which is true if and only if x appears (at least once) in sequence xs.

7.9 Define a predicate *sorted* which, given a sequence $xs \in seq\ X$ and an ordering relation $\leq_X \in X \leftrightarrow X$ (where $x \leq_X y$ is taken to mean that x is not larger than y), returns whether or not xs is sorted in ascending order.

7.10 Define a function *items*, which given a sequence xs, returns a multiset with the items in a finite sequence xs.

7.11 Use the solutions of the previous two exercises to define a relation *sorts* such that $(xs, xs') \in sorts$ if an only if xs is a sorted version of xs'.

7.12 Define the sequence equality relation $\stackrel{seq}{=}$.

7.13 The reverse of a sequence, written $reverse(xs)$, returns the original sequence but in reverse order. For example the reverse of $\langle a, b, c \rangle$ is $\langle c, b, a \rangle$. Just as we did with catenation, define *reverse* in two different ways: (i) using lambda notation, and (ii) using recursion.

7.3 Graph Theory

As we have seen time and time again, mathematics is about discovering common patterns, commonalities between problems. In this manner, we can prove properties of the general case, and apply the results to all instances. The notion of a sequence can be equally applied to arrays of information on a computer, queues of persons waiting to be served, and an ordered list of items a user has put in his virtual shopping cart. Once we prove a law about sequences, all these instances must obey it—for example, in all cases, the catenation of sequences must be associative.

Another notion that occurs frequently in different forms is that of a graph. Consider the following three diagrammatic representations of real-life concepts.

Air travel between cities: Consider a representation of flights between cities in Europe with a particular airline and the associated costs. The information can be represented using a table, or visually using a diagram, with arrows between cities to show the possibility of flying between those two cities (in the direction marked by the arrow), and with the cost of that flight marked on the arrow between the cities.

The following diagram uses the graphical representation:

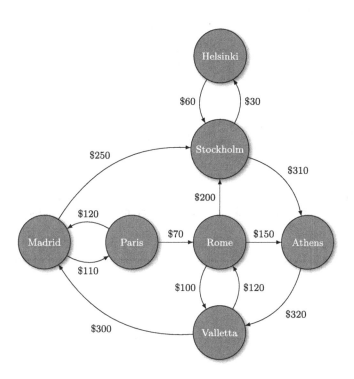

Through the use of such a representation, one can pose (and answer) questions such as 'Is there a way of travelling from a particular city c_1 to another city c_2 spending less than budget $\$B$?', or 'Starting from a particular city c, what budget is required to enable visiting all the cities?'

User authentication: The diagram on the next page gives a model of how user authentication is to take place on a particular system. The user is requested to input his or her credentials, and if successful, he or she is then allowed to read and write files until logging out. If a bad password is given, this is taken into account, so that after three consecutive bad passwords, the user's account is disabled. A successful login resets the count of wrong passwords to zero.

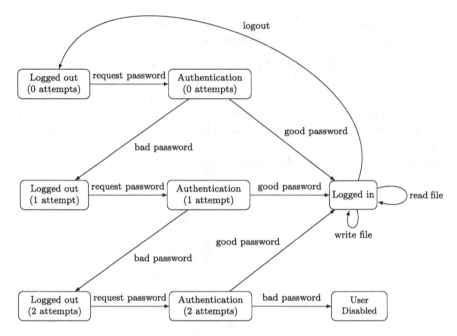

Apart from giving a concrete specification to be implemented, such a representation allows one to analyse the design proposed, for instance checking whether a user can read a file if the last login attempt had been with a wrong password, or whether the number of failed logins can ever be more than twice the number of good logins.

Playing tic-tac-toe: As a third model which lends itself to a diagrammatic approach, we will look at the game tic-tac-toe. Consider the diagram shown below:

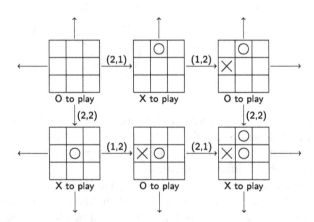

The diagram is a partial (since the depiction of all the game possibilities would be huge) representation of the game of tic-tac-toe. Each possible game position would be included in the full representation. Arrows show the result of playing at a certain coordinate. Although the representation of a simple game such as tic-tac-toe is massive, it provides us with a means of analysing the game and checking, for instance, whether the starting player has a guaranteed winning strategy.

Despite the radical differences between the problems, all three are represented by diagrams which share a number of common features: All three problems have been represented using (i) a number of nodes (cities, the state of the authentication procedure and positions in a game of tic-tac-toe), (ii) which are connected via arrows, (iii) each of which carries some relevant information (price of a flight between the cities, the operation performed and the move a player has performed). The objects are called nodes or vertices, the arrows are called edges, while the information is called a label. The whole construction is called a graph. To formalise a graph, we note that each edge can be represented as a triple (v, l, v') identifying the source vertex v, the destination vertex v' and the label on the edge l. The set of edges thus corresponds to a set of triples. In general, we can formalise a graph as follows:

Definition 7.10 *Given* (i) *a set of vertices (or nodes)* V, (ii) *a set of labels* L, *and* (iii) *a set of labelled edges between vertices* $E \subseteq V \times L \times V$, *the triple* $G = (V, L, E)$ *is called a graph.* ∎

Example The graph below is an insult generator—if we follow edges from vertex A to vertex E, the sequence of labels along the edges we take gives an insult. For instance, if we follow the vertices A, G, D, E, E (by taking the lower edge from E to E labelled with *vile*), E (now taking the upper edge labelled with *green*), F (via the lower edge from E to F), we will obtain the insult *You are a vile green turnip*. Such a graph can be used in different ways. One application would be to randomly follow a path through the graph, thus generating insults. Another application would be parsing—given a sentence, checking whether there is a path in the graph which matches that sentence, thus checking for validity of the insult.

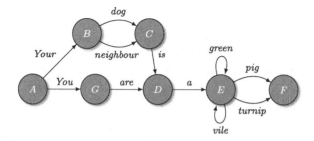

To formalise the graph G, we have to identify (i) the set of vertices, (ii) the set of labels and (iii) the set of edges:

$$V \overset{\mathrm{df}}{=} \{A, \ B, \ C, \ D, \ E, \ F, \ G\}$$

$$L \overset{\mathrm{df}}{=} \{You, \ Your, \ neighbour, \ dog, \ is, \ a, \ vile, \ green, \ pig, \ turnip\}$$

$$E \overset{\mathrm{df}}{=} \{(A, Your, B), \ (B, dog, C), \ (B, neighbour, C), \ (C, is, D),$$

$$(A, You, Z), \ (Z, are, D), \ (D, a, E), \ (E, vile, E),$$

$$(E, green, E), \ (E, pig, F), \ (E, turnip, F)\} \qquad \diamond$$

Using such a formalisation of graphs, we can define operators or predicates on them. For instance, a clique is a set of vertices such that any pair of different vertices in the clique are connected directly via an edge:

Definition 7.11 *A set of vertices V' forming part of a graph $G = (V, \ L, \ E)$ are said to be a clique if any pair of distinct vertices in V' are connected to each other via some edge in E: $\forall v, v' \in V' \cdot v \neq v' \Rightarrow \exists l : L \cdot (v, l, v') \in E$.* ∎

Note that every single vertex set is a trivial clique. In the insult-generator example, we find only trivial cliques. The air traveller's graph has various non-trivial cliques, such as $\{Rome, \ Valletta\}$. If we were to add flights from Athens to Rome and from Valletta to Athens, $\{Rome, \ Valletta, \ Athens\}$ would also be a clique.

As we have seen in the insult-generator example, we would like to be able to traverse the vertices of a graph along its edges, to study the sequences of vertices, which we call paths, or sequences of labels, which we call words, which can be followed in the graph. What such paths and words represent depends on what the graph depicts—in the last example, the words give text sequences which the insult generator can produce, while in the air travel example, a path would correspond to a sequence of cities which a traveller may visit in that exact order using only air travel.

Definition 7.12 *Given a graph $G = (V, \ L, \ E)$, a sequence of vertices $vs \in seq \ V$ is said to be a valid path in G if each consecutive pair of vertices in vs is joined via an edge in G:*

$$\forall i : \mathbb{N} \cdot i < length(vs) - 1 \Rightarrow \exists l : L \cdot (vs(i), l, vs(i + 1)) \in E$$

We write $paths(G)$ to denote the set of all valid paths in G.

A cycle in G is a valid path in G which terminates on the same vertex where it started. We write $cycles(G)$ to denote the set of all valid cycles in G:

$$cycles(G) \overset{\mathrm{df}}{=} \{vs : seq \ V \mid vs \in paths(G) \land head(vs) = last(vs)\}$$ ∎

Example Consider the authentication system graph. A valid path in the graph corresponds to a valid sequence of states the system may go through, such as:

$$\langle Logged\text{-}out\text{-}(1\text{-}attempt), \ Authentication\text{-}(1\text{-}attempt),$$
$$Logged\text{-}in, \ Logged\text{-}in, \ Logged\text{-}out\text{-}(0\text{-}attempts)\rangle$$

A cycle corresponds to a sequence of states which the system can loop through indefinitely often, such as:

⟨*Logged-out-(0-attempts)*, *Authentication-(0-attempts)*,
Logged-in, *Logged-in*, *Logged-out-(0-attempts)*⟩

Using these notions of paths and cycles, we can, for instance, specify that the system can never go through a cycle without the user logging in successfully at some point—*every cycle in the system must go through a logged-in state*:

$$\forall vs : seq\ V \cdot vs \in cycles(G) \Rightarrow Logged\text{-}in \in ran(vs)$$

A stronger property is that the system may not progress indefinitely without the user being eventually logged in. Using infinite sequences, our property can be stated as: *there is no infinite path in the system which does not go through the Logged-in state.* Since, we have only shown how to reason about finite paths, we have to express this property in a more intricate manner: *there is a limit on the length of paths through the system which do not go through Logged-in.* We can write this as: *there is a numeric value N such that any valid path beyond length N must go through the Logged-in state:*

$$\exists N : \mathbb{N} \cdot \forall vs : seq\ V \cdot vs \in paths(G) \wedge length(vs) > N \Rightarrow Logged\text{-}in \in ran(vs) \; \diamond$$

Some graphs are structurally simpler than others. For instance, if you try to draw the full graph for tic-tac-toe, you will discover that the graph has no cycles. Since every edge increases a symbol on the board, which cannot be removed, one can never go back to a vertex after leaving it. Such graphs are called directed acyclic graphs.

Definition 7.13 *A graph G is said to be a directed acyclic graph (DAG), if* $cycles(G) = \emptyset$. ∎

Example Trees are another class of graphs, which we will characterise as an example of how one approaches such a task. There are different ways which one can use to characterise trees. The approach we will use is to say that a tree is a graph with the constraint that each vertex may have no more than one predecessor (called the parent), and with only one vertex (called the root) having no predecessors. Also, every vertex should be reachable from the root vertex.

Before we characterise a tree, we need the notion of the predecessors of a vertex—the set of vertices which can reach v by traversing exactly one edge:

$$predecessors(v) \overset{df}{=} \{v' : V \mid \exists l : L \cdot (v', l, v) \in E\}$$

To simplify the specification of a tree, we will also defined the set of vertices *reachable* from a particular vertex:

$$reachable(v) \overset{df}{=}$$
$$\{v' : V \mid \exists vs : seq\ L \cdot vs \in paths(G) \wedge head(vs) = v \wedge last(vs) = v'\}$$

A graph $G = (V, L, E)$ is said to be a tree, if (i) vertices have no more than one predecessor: $\forall v : V \cdot \#(predecessors(v)) \leq 1$; (ii) exactly one vertex (which we call the root) which has no predecessors: $\exists_1 v : V \cdot \#(predecessors(v)) = 0$; and (iii) all vertices are reachable from the root: $\forall v : V \cdot \#(predecessors(v)) = 0 \Rightarrow reachable(v) = V$.

Note the notation $\# S$ to obtain the size of a finite set S. This notation will be formalised in Chap. 9. Based on this definition, one can prove that all trees are also DAGs. A rigorous proof can be rather intricate. However, the intuition behind the reasoning is as follows. We will assume that a tree has a cycle vs in order to reach a contradiction. Since every vertex in the cycle has one predecessor, the root r cannot be part of the cycle vs. However, by the definition of a tree, all the vertices, including the ones in vs, are reachable from the root r. This implies that some vertex in vs must have a predecessor other than the one appearing before it in the cycle, which contradicts the tree constraint that no vertex may have more than one predecessor. This contradiction allows us to conclude that there can be no cycles in the graph, thus making it a DAG.

There are other ways of characterising a tree, one of which is an exercise at the end of this section.   ◇

Although in some graphs we are interested in sequences of states which can be followed along the edges, in others we are more interested in the sequences of labels of edges which are followed. For example, in the insult generator, the insults are not sequences of vertices, but sequences of labels. We will define the set of such sequences similarly to what we did with paths.

Definition 7.14 *Given a graph $G = (V, L, E)$, a sequence of labels $ls \in seq\ L$ is said to be a* valid word *in G, if there is a path in G along which the sequence of labels appears on the edges traversed:*

$$\exists vs : paths(vs) \cdot length(vs) = length(ls) + 1\ \wedge$$
$$\forall i : \mathbb{N} \cdot i < length(vs) - 1 \Rightarrow (vs(i), l(i), vs(i+1)) \in E$$

The set of all valid words of a graph G will be referred to as $words(G)$.

We call such a word-path pair (ls, vs) a labelled path *of G. The set of all labelled paths of a graph G will be written as $lpaths(G)$.*   ∎

Example A well-known problem in computer science is the so-called *travelling salesman problem*. Many variants of the problem can be found, but the underlying problem is that of taking a graph with numeric costs on the edges, and asking whether a traveller can visit all the cities (vertices) such that the total cost (the sum of the costs of all the flights taken) does not exceed a particular budget. The problem can be specified rather easily using the notion of labelled paths: *Given a graph with numeric labels $G = (V, \mathbb{N}, E)$ and budget B, is the following predicate true?*

$$\exists (vs, ls) \in lpaths(G) \cdot ran(vs) = V \wedge sum(ls) \leq B$$

So, how does one go about programming a solution to this problem? The most obvious solution is to try out all possible sequence paths through the graph and check them individually. Unfortunately, the number of such paths can be exponential with respect to the number of cities—if the number of cities is c, then the number of paths can be up to 2^c. Exponential complexity is bad news since, adding one more city doubles the time taken by the program. For example, if we somehow manage to write a very efficient program which can solve an instance of the problem with 1,000,000 cities in just one second, solving it for ten additional cities will take over 1000 seconds, or over a quarter of an hour. Solving it for 1,000,021 cities will take over a year, eventually taking more than the age of the universe to solve it with 1,000,055 cities. So, the question is whether we can find a more efficient algorithm to solve this problem. Alas we know of no solution which is more efficient. Neither have computer scientists managed to prove that no more efficient algorithm exists. In fact there is a whole class of such problems, called *NP*-complete problems, which we know to be equivalent to each other—in that if we manage to solve one efficiently then the whole class of problems can be solved in an efficient manner, while if one is shown not to be possible to solve efficiently, then neither are any of the others. This class includes various practical problems, such as setting a timetable and running a number of jobs in parallel on different computers to minimise the time taken to produce the result.   ◇

We have already seen a wide spectrum of scenarios which can be modelled using a graph; however, there are other problems which may seem completely unrelated to graphs but which, with some insight, can be encoded as graph problems. Consider the following example:

Example Recall that in Chap. 2 we saw how different problems can be formulated using propositional logic. A propositional logic formula was said to be a contradiction if, no matter what values the propositional variables used take, the formula will always be false. A simple example of a contradiction is the formula $P \wedge \neg P$. No matter what value P has, the formula is false.

Consider the problem of checking whether or not a propositional formula is a contradiction. For simplicity, we will assume that the formula is in conjunctive normal form, which means that it is the conjunction of a number of clauses, each of which is a formula consisting of a disjunction of a number of literals, each of which is either just a variable or the negation of a variable. An example of a formula in conjunctive normal form is: $(P \vee Q \vee \neg P) \wedge (Q \vee \neg R)$. The formula is a conjunction over the clauses $(P \vee Q \vee \neg P)$ and $(Q \vee R)$, which, in turn, consist of a disjunction over literals. It can be shown that all propositional formulae can be rewritten in conjunctive normal form.

We will now see how we can pose the question of whether a propositional formula is a contradiction in terms of a question about graphs. Given a formula in conjunctive normal form, we will construct a graph in the following manner:

Vertices: For each literal—each instance of a variable or the negation of a variable in the formula—a different vertex will be created. For instance, $(P \vee Q \vee \neg R) \wedge (Q \vee R)$ will result in five vertices.

Labels: No labels are necessary for this transformation. However, since our graphs require labels, we will use a label 0 for each edge.

Edges: Two vertices will have an edge joining them if and only if (i) they come from different clauses; and (ii) they were not created from two complementary literals P and $\neg P$.

Consider the conjunctive normal form expression:

$$(P \vee Q \vee \neg R)$$
$$\wedge (\neg P \vee \neg R)$$
$$\wedge (P \vee R \vee \neg Q)$$

This will result in the following graph, where an edge with arrows on both ends is used to mark two edges going back and forth between the two vertices:

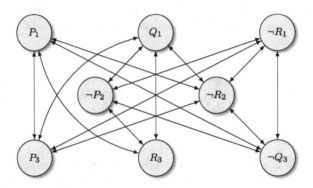

Despite the criss-crossed edges, making the graph difficult to read, it is very straight-forward to construct. Note, for instance, that P_1 is not connected to $\neg P_2$ because, despite the fact that they are from different clauses, they are complementary literals. Now consider the question of whether the resulting graph has a clique of size 3—the number of clauses in the original formula. One such clique is Q_1, $\neg P_2$ and R_3, and if we set variables Q and R to true and P to false, the whole formula is satisfied and is thus not a contradiction. Another clique we could have found is $\neg R_1$, $\neg R_2$ and $\neg Q_3$, and once again, if we set both Q and R to false, the formula is satisfied and hence is not a contradiction. If we find a clique with as many nodes as clauses, we are guaranteed that (i) we have picked up a vertex from each clause (we cannot choose two from the same clause since they would not be connected together), and (ii) the literals in the clique are not contradictory (since otherwise they would not be connected). Therefore, there is a way of making each clause true, thus making the whole formula true.

On the other hand, if the formula is not a contradiction, there is a way of making one literal of each clause true (to make each disjunction true, resulting in the formula being true). Furthermore, these vertices are all connected via edges since (i) they come from different clauses, and (ii) do not contradict each other. Hence, we must have a clique.

This leads to the observation that: *There is a clique the size of the number of clauses in the original formula if and only if the formula was not a contradiction.*

We have thus reduced checking whether a formula is a contradiction to checking whether a graph has a clique of a particular size. What this means is that checking for a contradiction can be no harder than (i) translating the expression into a graph, plus (ii) checking whether the graph has a clique of a particular size.

From this observation, two corollaries follow. The first is that, if we manage to write an efficient algorithm to check for cliques, then we can check for contradictions efficiently, since the translation into the graph is rather straightforward to perform. The second, less obvious, observation is that, if somehow we prove that there is no efficient algorithm to check for satisfiability of expressions in this format, then there can never be an efficient way of checking for cliques (otherwise we would use that algorithm to check for contradictions efficiently). Alas, we know that both checking for contradictions in propositional expressions and checking for the existence of cliques are as difficult as the travelling salesman problem. We thus still have no clue whether an efficient algorithm to solve these problems exists.   ◇

The graphs we have formalised are directed graphs with labelled edges. Directed because edges have a fixed direction, and can only be traversed in that direction, and labelled because each transition carries a label. As we have seen in the last example, in some cases, we do not need one or both of these features, in which case, we would have to formalise in a different manner. If, for example, we wanted to formalise unlabelled graphs, we would use a definition very similar to the definition of directed graphs which we have seen, but leave out the set of labels L, thus making the graph of the form (V, E), where the edges are a subset of $V \times V$ (since they now carry no label). If we needed undirected graphs, the formalism would be identical to the one we have used in this section, but the interpretation of the graph, for instance characterising what a valid path or the predecessors of a vertex are, would be different.

Exercises

7.14 A Hamiltonian path is a path which goes through each vertex exactly once. Define a predicate to check whether a given path is a Hamiltonian one.

7.15 A simple cycle in graph G is a valid cycle in G with no vertex repeated (other than that of the first and last item in the path). Define the set of simple cycles in graph G.

7.16 The diameter of a graph G is the length of the longest valid path in G which goes through no vertex more than once. Define the function *diameter*(G).

7.17 There are other ways one can define the class of trees. In this exercise, we will build the machinery to be able to characterise trees in a different way:

   (i) Given a graph G, we define the undirected version of G, written $\widehat{G}$, as a graph identical to G, but replicating each edge to be traversable in either direction; i.e. for every edge (v, l, v') in G, we also add edge (v', l, v) to $\widehat{G}$. Define $\widehat{G}$ in terms of G.

  (ii) A graph G is said to be weakly connected if there is a path joining any two nodes in $\widehat{G}$. Give a predicate stating this constraint.

 (iii) A graph is said to have shared structure if there are two vertices v and v' for which there are at least two distinct paths going from v to v'. Write a predicate which corresponds to this constraint.

 (iv) An alternative way of defining a tree is that it is a weakly connected DAG without shared structure. Express this alternative definition of a tree.

7.18 Show that $(P \vee \neg Q) \wedge (Q \vee P) \wedge P$ is not a contradiction by using the construction given and checking for cliques.

Chapter 8
Defining New Structured Types

Up till this point, we have constructed our structures on the basis of propositions and predicates, assuming the existence of various types which can be used in specifications. But where do these types come from? We have assumed that we have a type of counting numbers—the natural numbers $\mathbb{N}$, and the real numbers $\mathbb{R}$, and various other types as needs dictate, such as the type of persons PERSON and the type FILE of files in an operating system. Although we have already seen ways of how new types can be constructed from types we already know, using power sets and Cartesian products, we still need an approach to be able to define new basic types. In this chapter we will investigate how we can formally construct and reason about new types.

Let us start with an example. If we want to reason about black and white computer images, we can formalise an image as a table of pixels, with colour information about each location. We can describe the table as a sequence of rows, each of which is a sequence of colour information. For example, a 3×2 pixel image with all pixels being white except for the top-left- and bottom-right-hand corner ones, would be encoded as:[1]

Picture-BW $\overset{\mathrm{df}}{=}$ *seq* Row-BW

Row-BW $\overset{\mathrm{df}}{=}$ *seq* Colour-BW

pic3x2 $\in$ Picture-BW

pic3x2 $\overset{\mathrm{df}}{=}$ $\langle\langle Black,\ White,\ White\rangle,\ \langle White,\ White,\ Black\rangle\rangle$

---

[1]The description and formalisation we present here is not ideal, since it allows for rows of different length.

G.J. Pace, *Mathematics of Discrete Structures for Computer Science*,
DOI 10.1007/978-3-642-29840-0_8, © Springer-Verlag Berlin Heidelberg 2012

However, although we know how to reason about sequences, the problem is formulated using a colour type Colour-BW, about which we do not know how to reason. For instance, although we would expect the type to consist of exactly two items Black and White, we have no way of proving that Red is not of type Colour-BW. What are the axioms and rules of inference for this type?

If we axiomatise the new type from scratch, we would want to ensure that the axioms and rules of inference are (i) sufficiently strong to enable us to prove interesting statements about the type, and (ii) sound, in that they do not contradict each other. Ensuring this about a type even as simple as Colour-BW can be challenging. In this chapter, we explore a general approach to define and axiomatise new types of discrete structures, such as Colour-BW but also more complex ones such as lists, trees and natural numbers.

8.1 Notions and Notation

To start off, we will look at ways of intuitively declaring new structured types, and define functions on these types. We defer discussing how to formally reason about these types to a later section.

8.1.1 Simple Enumerated Types

Let us consider the type Colour-BW, consisting of only two distinct elements—Black and White. In such cases, when the items in a new type can be individually enumerated, we will write them as follows:

$$\text{Colour-BW} ::= \text{Black} \mid \text{White}$$

The type declaration says that we are defining a new type Colour-BW (named before the ::= symbol) consisting of objects matching one of a number of options separated by a vertical line | symbol. Later on, we will look at how we will ensure that the type works as expected—for instance, that it consists of exactly two distinct objects, and nothing else.

Since the number of items in the type is finite, we can define functions over these types by defining the result for each item separately. For example, if we want to define a *flip* function which toggles black to white, and vice versa, this could be done using patterns as follows:

$$\textit{flip} \in \text{Colour-BW} \rightarrow \text{Colour-BW}$$

$$\textit{flip}(\text{Black}) \stackrel{\text{df}}{=} \text{White}$$

$$\textit{flip}(\text{White}) \stackrel{\text{df}}{=} \text{Black}$$

As long as there is no overlap between the patterns on the left—only one line of the definition can be applied to a given input—such a definition is a valid one, and can be rewritten in a format, possibly more familiar from mathematical texts:

$$flip(x) \stackrel{df}{=} \begin{cases} \text{White} & \text{if } x = \text{Black} \\ \text{Black} & \text{if } x = \text{White} \end{cases}$$

Exercises

8.1 Define a type Compass which has the four cardinal points of the compass (north, south, east and west). Then define functions *turnLeft*, *turnRight* and *turnBack*, which given a direction give back a new direction faced after turning 90° to the left, 90° to the right and 180°, respectively.

8.2 Consider a type which describes the action of a computer game player facing two buttons—and who may (i) press neither button, (ii) press button A, (iii) press button B or (iv) press both buttons. Define a type ButtonsAction which encodes these four options. Imagine two players facing the same two buttons, both of whom can perform one of the four actions we identified. Define a function *twobuttons* $\in$ ButtonAction $\times$ ButtonAction $\rightarrow$ ButtonAction which defines the result of both players performing their action simultaneously. The function should combine the actions such that each button is pressed if at least one of the players is pressing it.

Define a similar function, *twoswitches* with the same type as *twobuttons*, but which treats the buttons as *switches*—if both players press the switch, it is assumed to return to the original state, hence being equivalent to not pressing the switch at all.

8.1.2 More Elaborate Types

In the previous section, we saw how one can declare types made up of a finite number of distinct objects. However, in some cases, some of these cases may depend on further information. For instance, consider a more elaborate colour definition scheme. There are various ways in which one may encode colours for computer processing. We have already seen a very simple scheme which allows for just black and white pixels. In this section we will encode two other schemes:

Greyscale: A greyscale image is made up of pixels, each of which is a shade of grey, ranging from black to white, but permitting many intermediate shades of grey. For instance, the colour of a pixel can be encoded as a natural number ranging from 0 to 100, with 0 corresponding to black, and 100 to white. In between, lie a further 99 shades of grey.[2]

---

[2]Due to the fact that computers work in binary, and organise their data in bytes or words, it would be more typical, for instance, to have the colour encoded using one byte, thus ranging from 0 to 255, or one 16-bit word, thus ranging from 0 to 65,535.

RGB: This representation allows for colour in images. Each colour is uniquely determined as a combination of red, green and blue (hence the name RGB) light, each set at different intensities.[3] If we use the 0 to 100 range again, we would represent each pixel as a triple of numbers (r, g, b) representing how intense the red, green and blue light should be used to obtain the desired colour. For example, $(0, 0, 0)$ would correspond to black, and $(100, 100, 100)$ to white. Pure red would be $(100, 0, 0)$ and similarly green and blue. Yellow, which is a mixture of equal amounts of red and green would be encoded as $(100, 100, 0)$, but if a darker shade is required, one could use $(50, 50, 0)$.

We can now encode the colour of a pixel using either its greyscale or its RGB value. In this example, we will also allow for a third possibility to specify that the colour is unknown. Notice that we now have three possibilities—Unknown, GS and RGB, however, unlike the example in the previous section, greyscale colours require one additional value, while RGB requires three further values. This can be written as follows:

$$\text{Colour} ::= \text{Unknown}$$
$$\quad | \quad \text{GS } \mathbb{N}_{100}$$
$$\quad | \quad \text{RGB } (\mathbb{N}_{100} \times \mathbb{N}_{100} \times \mathbb{N}_{100})$$

$\mathbb{N}_{100}$ is the type of natural numbers up to and including 100. Based on this definition, one can also define, as before, functions on this type by enumerating all the possibilities, but this time also using variables for the values of the parameters. Unknown, GS and RGB are called *constructors*, since they enable us to *construct* objects in our new type Colour.

One can now define functions to convert a pixel to RGB or to greyscale using patterns:

$$toRGB \in \text{Colour} \rightarrow \text{Colour}$$

$$toRGB(\text{Unknown}) \stackrel{\text{df}}{=} \text{Unknown}$$

$$toRGB(\text{RGB } (r, g, b)) \stackrel{\text{df}}{=} \text{RGB } (r, g, b)$$

$$toRGB(\text{GS } n) \stackrel{\text{df}}{=} \text{RGB } (n, n, n)$$

$$toGS \in \text{Colour} \rightarrow \text{Colour}$$

$$toGS(\text{Unknown}) \stackrel{\text{df}}{=} \text{Unknown}$$

$$toGS(\text{RGB } (r, g, b)) \stackrel{\text{df}}{=} \text{GS } ((r + g + b) \div 3)$$

$$toGS(\text{GS } n) \stackrel{\text{df}}{=} \text{GS } n$$

---

[3]This is called an *additive* colour representation, since we are specifying how much colour to add. In contrast, when mixing paints, we get a *subtractive* colour scheme, since each paint colour we add absorbs more colours from white light before reflecting it.

Thus, a greyscale of level n percent would be transformed to an RGB value with n percent red, green and blue. In the other direction, the average RGB value is taken as the greyscale value.[4] These definitions can also be easily transformed so as not to use patterns. For example the second function can be written as:

$$toGS(x) \stackrel{\mathrm{df}}{=} \begin{cases} \mathsf{Unknown} & \text{if } x = \mathsf{Unknown} \\ \mathsf{GS}\,((r + g + b) \div 3) & \text{if } x = \mathsf{RGB}\,(r, g, b) \\ \mathsf{GS}\,n & \text{if } x = \mathsf{GS}\,n \end{cases}$$

Exercises

8.3 Add a fourth possibility to the Colour type, to handle pure black and pure white, using the Colour-BW type defined earlier.

8.4 Coordinates on a plane surface can be described in two different ways—using Cartesian coordinates, with each point being described as (x, y), its distance along the x- and y-axis, or using polar coordinates, where a point is described as a pair (r, θ), with r being the distance from the origin, and θ the angle subtended by the line from the origin to the point with the x-axis.

   (i) Define a type Point which allows descriptions in either representation.
   (ii) Define functions *toPolar* and *toCartesian* to convert a given point into polar and Cartesian form, respectively.
   (iii) Rotating a point by a given angle around the origin can be easily done if the point is in polar coordinates. Define a function *rotate*, which given an angle and a point rotates the point around the origin by the given angle. Hint: Start by transforming into polar coordinates.

8.1.3 Self-Referential Types

The examples we have seen in the previous section enable us to build new objects in the new type from objects of types we already know how to handle. For example, given a natural number not larger than 100, we can construct an object of type Colour using the constructor GS.

Now consider a new colour type using the idea of paint mixing. The basic constructors, which take no parameters, will be the basic colours—red, yellow and blue. How do artists create different shades? They either lighten or darken a colour they already have, or mix two colours they already have. For instance, pink would be a lightened version of red, while orange would be a mixture of red and yellow.

We can similarly use three constructors: Lighten and Darken, which take one paint colour as a parameter, and Mix, which takes two paint colours as parameters.

---

[4]In practice, more elaborate formulae, taking into consideration colour intensity, are typically used to convert between colour spaces in a more photorealistic manner.

Note that the parameters passed to the constructors are of the same type as the one we are defining:

$$PaintColour ::= Red$$
$$|\ \ Yellow$$
$$|\ \ Blue$$
$$|\ \ Lighten\ PaintColour$$
$$|\ \ Darken\ PaintColour$$
$$|\ \ Mix\ (PaintColour \times PaintColour)$$

Notice that the definition of the type PaintColour is in terms of itself. To take the previous examples, pink would correspond to Lighten Red, while orange would correspond to Mix (Red, Yellow). Based on this type definition, we can define functions as before. For example, a function to identify which of the basic colours were used in a colour can be defined as follows:

$$required \in PaintColour \to \mathbb{P}(PaintColour)$$

$$required(Red) \stackrel{df}{=} \{Red\}$$

$$required(Yellow) \stackrel{df}{=} \{Yellow\}$$

$$required(Blue) \stackrel{df}{=} \{Blue\}$$

$$required(Lighten\ c) \stackrel{df}{=} required(c)$$

$$required(Darken\ c) \stackrel{df}{=} required(c)$$

$$required(Mix\ (c_1,\ c_2)) \stackrel{df}{=} required(c_1) \cup required(c_2)$$

Notice that the last three lines of the definition use the function $required$ themselves. Anyone with programming experience will find this familiar—it is a recursive definition of the function $required$. However, there is a snag—programmers will ask themselves whether 'calling' $required$ will always terminate and return a value, the equivalent worry of a mathematician, who would be worrying whether $required$ is a total and well-defined function.

To justify the recursive definition, we have to ensure that the function always take a smaller value on the right-hand side of the definition than the value on the left-hand side. In this case, all uses of $required$ on the right-hand side are applied to smaller parameters, in the sense that they remove one constructor. This guarantees that the definition is a safe one.

Exercises

8.5 Define a function to count how many units of red paint are required for a particular paint colour. You may assume that every use of Red requires one unit of paint.

8.6 Define a type to describe cocktails, made up of any combination of the following ingredients: whiskey, vodka, tequila, gin, lemonade and tonic. A cocktail may be shaken or stirred to give a new version of the cocktail, or may be mixed with another cocktail to give a new cocktail. Hint: The type will look very similar to the PaintColour type.

Define a function *taste*, which given a cocktail, tells you whether or not it is alcoholic.

8.7   Consider the following type which describes the lifetime of an amoeba:

$$\text{Amoeba} ::= \text{Die} \mid \text{Walk Amoeba} \mid \text{Split}(\text{Amoeba} \times \text{Amoeba})$$

The Walk constructor gives an amoeba behaviour which starts by walking, and then proceeding according to the behaviour given as a parameter. Split is given two behaviours, corresponding to the behaviour of the two amoebae resulting after the split. For instance, the following value corresponds to an amoeba which walks, then splits into two, one of which dies immediately, while the second walks once, then dies: Walk (Split (Die, Walk Die)).

Define a function which counts the total number of amoebae spawned in a given expression. For instance, in the example given, the result should be 1, since there was only one spawning of amoebae.

8.1.4 Parametrised Types

Self-referential types are very powerful. For instance, lists of numbers can be encoded directly using self-referential types. A list of numbers can be either (i) an empty list, or (ii) a number followed by another list. This can be encoded using two constructors as seen below:

$$\text{ListN} ::= \text{EmptyN} \mid \text{ConsN}(\mathbb{N} \times \text{ListN})$$

For instance, the list $\langle 3, 1, 4, 2 \rangle$ would be encoded as:

$$\text{ConsN}(3, \text{ConsN}(1, \text{ConsN}(4, \text{ConsN}(2, \text{EmptyN}))))$$

Various useful functions can be defined in a recursive manner of this type. For instance, the length of a list can be defined recursively by stating that the length of an empty list is zero, while the length of a list using the ConsN constructor would be one more than the length of the list given as the second parameter:

$$lengthN \in \text{ListN} \to \mathbb{N}$$

$$lengthN(\text{EmptyN}) \stackrel{\text{df}}{=} 0$$

$$lengthN(\text{ConsN}(n, xs)) \stackrel{\text{df}}{=} 1 + lengthN(xs)$$

To catenate two lists together, we can use a recursive definition on the first parameter. When the first list is empty, the result is simply the second list. On the other hand, if the first list to catenate is of the form $\mathsf{ConsN}(n, x)$, we start by catenating x to the second list (note that the recursion is on a smaller object, since we are removing one ConsN), and then adding n to the front of the list after we are done:

$$+\!\!\!+ \ \in\ \mathsf{ListN} \times \mathsf{ListN} \rightarrow \mathsf{ListN}$$

$$\mathsf{EmptyN} +\!\!\!+ ys \stackrel{df}{=} ys$$

$$\mathsf{ConsN}(n, xs) +\!\!\!+ ys \stackrel{df}{=} \mathsf{ConsN}(n, xs +\!\!\!+ ys)$$

Now consider the type of lists of colours. Similarly as was done in the case of lists of natural numbers, we would define the type as:

$$\mathsf{ColourList} ::= \mathsf{ColourEmpty} \mid \mathsf{ColourCons}\,(\mathsf{Colour} \times \mathsf{ColourList})$$

If we define functions to calculate the length of a list of colours, or a function to concatenate two lists of colours, the first thing we notice is that, other than the types, the definitions look identical. Worse than this, if we prove some properties of lists of numbers, the proofs would have to be redone for every new list type. This is something which we would like to avoid.

Parametrised type declarations allow us to do this. Rather than define the type of lists of numbers or lists of colours, we define a family of types:

$$\mathsf{List}\,\alpha ::= \mathsf{Empty} \mid \mathsf{Cons}\,(\alpha \times \mathsf{List}\,\alpha)$$

Given a type α, the type $\mathsf{List}\,\alpha$ would be the type of lists of objects of type α. Lists of natural numbers would be $\mathsf{List}\,\mathbb{N}$, while the type of lists of colours is $\mathsf{List}\,\mathsf{Colour}$. Note that the Cons constructor takes an object of type α and a list of objects of type α to produce a new list of objects of the same type. It is important to note that List without a type parameter is not a type.

Using this approach, we can also define functions on general lists, which would work for any type α:

$$length \in \mathsf{List}\,\alpha \rightarrow \mathbb{N}$$

$$length(\mathsf{Empty}) \stackrel{df}{=} 0$$

$$length(\mathsf{Cons}(v, x)) \stackrel{df}{=} 1 + length(x)$$

$$+\!\!\!+ \ \in\ \mathsf{List}\,\alpha \times \mathsf{List}\,\alpha \rightarrow \mathsf{List}\,\alpha$$

$$\mathsf{Empty} +\!\!\!+ y \stackrel{df}{=} y$$

$$\mathsf{Cons}(v, x) +\!\!\!+ y \stackrel{df}{=} \mathsf{Cons}(v, x +\!\!\!+ y)$$

Note that the patterns on the left-hand sides of the definitions are mutually exclusive. For instance, consider adding the following line to the concatenation definition:

$$x +\!\!\!+ \mathsf{Empty} \stackrel{df}{=} x$$

This introduces ambiguity in that $\mathsf{Cons}(x, xs) \mathbin{+\!\!+} \mathsf{Empty}$ can be reduced to either (i) $\mathsf{Cons}(x, xs \mathbin{+\!\!+} \mathsf{Empty})$ according to the second rule in the original definition, or (ii) $\mathsf{Cons}(x, xs)$ according to the new definition. Which of the two is correct? In this case, further applications of the definition rules would result in the same answer in both cases. However, this is not always the case, and unless the definition patterns are mutually exclusive, we would have to *prove* that a definition is consistent.

In the rest of the chapter, we will, however, allow for pattern-based definitions which may overlap. To avoid having to prove that the function definition is not open to ambiguous results, we apply the *first* definition line (in the order they are written) whose pattern matches with the input. For example, to resolve the value of $\mathsf{Cons}(x, xs) \mathbin{+\!\!+} \mathsf{Empty}$ we would apply the first line which matches. If the new definition line we saw above is added to the end of the list of pattern-based definitions, we would obtain the result $\mathsf{Cons}(x, xs \mathbin{+\!\!+} \mathsf{Empty})$, but if it is added before the rule for Cons, we would get $\mathsf{Cons}(x, xs)$.

Exercises

8.8 Define a function *reverse*, which reverses a list recursively.

8.9 Consider simplified versions of PaintColour and Cocktail:

$$\mathsf{PaintColour} ::= \mathsf{Red} \mid \mathsf{Blue} \mid \mathsf{Yellow}$$
$$\mid \quad \mathsf{Mix}\,(\mathsf{PaintColour} \times \mathsf{PaintColour})$$

$$\mathsf{Cocktail} \quad ::= \mathsf{Whiskey} \mid \mathsf{Vodka} \mid \mathsf{Lemonade} \mid \mathsf{Tonic}$$
$$\mid \quad \mathsf{Mixer}\,(\mathsf{Cocktail} \times \mathsf{Cocktail})$$

   (i) Define two types PrimaryColour and SimpleDrink, the first consisting only of basic colours and the second of basic drinks.
   (ii) Define a parametrised type Mixer α to be able to handle both paint colours and cocktails.

8.10 To describe a two-dimensional table of entries of type α, the following type has been proposed:

$$\mathsf{Table}\,\alpha ::= \mathsf{Entry}\,(\alpha \times \mathsf{Table}\,\alpha)$$
$$\mid \quad \mathsf{NewRow}\,(\mathsf{Table}\,\alpha)$$
$$\mid \quad \mathsf{EndOfTable}$$

The entries in a row are added in sequence using the Entry constructor, and NewRow is used to start describing the entries of the next row. Finally, EndOfTable is used to define the end of the last row (and hence the table).

   (i) Define a function *height*, which given a table, returns the number of rows in the table.
   (ii) Define *size*, which returns the total number of entries in a table.
   (iii) Define *linearise* which, given a table of type Table α, linearises it into a single list of type List α.

(iv) A table is essentially a list of lists. Show how these functions would have been defined if, instead of defining a new type, we would have used List (List α).

8.2 Reasoning About New Types

As we have seen in the previous section, being able to define new types can be very useful to formalise new concepts. However alluring these type definitions are, without a mathematical formalisation, they are useless in that we lose our original objective of being able to reason precisely about them. For instance, without a mathematical formalisation of the notion of types, there is no way in which we can verify whether the catenation operator ++ on lists is associative, or check whether converting a colour to RGB then to greyscale is equivalent to converting it directly to greyscale. In this section, we will show how we can axiomatise new types, thus giving us strong tools to reason about objects created in this manner.

8.2.1 Axiomatising New Types

Identifying the items in the type: We start by giving rules of inference to *identify* the items in a type T defined using a type declaration.

Basic items: The basic items in a type declaration—constructors which take no parameters—are rather straightforward to formalise. For each such object B, we simply add an axiom saying that B is an object of type T, which we write as $B : T$. For instance, based on the examples given earlier, we would have that Black : Colour-BW and that Empty : List α.

Constructors: Constructors which take parameters are different from the basic items in that they typically generate more than a single object. For instance in the Colour type, the constructor RGB takes a triple of numbers and in return gives an instance of the Colour type. For example we would expect pure green, written as RGB$(0, 100, 0)$, to be of type Colour—for a particular triple of numbers we would expect a colour to be associated. The constructor RGB applied to a triple of numbers, is thus guaranteed to be an item of Colour. If we can show that $t : \mathbb{N}_{100} \times \mathbb{N}_{100} \times \mathbb{N}_{100}$, then we can conclude that RGB(t) : Colour.

This approach also works for self-referential type constructors—constructors which take parameters of the same type they are constructing. For example, from the type specification, we can conclude that Darken Yellow : PaintColour, since we know that Yellow : PaintColour.

In general, for a constructor C appearing in the specification of a type T as $C\,X$ (where X is the type following the constructor name C), we obtain the rule $x : X \vdash C(x) : T$.

Equality: The idea behind these type definitions is that items in the type being defined can only be constructed using the basic items and constructors appearing in
the type declaration. This enables us to identify distinct items only in terms of their
structure—for example the colour Yellow and Darken Green are structurally distinct, as are Mix (Red, Blue) and Mix (Blue, Red).[5] Structural equivalence is crucial to enable us to define functions over these types using patterns. The approach
taken is to identify which pairs of objects are different from each other—any other
pair of objects being considered to be equal to each other:

Basic items are distinct: As we did before, we start by looking at the basic (pa-
rameterless) constructors in a type T. If B and B' are distinct basic constructors,
we will have an axiom saying that $B \neq B'$. In general, if the basic constructors
are B_1 to B_b, we will have axioms stating that, for any distinct i and j, $B_i \neq B_j$.

No overlap between basic items and constructors: We want to be sure that
none of the basic constructors are equivalent to an object constructed using a
parametrised constructor. For instance, we would want to ensure that Red is
not equal to any colour of the form Mix (x, y). We can write this as x, y :
PaintColour $\vdash$ Red $\neq$ Mix (x, y).
In general, if the basic items are B_1 to B_b, and the constructors (which take parameters) are C_1 to C_c respectively taking parameters of types X_1 to X_c, we will
have axioms to ensure that, for any values i and j, $x : X_j \vdash C_j(x) \neq B_i$.

Constructors are distinct: Distinct non-basic constructors also produce distinct
objects. For instance an object of the form Darken (x) should be distinct from
another object of the form Mix (y, z). Formally, we can write this as c, c_1, c_2 :
PaintColour $\vdash$ Darken$(c) \neq$ Mix(c_1, c_2).
In general, if the parametrised constructors are C_1 to C_c (taking parameters of
type X_1 to X_c respectively), we will need an axiom for each distinct i and j:
$x_i : X_i$ $x_j : X_j \vdash C_i(x_i) \neq C_j(x_j)$.

Distinct parameters, distinct objects: Apart from distinguishing between items
built using different constructors, we would also want to ensure that, for example, given two different colours c and c', Darken(c) should be different from
Darken(c'). Formally, we would write this as the axiom x, y : PaintColour $\vdash x \neq$
$y \Rightarrow$ Darken$(x) \neq$ Darken(y). In general, we have an axiom for each value i:
$x, y : X_i \vdash i \neq j \Rightarrow C_i(x) \neq C_i(y)$.

No superfluous items: We have shown how we can axiomatise which items are in
the new type we are defining, and how to compare items for structural equality.
However, we still lack means to show that the *only* items in the new type are the
ones created by the constructors. For example, we need to be able to show that
Lemur is not an object in PaintColour. By simply listing which items appear in
the new type, we are not excluding anything else from the type. Informally, what

---

[5]Note that this is structural equivalence, which cares about order of parameters. This is not a
limitation of the approach, since if we wanted to equate these two colours, we would be able to
separately define our own relation which is not purely structural.

we would like to say is that the type we desire is the *smallest* one which includes everything which can be constructed from the constructors, basic and otherwise.

Consider a structural definition of a type T, and a collection of objects X which satisfies all the axioms of T we have considered above (in that X includes all the basic constructors of T, and the objects generated by the parametrised constructors of T). What we would like to ensure is that either X is the same as T, or X contains all the objects in T and additionally some other objects. In other words, $\forall x : T \cdot x : X$.

In general, given a type T defined as follows:

$$T ::= B_1 \mid B_2 \mid \ldots \mid B_b \mid$$
$$C_1(X_1) \mid C_2(X_2) \mid \ldots \mid C_c(X_c)$$

The following axioms are used to enable reasoning about inclusion in type T and structural equality over T:

1. Basic constructors are in T: for every i, we have the axiom: $\vdash B_i : T$.
2. Parametrised constructors are in T: for every i we get the rule $x : X_i \vdash C_i(x_i) : T$.
3. Basic constructors are distinct: for every distinct i and j: $\vdash B_i \neq B_j$.
4. Parametrised constructors are distinct: for every distinct i and j: $x : X_i$, $y : X_j \vdash C_i(x) \neq C_j(y)$.
5. Parametrised constructors give distinct values when applied to distinct parameters: for every i, x, $y : X_i \vdash x \neq y \Rightarrow C_i(x) \neq C_i(y)$.
6. Basic and parametrised constructors are distinct: for every i and j: $x : X_j \vdash B_i \neq C_j(x)$.
7. T is the smallest possible type: $A_{1,2}(T') \vdash \forall t : T \cdot t : T'$, where $A_{1,2}(T)$ are axioms 1 and 2 written for type T'.

In addition, structural equality also has axioms stating that it is an equivalence relation—it is reflexive, symmetric and transitive.

Example The axioms we obtain to reason about a type may seem rather complex. To illustrate how this axiom schema can be used to obtain the axioms to reason about a concrete type, we will consider a simpler version of the type of paint colours:

> Paint ::= Black
> | White
> | Lighten PaintColour
> | Darken PaintColour

To reason about which objects are to be found in the type Paint, we have the following axioms:

$$\vdash \text{Black} : \text{Paint}$$

$$\vdash \text{White} : \text{Paint}$$

$$c : \text{Paint} \vdash \text{Lighten}(c) : \text{Paint}$$

$$c : \text{Paint} \vdash \text{Darken}(c) : \text{Paint}$$

To ensure that it is the smallest such type:

$$\vdash \text{Black} : T$$
$$\vdash \text{White} : T$$
$$c : T \vdash \text{Lighten}(c) : T$$
$$\underline{c : T \vdash \text{Darken}(c) : T}$$
$$\forall c : \text{Paint} \cdot c \in T$$

Finally, we also get axioms to state which objects are equal to which:

1. The basic values are distinct:

$$\vdash \text{Black} \neq \text{White}$$

$$\vdash \text{White} \neq \text{Black}$$

2. The parametrised constructors produce disjoint values:

$$c, \ c' : \text{Paint} \vdash \text{Lighten}(c) \neq \text{Darken}(c')$$

$$c, \ c' : \text{Paint} \vdash \text{Darken}(c) \neq \text{Lighten}(c')$$

3. Each parametrised constructor produces distinct items when applied to distinct parameters:

$$c, \ c' : \text{Paint} \vdash c \neq c' \Rightarrow \text{Lighten}(c) \neq \text{Lighten}(c')$$

$$c, \ c' : \text{Paint} \vdash c \neq c' \Rightarrow \text{Darken}(c) \neq \text{Darken}(c')$$

4. The parametrised constructors produce values distinct from the basic constructors:

$$c : \text{Paint} \vdash \text{Black} \neq \text{Lighten}(c)$$

$$c : \text{Paint} \vdash \text{White} \neq \text{Lighten}(c)$$

$$c : \text{Paint} \vdash \text{Black} \neq \text{Darken}(c)$$

$$c : \text{Paint} \vdash \text{White} \neq \text{Darken}(c)$$

These axioms and rules of inference also justify the use of pattern-based definitions, since we can now use structural equality given by the axioms to check whether a colour matches a particular pattern. For example, the condition corresponding to whether c is of the form $\text{Darken}(c')$, is: $\exists c' : \text{Paint} \cdot \text{Darken}(c') = c$.   ◇

The axiomatisation of these types may seem overly complex. However, we will now identify an important principle for these types, which will make reasoning about them more straightforward.

8.2.2 A General Inductive Principle

One interesting corollary of these axioms is that one is only allowed to create objects in the type with a finite number of instances of constructors. This fact can be used to prove properties about all items in a type. For instance, considering the black and white type Colour-BW, which consisted of just two basic constructors, Black and White, we can prove properties about the whole type by proving it separately for the two colours.

Recall the definition of the function to flip colours:

$$flip(\text{Black}) \stackrel{\text{df}}{=} \text{White}$$

$$flip(\text{White}) \stackrel{\text{df}}{=} \text{Black}$$

A seemingly reasonable property of this function would be that applying *flip* twice returns the original colour: $\forall c : \text{Colour-BW} \cdot flip(flip(c)) = c$. How can we prove this? Since we know that the only two items in the type Colour-BW are Black and White, we can prove the property independently for the two, from which we would like to conclude that the property holds for all items in Colour-BW:

$$\frac{\begin{array}{l} flip(flip(\text{Black})) = \text{Black} \\ flip(flip(\text{White})) = \text{White} \end{array}}{\forall c : \text{Colour-BW} \cdot flip(flip(c)) = c}$$

The individual proofs for Black and White are rather straightforward. The proof for Black would be the following:

$$
\begin{array}{l}
\quad \text{left-hand side of equality} \\
= flip(flip(\text{Black})) \\
= \{\text{using line 1 of the definition of } flip\} \\
\quad flip(\text{White}) \\
= \{\text{using line 2 of the definition of } flip\} \\
\quad \text{Black} \\
= \text{right-hand side of equality}
\end{array}
$$

The proof for the case of White is almost identical, and the two proofs appear to be sufficient to prove the original property.

This proof technique can be generalised for any property universally quantified over Colour-BW:

$$\frac{\begin{array}{l} \pi(\text{Black}) \\ \pi(\text{White}) \end{array}}{\forall c : \text{Colour-BW} \cdot \pi(c)}$$

With types which include parametrised constructors, this technique is essentially identical. For instance, recall the definition of the Colour type:

$$\text{Colour} ::= \text{Unknown}$$
$$| \quad \text{GS } \mathbb{N}_{100}$$
$$| \quad \text{RGB } (\mathbb{N}_{100} \times \mathbb{N}_{100} \times \mathbb{N}_{100})$$

Any item belonging to this type is in one of three forms. For the value Unknown, we just proceed as in the previous example. However, in the case of GS, we note that this is not a single possible value, but a whole family of them, one for each parameter. Therefore, we would have to prove the property for every possible value of the parameter. Similarly for RGB.

If we were to require a proof that transforming a colour to RGB and then to greyscale is the same as transforming it directly to greyscale, we would want to prove that:

$$\forall c : \text{Colour} \cdot toGS(toRGB(c)) = toGS(c)$$

We can prove this by looking at the instances separately, as we did before:

$$toGS(toRGB(\text{Unknown})) = toGS(\text{Unknown})$$
$$\forall (r, g, b) : \mathbb{N}_{100} \times \mathbb{N}_{100} \times \mathbb{N}_{100}.$$
$$toGS(toRGB(\text{RGB}(r, g, b))) = toGS(\text{RGB}(r, g, b))$$
$$\forall n : \mathbb{N}_{100} \cdot toGS(toRGB(\text{GS}(n))) = toGS(\text{GS}(n))$$

$$\overline{\forall c : \text{Colour} \cdot toGS(toRGB(c)) = toGS(c)}$$

The proof would thus consist of three proofs, one for each case:

Case (i): For the case of Unknown.

$$\text{left-hand side of equality}$$
$$= toGS(toRGB(\text{Unknown}))$$
$$= \{\text{using line 1 of the definition of } toRGB\}$$
$$toGS(\text{Unknown})$$
$$= \text{right-hand side of equality}$$

Case (ii): For the case of RGB.

$$\text{left-hand side of equality}$$
$$= toGS(toRGB(\text{RGB}(r, g, b)))$$
$$= \{\text{using line 2 of the definition of } toRGB\}$$
$$toGS(\text{RGB}(r, g, b))$$
$$= \text{right-hand side of equality}$$

Universal quantification can then be introduced on the equality proved here to obtain the desired result.

Case (iii): For the case of GS.

$$
\begin{aligned}
&\text{left-hand side of equality} \\
={}& toGS(toRGB(\mathsf{GS}(n))) \\
={}& \{\text{using line 3 of the definition of } toRGB\} \\
& toGS(\mathsf{RGB}(n, n, n)) \\
={}& \{\text{using line 2 of the definition of } toGS\} \\
& \mathsf{GS}((n + n + n) \div 3) \\
={}& \{\text{basic arithmetic}\} \\
& \mathsf{GS}(n) \\
={}& \{\text{using line 3 of the definition of } toGS\} \\
& toGS(\mathsf{GS}(n)) \\
={}& \text{right-hand side of equality}
\end{aligned}
$$

Note that in constructing the proof, one would proceed by applying the definitions on the left-hand side (from the top of the proof downwards), and on the right-hand side (from the bottom of the proof upwards), until the resulting values match.

As we did before in the case of Colour-BW, this proof technique can be generalised for any property universally quantified over Colour:

$$
\frac{\pi(\mathsf{Unknown}) \quad \forall(r, g, b) : \mathbb{N}_{100} \times \mathbb{N}_{100} \times \mathbb{N}_{100} \cdot \pi(\mathsf{RGB}(r, g, b)) \quad \forall n : \mathbb{N}_{100} \cdot \pi(\mathsf{GS}(n))}{\forall c : \mathsf{Colour} \cdot \pi(c)}
$$

This approach to proving properties using the structure of the type can also be applied to self-referential types. However, the approach rarely works in practice for interesting properties. Consider, for instance, trying to prove a property about lists. Recall that the type of general lists is:

$$
\mathsf{List}\,\alpha ::= \mathsf{Empty} \mid \mathsf{Cons}\,(\alpha \times \mathsf{List}\,\alpha)
$$

A naïve rule we would extract for this type would be:

$$
\frac{\pi(\mathsf{Empty}) \quad \forall(k, ks) : \alpha \times \mathsf{List}\,\alpha \cdot \pi(\mathsf{Cons}(k, ks))}{\forall xs : \mathsf{List}\,\alpha \cdot \pi(xs)}
$$

This rule is rather weak. Consider, for instance, if we try to prove that catenating an empty list after a list results in the original list: $\forall xs : \mathsf{List}\,\alpha \cdot xs \mathbin{+\!\!+} \mathsf{Empty} = xs$.[6]

---

[6]Note that this does not follow immediately from the definition of concatenation, which only states that catenating the empty list as the first parameter to a list leaves the list unchanged. Simply adding another line to the definition introduces overlapping rules, or increases the complexity of the definition unnecessarily.

The first proof, for the case of Empty, goes through without any problems:

> left-hand side of equality
> = Empty ++ Empty
> = {using line 1 of the definition of ++}
>   Empty
> = right-hand side of equality

The other case, however, gets stuck halfway through:

> left-hand side of equality
> = Cons(x, xs) ++ Empty
> = {using line 2 of the definition of ++}
>   Cons(x, xs ++ Empty)
> = {...}
>   Cons(x, xs)
> = right-hand side of equality

To complete the proof, we need to show that xs ++ Empty $= xs$—in other words, we need to know that the result we are trying to prove is true. How can we escape this vicious circle? If we knew that the property holds for xs, the proof would be complete. The second proof obligation will be that, whenever the property holds for a list xs, it must also hold if we add a *single* Cons to it:

$$\frac{\pi(\text{Empty}) \quad \forall k : \alpha \cdot \forall ks : \text{List } \alpha \cdot \pi(ks) \Rightarrow \pi(\text{Cons}(k, ks))}{\forall xs : \text{List } \alpha \cdot \pi(xs)}$$

For instance, consider the list: ConsN(3, ConsN(1, ConsN(4, EmptyN))). Since property π holds for Empty (by the first line of the rule) then, we can conclude that it also holds for ConsN(4, EmptyN) (by the second line of the rule). Using the second line of the rule again, we can conclude that π holds for ConsN(1, ConsN(4, EmptyN)) and, with yet another application of line 2, also for ConsN(3, ConsN(1, ConsN(4, EmptyN))). From this example, it should be clear that intuitively π should hold for all items of List α. This rule is called *structural induction*.

Applying this new rule allows us to prove that $\forall xs : \text{List } \alpha \cdot xs$ ++ Empty $= xs$. The first line is the same as the one we previously proved. To prove the second line, we assume that the property holds for some sequence ks: ks ++ Empty $= ks$ (which we call the *inductive hypothesis*). From this we try to prove that the property also holds for $\mathsf{Cons}(k, ks)$:

$$
\begin{aligned}
&\text{left-hand side of equality}\\
&= \mathsf{Cons}(k, ks) + \mathsf{Empty}\\
&= \{\text{using line 2 of the definition of } +\!\!+\}\\
&\quad \mathsf{Cons}(k, ks + \mathsf{Empty})\\
&= \{\text{using the inductive hypothesis}\}\\
&\quad \mathsf{Cons}(k, ks)\\
&= \text{right-hand side of equality}
\end{aligned}
$$

Using implication introduction, and universal quantification introduction, the second line of the structural induction rule is proved, hence allowing us to conclude that: $\forall xs : \mathsf{List}\,\alpha \cdot xs + \mathsf{Empty} = xs$.

The rule of structural induction obtained from recursive types can be a very powerful tool to prove properties. It can also be used on types which use more than one constructor which refers to the type itself, or constructors which refer to multiple instances of the type itself. Recall the definition of PaintColour:

$$
\begin{aligned}
\mathsf{PaintColour} ::= \;& \mathsf{Red}\\
\mid\;& \mathsf{Yellow}\\
\mid\;& \mathsf{Blue}\\
\mid\;& \mathsf{Lighten\ PaintColour}\\
\mid\;& \mathsf{Darken\ PaintColour}\\
\mid\;& \mathsf{Mix}\,(\mathsf{PaintColour} \times \mathsf{PaintColour})
\end{aligned}
$$

The inductive rule for this type contains rules for Lighten and Darken, which are similar to the ones we have seen for List α. In the case of Mix, however, the constructor has two references to the type being defined, and we will assume that the property holds for each instance:

$$
\begin{array}{l}
\pi(\mathsf{Red})\\
\pi(\mathsf{Yellow})\\
\pi(\mathsf{Blue})\\
\forall k : \mathsf{PaintColour} \cdot \pi(k) \Rightarrow \pi(\mathsf{Lighten}(k))\\
\forall k : \mathsf{PaintColour} \cdot \pi(k) \Rightarrow \pi(\mathsf{Darken}(k))\\
\forall k, k' : \mathsf{PaintColour} \cdot \pi(k) \wedge \pi(k') \Rightarrow \pi(\mathsf{Mix}(k, k'))\\
\hline
\forall c : \mathsf{PaintColour} \cdot \pi(c)
\end{array}
$$

We will see applications of structural induction on various types in the coming sections.

8.2.3 Structural Induction

Structural induction allows us to prove properties of the form $\forall t : T \cdot \pi(t)$ for a structurally defined type T. To conclude this using structural induction we must provide a proof for each constructor of the type:

Basic constructors: For each basic constructor C, we have to prove that $\pi(C)$ holds.

Non-self-referential constructors: For every constructor of the form $C(X)$, where X does not include type T, we require to prove that $\forall x : X \cdot \pi(C(x))$.

Self-referential constructors: For each constructor of type T, of the form $C(X, T, T, \ldots T)$ (with T appearing in the parameters n times), we prove:

$$\forall x : X, \ k_1, \ldots k_n : T \cdot \pi(k_1) \wedge \cdots \pi(k_n) \Rightarrow \pi(C(x, k_1, \ldots k_n))$$

If we manage to complete these proofs (one for each of the constructors of the type), we can conclude using structural induction that π holds on all elements of T. The first two proof obligations are usually referred to as the *base cases* of the proof, while the proof obligations coming from the third case are called the *inductive cases*, each of which assumes a number of *inductive hypotheses* (one for each instance of T in the parameters of the constructor) and from which we try to prove the result for the constructor in question.

One interesting thing to note is that structural induction works thanks to the axioms we saw earlier, which ensure that one is only allowed to create objects in the type with a finite number of instances of constructors.

Exercises

8.11 State the rules of structural induction for the types Amoeba, Cocktail and Table α.

8.12 Consider the following type:

$$
\begin{aligned}
\mathsf{Path} ::= \ &\mathsf{Stop} \\
\mid \ &\mathsf{Left\ Path} \\
\mid \ &\mathsf{Right\ Path}
\end{aligned}
$$

(i)   State the rule of structural induction for this type.

(ii) Define the function *invert*, which switches every left with a right and vice versa.

(iii) State the proof obligations required to prove the following statement using structural induction:

$$\forall p : \mathsf{Path} \cdot invert(invert(p)) = p$$

(iv) Complete the proof.

8.3 Using Structured Types

In this section, we will look at a number of examples of uses of structured types
and how one can prove useful properties about functions on these types using the
principle of structural induction. We will also be investigating more thoroughly two
particularly interesting types in the coming chapters.

8.3.1 Three-Valued Logic

Sometimes, when using logic to reason about propositions, we need to reason about
the possibility of unknown truth values. Whether we are modelling knowledge of an
agent which may not always know whether a particular proposition is true or false,
or whether we are modelling electronic circuits at a physical level, in which a wire
may be carrying a high (true) or low (false) value, or not be driven to either value,
the need for a third value is frequently encountered. Typically, if we want to reason using a three-valued logic, we would axiomatise such a logic and its operators.
However, in this section, we will investigate how we can encode such a logic using
structured types.

We start by encoding the type of three values Logic3, with T representing truth,
F representing false, and X for the third 'unknown' value:

$$Logic3 ::= T \mid F \mid X$$

The proof principle for this type is quite simple—since the type is not recursively
defined, it does not use induction, but simply requires proving that the property
holds for all three basic constructors:

$$\frac{\pi(T) \\ \pi(F) \\ \pi(X)}{\forall x : Logic3 \cdot \pi(x)}$$

We can now define a few functions over this type, and prove properties about them.
Negation in three-valued logic is rather straightforward: if the value is known, return
its negation, if it is the unknown value X, then return X.

$$not_3(T) \stackrel{\mathrm{df}}{=} F$$

$$not_3(F) \stackrel{\mathrm{df}}{=} T$$

$$not_3(X) \stackrel{\mathrm{df}}{=} X$$

Let us now try to prove a familiar law from propositional logic—applying negation
twice returns the original value: $\forall x : Logic3 \cdot not_3(not_3(x)) = x$. Does it hold?

Proposition 8.1 *Negating twice a three-valued logic value leaves it intact:* $\forall x :$ Logic3 $\cdot$ $not_3(not_3(x)) = x$.

Proof To prove this property, we need to prove it holds for three cases: T, F and X.

Case (i): $x = $ T. We are required to prove that $not_3(not_3(T)) = $ T.

> left-hand side of equality
> $= not_3(not_3(T))$
> $= \{$using line 1 of the definition of $not_3\}$
> $not_3(F)$
> $= \{$using line 2 of the definition of $not_3\}$
>   T
> $= $ right-hand side of equality

Case (ii): $x = $ F. We are required to prove that $not_3(not_3(F)) = $ F.
This proof is almost identical to the previous one.
Case (iii): $x = $ X. We are required to prove that $not_3(not_3(X)) = $ X.

> left-hand side of equality
> $= not_3(not_3(X))$
> $= \{$using line 3 of the definition of $not_3\}$
> $not_3(X)$
> $= \{$using line 3 of the definition of $not_3\}$
>   X
> $= $ right-hand side of equality

This completes the required proofs to be able to conclude that the law holds for all values ranging over Logic3.   ∎

Conjunction is slightly more elaborate than negation. If both values are known, then we give the same value conjunction would give in Boolean logic. If both are unknown values, then the result should also be X. The question is what to do if one value is known and one is unknown. If the known value is false, then no matter what the other value is, the conjunction should return false. If the known value is true, then the result cannot be determined and would have to be X:[7]

$$and_3(T, T) \overset{df}{=} T$$

$$and_3(F, x) \overset{df}{=} F$$

$$and_3(x, F) \overset{df}{=} F$$

$$and_3(x, y) \overset{df}{=} X$$

---

[7]This is a choice—we could also have chosen, for instance, to define the conjunction of X with any value to return X.

Note that, to avoid having to write nine separate definitions, we used overlapping definitions (both the second and third line match if the input is two false values, while the fourth line overlaps with all the others). However, we will assume that the function definition is defined to match the first line which matches the input. If we were to define this using a conditional definition, we would write it in the following way, assuming that the value on the first line whose condition is true is the one returned.

$$and_3(x, y) \stackrel{df}{=} \begin{cases} \mathsf{T} & \text{if } x = y = \mathsf{T} \\ \mathsf{F} & \text{if } x = \mathsf{F} \\ \mathsf{F} & \text{if } y = \mathsf{F} \\ \mathsf{X} & \text{otherwise} \end{cases}$$

It is interesting to note that some familiar laws no longer hold. For instance, it may seem reasonable to assume that $\forall x : \mathsf{Logic3} \cdot and_3(x, not_3(x)) = \mathsf{F}$. However, this is not true for all cases. Consider the unknown value X:

$$and_3(\mathsf{X}, not_3(\mathsf{X}))$$
$= \{\text{using line 3 of the definition of } not_3\}$
$$and_3(\mathsf{X}, \mathsf{X})$$
$= \{\text{using line 4 of the definition of } and_3\}$
$$\mathsf{X}$$

So, the equation does not hold for all values of $\mathsf{Logic3}$. However, other useful laws still hold. Let us consider the law of commutativity of conjunction:

$$\forall x, y : \mathsf{Logic3} \cdot and_3(x, y) = and_3(y, x)$$

To prove this, we note that we have two universally quantified variables, namely x and y. We have to choose on which variable we will be applying the proof rule for $\mathsf{Logic3}$. In practice, in this case, both options work out well. Let us prove it on variable x. We thus have to consider the three possible values $x = \mathsf{T}$, $x = \mathsf{F}$ and $x = \mathsf{X}$.

Case (i): $x = \mathsf{T}$. We are required to prove that $\forall y : \mathsf{Logic3} \cdot and_3(\mathsf{T}, y) = and_3(y, \mathsf{T})$. We will prove the equality, after which the universal quantification can be introduced.

left-hand side of equality
$= and_3(\mathsf{T}, y)$
$= \{\text{using lines 1, 3 and 4 of the definition of } and_3\}$
y
$= \{\text{using lines 1, 2 and 4 of the definition of } and_3\}$
$and_3(y, \mathsf{T})$
$= \text{right-hand side of equality}$

Case (ii): $x = F$. We are required to prove that $\forall y : \text{Logic3} \cdot and_3(F, y) = and_3(y, F)$.

> left-hand side of equality
> $= and_3(F, y)$
> $= \{\text{using line 2 of the definition of } and_3\}$
>   F
> $= \{\text{using line 3 of the definition of } and_3\}$
> $and_3(y, F)$
> $= \text{right-hand side of equality}$

Case (iii): $x = X$. We are required to prove that $\forall y : \text{Logic3} \cdot and_3(X, y) = and_3(y, X)$.
A direct proof of this cannot be achieved based on what we have already proved. We can, however, prove the result using the proof rule of Logic3 on variable y, to prove that: $\forall y : \text{Logic3} \cdot and_3(X, y) = and_3(y, X)$. We have three cases to prove:

Case (a): $y = T$.

> left-hand side of equality
> $= and_3(X, T)$
> $= \{\text{using line 4 of the definition of } and_3\}$
>   X
> $= \{\text{using line 4 of the definition of } and_3\}$
> $and_3(T, X)$
> $= \text{right-hand side of equality}$

Case (b): $y = F$.

> left-hand side of equality
> $= and_3(X, F)$
> $= \{\text{using line 3 of the definition of } and_3\}$
>   F
> $= \{\text{using line 2 of the definition of } and_3\}$
> $and_3(F, X)$
> $= \text{right-hand side of equality}$

Case (c): $y = X$.

> left-hand side of equality
> $= and_3(X, X)$
> $= \{\text{using line 4 of the definition of } and_3\}$
>   X
> $= \{\text{using line 4 of the definition of } and_3\}$
> $and_3(X, X)$
> $= \text{right-hand side of equality}$

This completes the proof for $\forall y : \text{Logic3} \cdot and_3(X, y) = and_3(y, X)$.

This proof required the use of the proof rule for Logic3 applied in a nested manner—in the third case of the rule applied to x, we had to apply the rule again for the other variable y.

Exercises

8.13 In the first two cases of the proof of commutativity of conjunction, you may have noticed that we assumed that $\forall x : \text{Logic3} \cdot and_3(\text{T}, x) = x$ and that $\forall x : \text{Logic3} \cdot and_3(x, \text{T}) = x$. Prove these two laws.

8.14 Define disjunction for Logic3 in a similar manner to the definition of conjunction.

8.15 Show that the law of the excluded middle: $\forall x : \text{Logic3} \cdot or_3(x, not_3(x)) = \text{T}$ does not always hold in three-valued logic.

8.16 Do De Morgan's laws hold in three-valued logic?[8] Give counterexamples or proofs to support your answers.

8.3.2 Processing Data

In this example, we will look at the use of structured types for reasoning about data processing models. The type we will define will allow us to describe a data-processing pipeline such as the one below:

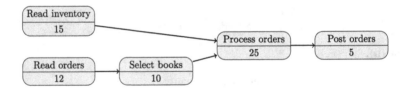

The diagram shows a pipeline handling online orders. The buyers' orders are restricted to ones for books, and processed together with the shop's inventory, after which they are posted. Each box represents a job with the amount of time units required to perform it shown in the bottom half of the box. The arrows tell us the jobs whose results are needed to be able to perform another job. For instance, the job *process orders* requires the results obtained from the job *read inventory* and the job *select books*, while the job *read orders* requires no information from other jobs. Notice that there are three types of jobs in this diagram: ones which require no input, ones which require a result obtained from one job and ones which require results obtained from two jobs. We can describe such data-processing pipelines using a structured type with three corresponding constructors:

$$\text{Job} ::= \text{Basic } \mathbb{N} \mid \text{Use } (\mathbb{N} \times \text{Job}) \mid \text{Combine } (\mathbb{N} \times \text{Job} \times \text{Job})$$

---

[8]Please answer in two-valued logic—replying 'Unknown' is not acceptable.

Each constructor has a natural number (corresponding to how much time is required to perform that job). For simplicity, the names of the jobs are not part of the data, but can easily be added. The online store example can be encoded in this type as: Use(5, Combine(25, Basic(15), Use(10, Basic(12)))).

The structural induction principle for the type of data-processing pipelines has one base case (for Basic) and two inductive cases (for Use and Combine):

$$\forall n : \mathbb{N} \cdot \pi(\text{Basic } n)$$
$$\forall n : \mathbb{N}, \; j : \text{Job} \cdot \pi(j) \Rightarrow \pi(\text{Use}(n, j))$$
$$\frac{\forall n : \mathbb{N}, \; j_1, j_2 : \text{Job} \cdot \pi(j_1) \land \pi(j_2) \Rightarrow \pi(\text{Combine}(n, j_1, j_2))}{\forall j : \text{Job} \cdot \pi(j)}$$

We can now write functions to process these pipelines. Since we have information as to how long each job takes, we can calculate the total time required for a pipeline to be completed. The simplest approach to this is to assume that all jobs are done sequentially. For instance, in the order processing job, we will assume that reading the inventory (taking 15 time units) is done before or after retrieving and filtering the orders (taking 22 time units). Therefore, before the orders are processed, 37 time units will already have elapsed. In total, the complete job would take 52 units, the sum of the times of all the jobs.

The function *sequential* calculates the time required to perform a job in this manner:

$$sequential(\text{Basic}(n)) \overset{\text{df}}{=} n$$

$$sequential(\text{Use}(n, j)) \overset{\text{df}}{=} n + sequential(j)$$

$$sequential(\text{Combine}(n, j_1, j_2)) \overset{\text{df}}{=} n + sequential(j_1) + sequential(j_2)$$

However, this approach is very wasteful of resources. In practice, if we were to do the job manually and we were to have sufficient human resources, we would send someone to get the inventory listing at the same time as someone else is processing the orders. The time elapsed before the orders can be processed would now be 22 time units as opposed to 37. Therefore, whenever a job requires information from two other jobs, we can perform them independently and concurrently. When programming the automation of this pipeline, we can perform the two branches in parallel. The cost of combining two jobs with costs c_1 and c_2 is thus reduced from $c_1 + c_2$ to $max(c_1, c_2)$:

$$parallel(\text{Basic}(n)) \overset{\text{df}}{=} n$$

$$parallel(\text{Use}(n, j)) \overset{\text{df}}{=} n + parallel(j)$$

$$parallel(\text{Combine}(n, j_1, j_2)) \overset{\text{df}}{=} n + max(parallel(j_1), parallel(j_2))$$

Our intuition tells us that performing the jobs concurrently can save time and is never worse than performing them sequentially. Can we prove this? Formally, the property we would like to prove is:

$$\forall j : \textsf{Job} \cdot parallel(j) \leq sequential(j)$$

We will prove this using structural induction on variable j. From the rule of induction, we have three results to prove:

Base case: For the only base case, with constructor Basic, we have to prove that $\forall n : \mathbb{N} \cdot parallel(\textsf{Basic}(n)) \leq sequential(\textsf{Basic}(n))$. To prove the inequality, we just have to use the definitions of the different execution strategies:

$$\begin{aligned}
&parallel(\textsf{Basic}(n)) \\
=\ & \{\text{line 1 of the definition of } parallel\} \\
&n \\
\leq\ &n \\
=\ & \{\text{line 1 of the definition of } sequential\} \\
&sequential(\textsf{Basic}(n))
\end{aligned}$$

Inductive case 1: For the first inductive case, for the constructor of Use, we have to prove that:

$$\forall n : \mathbb{N},\ j : \textsf{Job} \cdot parallel(j) \leq sequential(j) \Rightarrow$$
$$parallel(\textsf{Use}(n, j)) \leq sequential(\textsf{Use}(n, j))$$

We have one inductive hypothesis: $parallel(j) \leq sequential(j)$, from which we prove that: $parallel(\textsf{Use}(n, j)) \leq sequential(\textsf{Use}(n, j))$.

$$\begin{aligned}
&parallel(\textsf{Use}(n, j)) \\
=\ & \{\text{line 2 of the definition of } parallel\} \\
&n + parallel(j) \\
\leq\ & \{\text{using the inductive hypothesis}\} \\
&n + sequential(j) \\
=\ & \{\text{line 2 of the definition of } sequential\} \\
&sequential(\textsf{Use}(n, j))
\end{aligned}$$

Inductive case 2: The inductive case for the constructor Combine is slightly more elaborate. What we require to prove is that:

$$\forall n : \mathbb{N},\ j_1, j_2 : \textsf{Job} \cdot$$
$$parallel(j_1) \leq sequential(j_1) \wedge parallel(j_2) \leq sequential(j_2) \Rightarrow$$
$$parallel(\textsf{Combine}(n, j_1, j_2)) \leq sequential(\textsf{Combine}(n, j_1, j_2))$$

We thus have *two* inductive hypotheses: (i) $parallel(j_1) \leq sequential(j_1)$, and (ii) $parallel(j_2) \leq sequential(j_2)$.

Assuming these two hypotheses, we have to prove that the result holds when we combine the jobs using Combine:

$$parallel(\textsf{Combine}(n, j_1, j_2)) \leq sequential(\textsf{Combine}(n, j_1, j_2))$$

The proof follows directly from the inductive cases:

$$parallel(\textsf{Combine}(n, j_1, j_2))$$
$$= \{\text{line 3 of the definition of } parallel\}$$
$$n + max(parallel(j_1), parallel(j_2))$$
$$\leq \{\text{using arithmetic}\}$$
$$n + parallel(j_1) + parallel(j_2)$$
$$\leq \{\text{using the inductive hypotheses}\}$$
$$n + sequential(j_1) + sequential(j_2)$$
$$= \{\text{line 3 of the definition of } sequential\}$$
$$sequential(\textsf{Combine}(n, j_1, j_2))$$

With these three proofs, we can apply the rule of induction to prove that computing the branches in parallel never yields worse results than when computing them sequentially: $\forall j : \textsf{Job} \cdot parallel(j) \leq sequential(j)$.

In this section, we have seen how one can use structured types to reason about computation pipelines. The model we have used is rather simple, for example, not catering for jobs requiring information from more than two other jobs, or allowing an output to be shared between two jobs. It does illustrate, however, how mathematics can be used to reason about notions of computation, and the model can be extended to cater for more complex scenarios.

Exercises

8.17 Add another constructor to the type Job to be able to reason about jobs which use data from three other jobs. Extend the definitions and proofs in this section to cater for the additional case.

8.18 Define the following functions on the type Job:

- The function *unit*, which takes a job description and returns an identical job, but with the cost of each processing unit to 1. Define *unit*.
- The function *depth* takes a job and returns the number of jobs which have to run in sequence even if we adopt a parallel strategy. For instance, the online store example would return 4, since the output of the job has to go through *read orders*, *select books*, *process orders* and *post orders* in sequence. Define *depth*.

Now prove that $\forall j : \textsf{Job} \cdot depth(j) = parallel(unit(j))$.

8.19 Another constructor we can introduce to the type Job is: $\textsf{Choice}(n, j_1, j_2)$. This case is very similar to the Combine constructor, but with one key difference: the data from one of the two jobs is sufficient for the computation to continue. For example, one can have two jobs trying to authenticate a user against two different databases, and whichever value is returned first is used. When computing this in parallel, the choice of two jobs takes as long as the faster of the two (the one which takes least time), but when executed in sequence, the strategy will be just to run the first job (since we assume that at runtime we do not know how long each job will take). Extend the type Job and definitions to cater for Choice, and prove that parallel composition is still the better option.

8.20 Extend the type of Job to handle jobs which depend on data from any number of other jobs.

8.3.3 Lists

We have already seen how one can reason about sequences using functions and natural numbers. In the beginning of this chapter, we have seen an alternative approach, defining sequences using structured types. We will now look at how we can reason about sequences using such a representation. Recall the definition of List α, which may be either empty, or consist of a first object followed by another list:

$$\text{List } \alpha ::= \text{Empty} \mid \text{Cons}(\alpha \times \text{List } \alpha)$$

The inductive rule for lists is the following:

$$\frac{\pi(\text{Empty}) \qquad \forall k : \alpha, \; ks : \text{List } \alpha \cdot \pi(ks) \Rightarrow \pi(\text{Cons}(k, ks))}{\forall xs : \text{List } \alpha \cdot \pi(xs)}$$

We have already seen how one can define functions to calculate the length of a list and the catenation of two lists using recursion and input value patterns:

$$length \in \text{List } \alpha \to \mathbb{N}$$

$$length(\text{Empty}) \overset{\text{df}}{=} 0$$

$$length(\text{Cons}(x, xs)) \overset{\text{df}}{=} 1 + length(xs)$$

$$\mathbin{+\!\!+} \in \text{List } \alpha \times \text{List } \alpha \to \text{List } \alpha$$

$$\text{Empty} \mathbin{+\!\!+} xs \overset{\text{df}}{=} xs$$

$$\text{Cons}(x, xs) \mathbin{+\!\!+} ys \overset{\text{df}}{=} \text{Cons}(x, xs \mathbin{+\!\!+} ys)$$

In a similar manner, we can define a function to reverse a list—the basic case is trivial, since the reverse of an empty list is itself empty, while for the case of a list of the form $\text{Cons}(x, xs)$ we note that we can reverse the rest of the list xs, and append x at the end of the list:

$$reverse \in \text{List } \alpha \to \text{List } \alpha$$

$$reverse(\text{Empty}) \overset{\text{df}}{=} \text{Empty}$$

$$reverse(\text{Cons}(x, xs)) \overset{\text{df}}{=} reverse(xs) \mathbin{+\!\!+} \text{Cons}(x, \text{Empty})$$

Recall that we have already proved that catenating an empty list to the end of a list leaves the original list unchanged: $\forall xs : \text{List } \alpha \cdot xs \mathbin{+\!\!+} \text{Empty} = xs$.

We will now prove a number of other results about these definitions. Let us start by proving that catenation is associative:

$$\forall xs, ys, zs : \text{List } \alpha \cdot xs \mathbin{+\!\!+} (ys \mathbin{+\!\!+} zs) = (xs \mathbin{+\!\!+} ys) \mathbin{+\!\!+} zs$$

We have to start by deciding on which variable to apply induction. We have three options: xs, ys and zs from which we choose to apply it on xs.[9]

To apply structural induction on variable xs, we need to prove the two proof obligations of the inductive principle.

Base case: For the base case, when $xs = \text{Empty}$, we have to prove that:

$$\forall ys, zs : \text{List } \alpha \cdot \text{Empty} \mathbin{+\!\!+} (ys \mathbin{+\!\!+} zs) = (\text{Empty} \mathbin{+\!\!+} ys) \mathbin{+\!\!+} zs$$

The proof is the following:

$$
\begin{aligned}
&\quad \text{left-hand side of equality} \\
&= \text{Empty} \mathbin{+\!\!+} (ys \mathbin{+\!\!+} zs) \\
&= \{\text{using line 1 of the definition of } \mathbin{+\!\!+}\} \\
&\quad ys \mathbin{+\!\!+} zs \\
&= \{\text{using line 1 of the definition of } \mathbin{+\!\!+}\} \\
&= (\text{Empty} \mathbin{+\!\!+} ys) \mathbin{+\!\!+} zs \\
&= \text{right-hand side of equality}
\end{aligned}
$$

Inductive case: For the only inductive case, we assume that the property holds for some list ks:

$$\forall ys, zs : \text{List } \alpha \cdot ks \mathbin{+\!\!+} (ys \mathbin{+\!\!+} zs) = (ks \mathbin{+\!\!+} ys) \mathbin{+\!\!+} zs$$

Based on this inductive hypothesis, we prove the property for the list $\text{Cons}(k, ks)$, namely that:

$$\forall ys, zs : \text{List } \alpha \cdot \text{Cons}(k, ks) \mathbin{+\!\!+} (ys \mathbin{+\!\!+} zs) = (\text{Cons}(k, ks) \mathbin{+\!\!+} ys) \mathbin{+\!\!+} zs$$

---

[9]The intuition, once again, comes from the fact that the recursive definition of catenation pattern-matches on the first parameter, and variable xs appears as the first parameter of the catenation operators in the theorem.

        left-hand side of equality
$= \mathsf{Cons}(k, ks) \mathbin{+\!\!+} (ys \mathbin{+\!\!+} zs)$
$= \{\text{using line 2 of the definition of } \mathbin{+\!\!+}\}$
    $\mathsf{Cons}(k, ks \mathbin{+\!\!+} (ys \mathbin{+\!\!+} zs))$
$= \{\text{using the inductive hypothesis}\}$
    $\mathsf{Cons}(k, (ks \mathbin{+\!\!+} ys) \mathbin{+\!\!+} zs)$
$= \{\text{using line 2 of the definition of } \mathbin{+\!\!+}\}$
    $\mathsf{Cons}(k, ks \mathbin{+\!\!+} ys) \mathbin{+\!\!+} zs$
$= \{\text{using line 2 of the definition of } \mathbin{+\!\!+}\}$
    $(\mathsf{Cons}(k, ks) \mathbin{+\!\!+} ys) \mathbin{+\!\!+} zs$
$= \text{right-hand side of equality}$

This completes the two proof obligations, and therefore, we can conclude using structural induction that catenation is associative.

We can also prove a property relating catenation and the reversing of a list. Reversing the catenation of two lists can be done by first reversing the lists and then catenating them in reverse order. For instance, to reverse a list $\langle a, b, c \rangle \mathbin{+\!\!+} \langle d, e, f \rangle$, we can reverse the two lists and then catenate them in reverse order: $\langle f, e, d \rangle \mathbin{+\!\!+} \langle c, b, a \rangle$. We will prove that this statement works for any two lists:

$$\forall xs, ys : \mathsf{List}\,\alpha \cdot reverse(xs \mathbin{+\!\!+} ys) = reverse(ys) \mathbin{+\!\!+} reverse(xs)$$

Once again, we have to choose whether to apply induction on xs or on ys, and as before, we choose xs, since it appears in the parameter of the catenation operator on which the patterns are used in the definition.

Base case: For the base case, when $xs = \mathsf{Empty}$, we have to prove that:

$$\forall ys : \mathsf{List}\,\alpha \cdot reverse(\mathsf{Empty} \mathbin{+\!\!+} ys) = reverse(ys) \mathbin{+\!\!+} reverse(\mathsf{Empty})$$

The proof is the following:

        left-hand side of equality
$= reverse(\mathsf{Empty} \mathbin{+\!\!+} ys)$
$= \{\text{using line 1 of the definition of } \mathbin{+\!\!+}\}$
    $reverse(ys)$
$= \{\text{using the theorem we proved about } \mathbin{+\!\!+}\}$
    $reverse(ys) \mathbin{+\!\!+} \mathsf{Empty}$
$= \{\text{using line 1 of the definition of } \mathbin{+\!\!+}\}$
$= reverse(ys) \mathbin{+\!\!+} reverse(\mathsf{Empty})$
$= \text{right-hand side of equality}$

Inductive case: For the only inductive case, we assume that the property holds for some list ks:

$$\forall ys : \mathsf{List}\,\alpha \cdot reverse(ks \mathbin{+\!\!+} ys) = reverse(ys) \mathbin{+\!\!+} reverse(ks)$$

Using this inductive hypothesis, we will prove the property for the list $\mathsf{Cons}(k, ks)$:

$\forall k : \alpha, \ ys : \mathsf{List}\, \alpha \cdot$
$\quad reverse(\mathsf{Cons}(k, ks) \mathbin{+\!\!+} ys) = reverse(ys) \mathbin{+\!\!+} reverse(\mathsf{Cons}(k, ks))$

> left-hand side of equality
> $= reverse(\mathsf{Cons}(k, ks) \mathbin{+\!\!+} ys)$
> $= \{\text{using line 2 of the definition of } \mathbin{+\!\!+}\}$
> $\quad reverse(\mathsf{Cons}(k, ks \mathbin{+\!\!+} ys))$
> $= \{\text{using line 2 of the definition of } reverse\}$
> $\quad reverse(ks \mathbin{+\!\!+} ys) \mathbin{+\!\!+} \mathsf{Cons}(k, \mathsf{Empty})$
> $= \{\text{using the inductive hypothesis}\}$
> $\quad (reverse(ys) \mathbin{+\!\!+} reverse(ys)) \mathbin{+\!\!+} \mathsf{Cons}(k, \mathsf{Empty})$
> $= \{\text{using associativity of concatenation}\}$
> $\quad reverse(ys) \mathbin{+\!\!+} (reverse(ks) \mathbin{+\!\!+} \mathsf{Cons}(k, \mathsf{Empty}))$
> $= \{\text{using line 2 of the definition } reverse\}$
> $\quad reverse(ys) \mathbin{+\!\!+} reverse(\mathsf{Cons}(k, ks))$
> $= \text{right-hand side of equality}$

These two cases allow us to complete the proof by induction.

In this section, we have seen how one can reason about definitions of functions on structured types. In practice, this allows us not only to confirm that our definitions correspond to what we had in mind when we defined them, but also, for instance, to apply optimisations when designing systems.

Exercises

8.21 Show that $reverse(ys \mathbin{+\!\!+} \mathsf{Cons}(x, \mathsf{Empty})) = \mathsf{Cons}(x, reverse(ys))$ really does follow from the last theorem which relates catenation and list reversal. No induction is necessary—just apply the theorem and the definitions of catenation and reversal.

8.22 Prove that reversing a list twice returns the original list: $\forall xs : \mathsf{List}\, \alpha \cdot reverse(reverse(xs)) = xs$.

8.23 Prove that the length of the catenation of two lists is the sum of the lengths of the lists: $\forall xs, ys : \mathsf{List}\, \alpha \cdot length(xs \mathbin{+\!\!+} ys) = length(xs) + length(ys)$.

8.24 Prove that reversing a list leaves its length unchanged: $\forall xs : \mathsf{List}\, \alpha \cdot length(xs) = length(reverse(xs))$.

8.25 Consider the following function $double$, duplicating each item in a list:

$$double(\mathsf{Empty}) \stackrel{\mathrm{df}}{=} \mathsf{Empty}$$
$$double(\mathsf{Cons}(x, xs)) \stackrel{\mathrm{df}}{=} \mathsf{Cons}(x, \mathsf{Cons}(x, double(xs)))$$

Formally state and prove the property that the length of a doubled list is twice the length the original list.

8.3.4 Binary Trees

As a final example of the use of structured types, we will look at their use in reasoning about binary trees. Extensively used in computer science, trees are a special case of directed graphs—(i) consisting of a special vertex with no incoming edges, called the root; (ii) such that the graph has no loops; and (iii) starting from any vertex, the parts of the graph reachable from different edges are disjoint. Vertices with no outgoing edges are called leaves. Typically data is stored in each vertex. Binary trees are a special case, where each vertex can have no more than two outgoing edges. Below is an example of a binary tree, with 17 being at the root and the leaves marked using double circles:

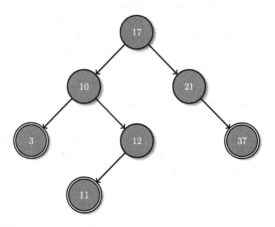

When using and reasoning about trees, we usually care whether a successor of a vertex is the one on the left or on the right. Thus, flipping the position of the two subtrees of the root, obtaining the tree starting from the vertex containing 21 on the left and the one starting from 10 on the right, would result in a different tree. Similarly, placing the vertex containing 37 as the left successor of the one containing 21 would be considered as changing the tree.

One possible way of describing trees using a structured type is to have a basic constructor for leaves, one for vertices with only a subtree to the left, one for those with only a subtree to the right, and those with two subtrees. Note that, since we do not want to limit ourselves to trees with natural numbers in the vertices, we use parametrised type just as we did with lists:

$$
\begin{aligned}
\text{TreeAttempt}\,\alpha ::= \ &\text{Leaf}\,\alpha \\
| \ &\text{NodeL}\,(\alpha \times \text{TreeAttempt}\,\alpha) \\
| \ &\text{NodeR}\,(\alpha \times \text{TreeAttempt}\,\alpha) \\
| \ &\text{NodeLR}\,(\alpha \times \text{TreeAttempt}\,\alpha \times \text{TreeAttempt}\,\alpha)
\end{aligned}
$$

This approach may seem reasonable until we realise that the empty tree is not part of this type. Just as the notion of zero is important when reasoning about numbers, so is that of the empty tree. One solution is to add a new basic constructor Nil, which can be used to model this value. However, now consider the following two structurally distinct objects: Leaf(10) and NodeLR(10, Nil, Nil). Although structurally different, the two objects are conceptually the same—a single vertex with no successors. Similarly, NodeL(10, t) is conceptually identical to NodeLR(10, t, Nil). To avoid having multiple ways of describing the same tree, we can simplify the type to the following one:

$$\text{Tree}\,\alpha ::= \text{Nil} \mid \text{Node}\,(\alpha \times \text{Tree}\,\alpha \times \text{Tree}\,\alpha)$$

All nodes now have exactly two outgoing edges, some of which, however can lead to the empty tree. The tree we saw earlier would be modelled in the following manner, with the small black circles marking the Nil values:

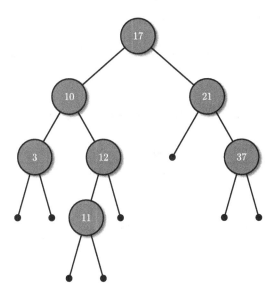

The rule of induction for trees contains one base case (for Nil) and one inductive case (for Node):

$$\frac{\pi(\text{Nil}) \qquad \forall x : \alpha,\ lt, rt : \text{Tree}\,\alpha \cdot \pi(lt) \wedge \pi(rt) \Rightarrow \pi(\text{Node}(x, lt, rt))}{\forall t : \text{Tree}\,\alpha \cdot \pi(t)}$$

We can define a function to measure the number of nodes in a tree by recursively going through the whole structure:

$$size \in \text{Tree } \alpha \to \mathbb{N}$$

$$size(\text{Nil}) \stackrel{\text{df}}{=} 0$$

$$size(\text{Node}(x, lt, rt)) \stackrel{\text{df}}{=} 1 + size(lt) + size(rt)$$

A tree can be flattened into a list in various ways. The following approach, called the *infix traversal* of a tree, flattens the left subtree of a vertex, then catenates the value at the top node, and finally flattens and concatenates the flattened right subtree:

$$flatten \in \text{Tree } \alpha \to \text{List } \alpha$$

$$flatten(\text{Nil}) \stackrel{\text{df}}{=} \text{Empty}$$

$$flatten(\text{Node}(x, lt, rt)) \stackrel{\text{df}}{=} flatten(lt) \mathbin{+\mkern-8mu+} \text{Cons}(x, flatten(rt))$$

For instance, the tree shown in the example flattens to $\langle 3, 10, 11, 12, 17, 21, 37 \rangle$. One thing we notice is that the number of items in the list should be equal to the number of vertices in the tree. Can we prove this?

The formal version of this law would be: $\forall t : \text{Tree } \alpha \cdot size(t) = length(flatten(t))$. Note that the property is about trees, so we would need to use the structural induction rule for trees on variable t.

Base case: We require to prove that $size(\text{Nil}) = length(flatten(\text{Nil}))$:

$$size(\text{Nil})$$
$$= \{\text{line 1 of the definition of } size\}$$
$$0$$
$$= \{\text{line 1 of the definition of } length\}$$
$$length(\text{Empty})$$
$$= \{\text{line 1 of the definition of } flatten\}$$
$$length(flatten(\text{Nil}))$$

Inductive case: We need to prove that, if the property holds for two trees lt and rt, it should also hold for $\text{Node}(x, lt, rt)$ whatever the value of x:

$$\forall x : \alpha, \ \forall lt, rt : \text{Tree } \alpha \cdot$$
$$size(lt) = length(flatten(lt)) \land size(rt) = length(flatten(rt)) \Rightarrow$$
$$size(\text{Node}(x, lt, rt)) = length(flatten(\text{Node}(x, lt, rt)))$$

We thus have two inductive hypotheses: (i) $size(lt) = length(flatten(lt))$, and (ii) $size(rt) = length(flatten(rt))$.

The proof of the inductive case can now be completed:

$$size(\mathsf{Node}(x, lt, rt))$$
$$= \{\text{line 2 of the definition of } size\}$$
$$1 + size(lt) + size(rt)$$
$$= \{\text{applying the inductive hypotheses}\}$$
$$1 + length(\mathit{flatten}(lt)) + length(\mathit{flatten}(lt))$$
$$= \{\text{basic arithmetic}\}$$
$$length(\mathit{flatten}(lt)) + (1 + length(\mathit{flatten}(rt)))$$
$$= \{\text{line 2 of the definition of } length\}$$
$$length(\mathit{flatten}(lt)) + length(\mathsf{Cons}(x, \mathit{flatten}(rt)))$$
$$= \{\text{property relating } length \text{ of lists and catenation}\}$$
$$length(\mathit{flatten}(lt) \mathbin{+\!\!+} \mathsf{Cons}(x, \mathit{flatten}(rt)))$$
$$= \{\text{line 2 of the definition of } \mathit{flatten}\}$$
$$length(\mathit{flatten}(\mathsf{Node}(x, lt, rt)))$$

This allows us to complete the proof using structural induction.

This shows how properties can be proved about trees and functions operating on trees using structural induction. In a later chapter, we will look at more elaborate proofs about trees.

Exercises

8.26 Consider the following function, which switches the left and right subtrees of a tree recursively:

$$reflect \in \mathsf{Tree}\ \alpha \to \mathsf{Tree}\ \alpha$$

$$reflect(\mathsf{Nil}) \stackrel{\mathrm{df}}{=} \mathsf{Nil}$$

$$reflect(\mathsf{Node}(x, lt, rt)) \stackrel{\mathrm{df}}{=} \mathsf{Node}(x, reflect(rt),\ reflect(lt))$$

(i) Draw the tree resulting from applying *reflect* to the tree shown in the diagram.

(ii) Prove that reflecting a tree does not change its size: $\forall t : \mathsf{Tree}\ \alpha \cdot size(t) = size(reflect(t))$.

(iii) Prove that reflecting, then flattening is the same as flattening and then reversing: $\forall t : \mathsf{Tree}\ \alpha \cdot flatten(reflect(t)) = reverse(flatten(t))$.

8.27 The *flatten* function, as given, is called *an infix traversal* of the tree, since the value of the root is placed between the flattened versions of the left and the right subtree.

Two other variants are prefix traversal, where the value of the root is written before the left subtree, and postfix traversal, where the value of the root is written after the right subtree:

$$prefix(\text{Nil}) \stackrel{\text{df}}{=} \text{Empty}$$

$$prefix(\text{Node}(x, lt, rt)) \stackrel{\text{df}}{=} \text{Cons}(x, prefix(lt) \mathbin{+\!\!+} prefix(rt))$$

$$postfix(\text{Nil}) \stackrel{\text{df}}{=} \text{Empty}$$

$$postfix(\text{Node}(x, lt, rt)) \stackrel{\text{df}}{=} postfix(lt) \mathbin{+\!\!+} postfix(rt) \mathbin{+\!\!+} \text{Cons}(x, \text{Nil})$$

Prove that a postfix traversal of a tree returns the reverse of the prefix traversal:
$\forall t : \text{Tree } \alpha \cdot postfix(t) = reverse(prefix(t))$.

8.4 Summary

In this chapter we have seen how one can introduce new types and how to reason about them. The notion of structured types turns out to be surprisingly useful, and the principle of structural induction is indispensable when proving properties of various discrete structures. In the next chapter, we will see how these new types can be used to reason about numbers. Although we have already seen useful applications of structured types in this chapter, other, more advanced, uses in computer science will be explored in Chap. 10.

In practice, structured types appear in an almost identical fashion in programming. Languages such as Scala, Haskell, ML and F♯ allow for the definition of such types directly, including definitions which use pattern matching. The definitions we saw in this chapter can be used almost verbatim as code in such languages. In object-oriented languages, one can encode the type constructors as different object constructors or using inheritance for the different subclasses of the type. The rules using pattern matching would mostly have to be encoded using case analysis, not unlike the conditional definitions typically used in mathematics. As we have seen, the translation of pattern-matching definitions into ones using conditionals is straightforward. Finally, in languages such as Pascal and C, one could emulate these types using enumerated types and variant records. Structured data types thus also give us a clean way of representing our data in our programs.

Chapter 9
Numbers

We have already reached Chap. 9 in a book about mathematics without having yet discussed numbers. Although we have already used numbers in previous chapters to encode notions such as multisets and sequences, unlike the formal treatment we gave the other concepts we have explored, we have brushed aside parts of proofs which use properties of numbers simply by saying that 'this follows from basic laws of arithmetic'. It is now time to show how we can encode the notions of numbers, addition and other operators in a formal manner and prove these basic laws of arithmetic once and for all.

9.1 Natural Numbers

The notion of numbers and quantities, which we use on a daily basis, is incredibly general. The concept of a number, say eight, is an abstraction of collections of objects at some point in time (there are eight apples in the basket), but also across time (I have been to Paris eight times). Identifying and naming the abstract notion of *eight*, encompassing these uses, allows us to show properties across all concrete uses of the notion. For example, if we show that *eight* can be split into two equal parts (*four* and *four*), we can apply this to apples in a basket, times I have been to Paris and the price of a *cortado* in Buenos Aires.

Mathematicians have refined this general notion of quantities, identifying different types of numbers useful for different applications—the basic counting numbers (useful to count sheep), extended with negative numbers (useful to keep track of a bank account balance), fractions (useful to split cakes and divide an inheritance), etc. The most basic of these notions, and the underlying notion behind numbers, is the counting, or so-called *natural numbers*—denoted by the symbol $\mathbb{N}$, and containing all theoretically possible answers you can give to the question 'how many angels fit on the head of a pin?' or 'how many sheep do you keep in your kitchen?'[1] This allows for answers ranging over finite whole numbers. Note that zero is a nat-

---

[1] Theoretical, because we do not limit the answer depending on the size of your kitchen or the number of angels or sheep that exist in the universe.

ural number, since I have no sheep in my kitchen. Some textbooks take the natural numbers to start from one, which can be convenient in some contexts, but is not so *natural*, if we want to use these numbers to count.

9.1.1 Defining the Counting Numbers

Based on this informal explanation of what the natural numbers are, we can conceptually construct them by starting with zero corresponding to an empty kitchen. We then let sheep into the kitchen, one at a time, each time declaring the resulting number of sheep in the kitchen to be a natural number. This works, since when given a natural number we can take its successor (increment it) to produce a new number. This gives us a way of encoding the natural numbers using structured types:

$$\mathbb{N} ::= \text{Zero} \mid \text{Successor } \mathbb{N}$$

Note that the axioms for structured types which we saw in the previous chapter give us structural equality over natural numbers, which corresponds to our everyday notion of equality of numbers. For instance, zero is not equal to the successor of any natural number, and if two numbers are equal, so are their successors. Similarly, the axiom guaranteeing minimality of the type guarantees that, for instance, negative numbers and fractions are not included in the type.

The basis for this axiomatisation of the natural numbers was originally formulated by Giuseppe Peano in the 19th century, and they are usually referred to as *Peano's axioms*. Here we have started with a more general formalisation for structured types and taken a concrete instance to obtain the natural numbers. The inductive principle for this structured type is nothing but the familiar rule of numeric induction:

$$\frac{\pi(\text{Zero}) \qquad \forall k : \mathbb{N} \cdot \pi(k) \Rightarrow \pi(\text{Successor}(k))}{\forall n : \mathbb{N} \cdot \pi(n)}$$

This corresponds to the principle of numeric induction with which you are probably already familiar. If we prove a property for the value of 0 and we prove that when it holds for k it also holds for $k + 1$ (the successor of k), then the property must hold for all the natural numbers. We will see this principle applied to various operators in the rest of this chapter.

9.1.2 Defining the Arithmetic Operators

As we have seen in the previous chapter, the notion of structural equality allows us to define functions using constructor patterns.

9.1.2.1 Comparing Natural Numbers

We will use the notion of structural equality for numeric equality, which we write as $n = m$. We will also write $n \neq m$ to denote that n is not equal to m, which is defined as $\neg(n = m)$. In addition to equality and non-equality, we will define inequality relations less-than ($<$), greater-than ($>$), less-than-or-equal-to ($\leq$) and greater-than-or-equal-to ($\geq$). We start by recursively defining less-than-or-equal-to as a function returning a Boolean value.

$$\text{Zero} \leq m \overset{\text{df}}{=} true$$

$$\text{Successor}(n) \leq \text{Zero} \overset{\text{df}}{=} false$$

$$\text{Successor}(n) \leq \text{Successor}(m) \overset{\text{df}}{=} n \leq m$$

Based on this definition, we can prove properties of $\leq$. We start by proving that it is transitive.

Theorem 9.1 *The relation $\leq$ is transitive*: $\forall l, m, n : \mathbb{N} \cdot l \leq m \wedge m \leq n \Rightarrow l \leq n$.

Proof The proof follows by induction on variable l.

Base case: For $l = \text{Zero}$. We have to prove that $\forall m, n : \mathbb{N} \cdot \text{Zero} \leq m \wedge m \leq n \Rightarrow \text{Zero} \leq n$. This clearly holds, since $\text{Zero} \leq n$ for any value of n.
Inductive case: For the inductive case, we assume that the result holds for $l = k$:

$$\forall m, n : \mathbb{N} \cdot k \leq m \wedge m \leq n \Rightarrow k \leq n$$

On the basis of this, we have to prove that it holds for $\text{Successor}(k)$. We thus have to prove that:

$$\forall m, n : \mathbb{N} \cdot \text{Successor}(k) \leq m \wedge m \leq n \Rightarrow \text{Successor}(k) \leq n$$

$$\text{Successor}(k) \leq m \wedge m \leq n$$
$\Rightarrow$ {from the definition of $\leq$, we can identify m'
          such that $m = \text{Successor}(m')$}
$$\text{Successor}(k) \leq \text{Successor}(m') \wedge \text{Successor}(m') \leq n$$
$\Rightarrow$ {from the definition of $\leq$, we can identify n'
          such that $n = \text{Successor}(n')$}
$$\text{Successor}(k) \leq \text{Successor}(m') \wedge$$
$$\text{Successor}(m') \leq \text{Successor}(n')$$
$\Leftrightarrow$ {applying line 3 of the definition of $\leq$ twice}
$$k \leq m' \wedge m' \leq n'$$
$\Rightarrow$ {by the inductive hypothesis}
$$k \leq n'$$
$\Leftrightarrow$ {by line 3 of the definition of $\leq$}
$$\text{Successor}(k) \leq \text{Successor}(n')$$
$\Leftrightarrow$ {since $n = \text{Successor}(n')$}
$$\text{Successor}(k) \leq n$$

It thus follows that: $\text{Successor}(k) \leq m \wedge m \leq n \Rightarrow \text{Successor}(k) \leq n$, which allows us to complete the proof of the inductive case.

The result thus follows by the inductive principle.   ■

We will show another property of $\leq$, in which we use the structural equality axioms.

Proposition 9.2 *Less-than-or-equal-to is reflexive—it relates pairs of structurally equivalent numbers*: $\forall n, m : \mathbb{N} \cdot n = m \Rightarrow n \leq m$.

Proof The proof uses induction on variable n.

Base case: For $n = $ Zero—we thus have to prove that $\forall m : \mathbb{N} \cdot$ Zero $= m \Rightarrow$ Zero $\leq$ m. However, by line 1 of the definition of $\leq$, it follows that Zero $\leq m$ for any value of m, which is sufficient to prove the base case.

Inductive case: We will assume that the property holds for $n = k$:

$$\forall m : \mathbb{N} \cdot k = m \Rightarrow k \leq m$$

We now prove that it also holds for Successor(k): $\forall m : \mathbb{N} \cdot$ Successor(k) $= m \Rightarrow$ Successor(k) $\leq m$.

$$
\begin{aligned}
&\text{Successor}(k) = m \\
\Rightarrow\ &\{\text{by definition of structural equality,} \\
&\qquad m = \text{Successor}(m') \text{ for some value } m'\} \\
&\text{Successor}(k) = \text{Successor}(m') \\
\Rightarrow\ &\{\text{by definition of structural equality}\} \\
&k = m' \\
\Rightarrow\ &\{\text{using the inductive hypothesis}\} \\
&k \leq m' \\
\Leftrightarrow\ &\{\text{by line 3 of the definition of } \leq\} \\
&\text{Successor}(k) \leq \text{Successor}(m') \\
\Leftrightarrow\ &\{\text{since } m = \text{Successor}(m')\} \\
&\text{Successor}(k) \leq m
\end{aligned}
$$

This completes the proof for the inductive case.

The result follows using the principle of induction.   ■

The other inequality relations can be similarly defined using constructor matching. However, another approach is to define them in terms of $\leq$:

$$n \geq m \stackrel{\text{df}}{=} m \leq n$$

$$n < m \stackrel{\text{df}}{=} n \leq m \wedge n \neq m$$

$$n > m \stackrel{\text{df}}{=} n \geq m \wedge n \neq m$$

This allows us to reduce all proofs about these operators to use the definition of $\leq$.

Exercises

9.1 Prove that $\forall n, m : \mathbb{N} \cdot n \leq m \wedge m \leq n \Rightarrow n = m$.

9.2 Prove that $>$ is transitive.

9.3 Prove that $\leq$ is a total order: $\forall n, m : \mathbb{N} \cdot n \leq m \vee m \leq n$.

9.4 Define $\leq$ as a relation over the natural numbers, using the function defined in this section.

9.5 Define $>$ using a pattern-matching approach.

9.1.2.2 Addition

Addition conceptually corresponds to the combination of two collections of things together—given two bags of apples their sum is the result of transferring them all into a single bag. One way of performing this operation is to repeatedly take one apple from the first bag and put it into the second until the first bag empties completely. The second bag is the result of the addition. This is the spirit of the formal definition of addition we use:

$$\mathsf{Zero} + n \stackrel{\mathrm{df}}{=} n$$

$$\mathsf{Successor}(n) + m \stackrel{\mathrm{df}}{=} n + \mathsf{Successor}(m)$$

For instance, to calculate the result of adding Successor(Zero) (which we usually call 1) and Successor(Successor(Zero)) (which we usually call 2) the definition gives us:

$$\mathsf{Successor}(\mathsf{Zero}) + \mathsf{Successor}(\mathsf{Successor}(\mathsf{Zero}))$$
$$= \{\text{line 2 of the definition of addition}\}$$
$$\mathsf{Zero} + \mathsf{Successor}(\mathsf{Successor}(\mathsf{Successor}(\mathsf{Zero})))$$
$$= \{\text{line 1 of the definition of addition}\}$$
$$\mathsf{Successor}(\mathsf{Successor}(\mathsf{Successor}(\mathsf{Zero})))$$

This gives the result Successor(Successor(Successor(Zero))), which we usually call 3. Congratulations, you have just followed a proof that $1 + 2 = 3$.

Note that the first operand of addition becomes smaller with each recursive call, thus guaranteeing that the definition is a sound one. Based on this definition, we can prove various properties of addition, such as commutativity and associativity. There are two ways of viewing these proofs. The first is that they are verification that addition satisfies these properties, while the second is that if we take the commutativity and associativity of addition as given, then the proofs are confirmation that the definition of addition makes sense.

Sometimes, a different definition of addition is used, replacing the second line equating $\mathsf{Successor}(n) + m$ with $\mathsf{Successor}(n + m)$. The intuition is that, rather than moving apples from the first to the second bag and then adding the two bags together, we put aside one apple from the first bag, add the bags, and then add the apple which we put aside to the result. We will start by proving that this equality can be proved to follow from our definition of addition.

Lemma 9.3 $\forall n : \mathbb{N} \cdot \forall m : \mathbb{N} \cdot \mathsf{Successor}(n) + m = \mathsf{Successor}(n + m)$.

Proof The proof will use induction on variable n.

Base case: We start by proving the property for $n = \mathsf{Zero}$. We have to prove that
$\forall m : \mathbb{N} \cdot \mathsf{Successor}(\mathsf{Zero}) + m = \mathsf{Successor}(\mathsf{Zero} + m)$.

$$
\begin{aligned}
&\text{left-hand side} \\
={}& \mathsf{Successor}(\mathsf{Zero}) + m \\
={}& \{\text{line 2 of the definition of addition}\} \\
& \mathsf{Zero} + \mathsf{Successor}(m) \\
={}& \{\text{line 1 of the definition of addition}\} \\
& \mathsf{Successor}(m) \\
={}& \{\text{line 1 of the definition of addition}\} \\
& \mathsf{Successor}(\mathsf{Zero} + m) \\
={}& \text{right-hand side}
\end{aligned}
$$

Inductive case: We now assume that the property holds for a particular value of n, with $n = k$:

$$\forall m : \mathbb{N} \cdot \mathsf{Successor}(k) + m = \mathsf{Successor}(k + m)$$

We now prove that the property also holds for $\mathsf{Successor}(k)$:

$$\forall m : \mathbb{N} \cdot \mathsf{Successor}(\mathsf{Successor}(k)) + m = \mathsf{Successor}(\mathsf{Successor}(k) + m)$$

The proof of this statement follows:

$$
\begin{aligned}
&\text{left-hand side} \\
={}& \mathsf{Successor}(\mathsf{Successor}(k)) + m \\
={}& \{\text{line 2 of the definition of addition}\} \\
& \mathsf{Successor}(k) + \mathsf{Successor}(m) \\
={}& \{\text{applying the inductive hypothesis}\} \\
& \mathsf{Successor}(k + \mathsf{Successor}(m)) \\
={}& \{\text{line 2 of the definition of addition}\} \\
& \mathsf{Successor}(\mathsf{Successor}(k) + m) \\
={}& \text{right-hand side}
\end{aligned}
$$

This completes the proof for the inductive case, thus allowing us to conclude the general result using induction. ∎

This result allows us to prove various other properties of addition. We will show that addition is commutative, confirming what you have been told in school since you were very young.

Theorem 9.4 *Addition is commutative*: $\forall n : \mathbb{N} \cdot \forall m : \mathbb{N} \cdot n + m = m + n$.

Proof The proof uses induction on variable n.

Base case: For $n =$ Zero. This part of the proof uses a result we have not proved, which says that zero is the right identity of addition. The proof of this conjecture will be left as an exercise to the reader.

$$
\begin{aligned}
&\text{left-hand side} \\
={}& \text{Zero} + m \\
={}& \{\text{line 1 of the definition of addition}\} \\
& m \\
={}& \{\text{conjecture that } m + \text{Zero} = m\} \\
& m + \text{Zero} \\
={}& \text{right-hand side}
\end{aligned}
$$

Inductive case: We assume that the result holds for some particular value k of n: $\forall m : \mathbb{N} \cdot k + m = m + k$. Based on this, we prove that it also holds for $n = \text{Successor}(k)$:

$$
\begin{aligned}
&\text{left-hand side} \\
={}& \text{Successor}(k) + m \\
={}& \{\text{Lemma 9.3}\} \\
& \text{Successor}(k + m) \\
={}& \{\text{inductive hypothesis}\} \\
& \text{Successor}(m + k) \\
={}& \{\text{Lemma 9.3}\} \\
& \text{Successor}(m) + k \\
={}& \{\text{line 2 of the definition of addition}\} \\
& m + \text{Successor}(k) \\
={}& \text{right-hand side}
\end{aligned}
$$

This concludes the inductive case, which allows us to conclude the desired result using the principle of induction.   ∎

We have shown how addition can be defined on the type of natural numbers, and how we can prove general properties of addition. In the rest of the chapter we will continue with this approach of defining new operators on the naturals and proving properties of the operators.

Exercises

9.6 Prove that zero is the right identity of addition: $\forall m : \mathbb{N} \cdot m + \text{Zero} = m$.

9.7 Prove that addition is associative: $\forall l, m, n : \mathbb{N} \cdot l + (m + n) = (l + m) + n$. Hint: use induction on variable l.

9.8 Prove that addition preserves $>$: if $n > m$ then $l + n > l + m$.

9.9 In the first example about addition, we proved that, from our definition, it follows that $\text{Successor}(n) + m = \text{Successor}(n + m)$. Now consider the opposite situation—let us take the following to be our definition of addition:

$$\text{Zero} + n \stackrel{\text{df}}{=} n$$

$$\text{Successor}(n) + m \stackrel{\text{df}}{=} \text{Successor}(n + m)$$

Based on this definition, prove that $\text{Successor}(n) + m = n + \text{Successor}(m)$. In this manner, we will have shown that the two definitions are, in fact, equivalent.

9.1.2.3 Subtraction

It is worthwhile to start by noting that subtraction on the natural numbers is a partial function, in that $n - m$ is not defined when m is larger than n. There are two intuitive ways of defining subtraction:

Property-based definition: One can look at subtraction as the operation which acts as the opposite of addition, and which should thus satisfy: (i) adding a number and then subtracting it should result in the original value: $\forall n, m : \mathbb{N} \cdot (n + m) - m = n$; and (ii) subtracting a number from one which is not smaller and then adding it should result in the original value: $\forall n, m : \mathbb{N} \cdot n \geq m \Rightarrow (n - m) + m = n$.

Algorithm-based definition: Subtraction can be seen as the operation which removes an item in the first bag for each item in the second, thus corresponding to the following pattern-based definition:

$$n - \text{Zero} \stackrel{\text{df}}{=} n$$

$$\text{Successor}(n) - \text{Successor}(m) \stackrel{\text{df}}{=} n - m$$

Note that, unlike the property-based approach, the equations give us an algorithm to *calculate* the subtraction of two numbers.

Which definition to adopt is largely a matter of style, because either one can be proved as a consequence of the other, thus making them equivalent formulations. In computer science texts, one tends to find more algorithmic (sometimes also called *constructive*) definitions, which tell us how to compute the result from which they then prove properties of the operators, whereas mathematical texts tend to prefer to define operators in terms of their properties, and then prove that a constructive approach gives a solution of the specification. Here, we will adopt the constructive approach, and prove the properties as consequences of the definitions.

Theorem 9.5 *Addition nullifies the effect of subtraction*: $\forall n, m : \mathbb{N} \cdot n \geq m \Rightarrow (n - m) + m = n$.

Proof The proof will follow by induction. Note that subtraction uses pattern-matching on the second operator, and thus induction on variable m will be applied.

Base case: For $m = \mathsf{Zero}$. We have to prove that $(n - \mathsf{Zero}) + \mathsf{Zero} = \mathsf{Zero}$. The result follows by applying line 1 of the definition of subtraction, followed by line 1 of the definition of addition.

Inductive case: We start by assuming that the property holds for $m = k$: $\forall n : \mathbb{N} \cdot n \geq k \Rightarrow (n - k) + k = n$, based on which we will prove that the property also holds for $m = \mathsf{Successor}(k)$:

$$\forall n : \mathbb{N} \cdot n \geq \mathsf{Successor}(k) \Rightarrow (n + \mathsf{Successor}(k)) - \mathsf{Successor}(k) = n$$

We thus assume that $n \geq \mathsf{Successor}(k)$, and prove the equality:

$$(n - \mathsf{Successor}(k)) + s(k)$$
$= \{$since $n \geq \mathsf{Successor}(k)$, by the definition of $\geq$
$\qquad\qquad n$ has to be of the form $\mathsf{Successor}(n')\}$
$$(\mathsf{Successor}(n') - \mathsf{Successor}(k)) + \mathsf{Successor}(k)$$
$= \{$line 2 of the definition of subtraction$\}$
$$(n' - k) + \mathsf{Successor}(k)$$
$= \{$Lemma 9.3 and commutativity of addition$\}$
$$\mathsf{Successor}((n' - k) + k)$$
$= \{$by the inductive hypothesis and that $n' \geq k$
$\qquad\qquad$ (since $\mathsf{Successor}(n') \geq \mathsf{Successor}(k))\}$
$$\mathsf{Successor}(n')$$
$= \{$since $n = \mathsf{Successor}(n')\}$
$$n$$

This completes the proof by numeric induction.   ■

The proof of the second property is left as an exercise. If we were to complete a proof of equivalence, we would have to prove that (i) the property-based approach defines a function, with at most one solution for $n - m$; and (ii) whenever the property-based approach provides a solution, the algorithmic approach always terminates, and gives an answer. At this stage we will not dwell on this further, but we will be discussing these issues later on in Chap. 10.

Exercises

9.10 Prove that $\forall n : \mathbb{N} \cdot n - n = \mathsf{Zero}$.

9.11 Prove that subtraction nullifies addition: $\forall n, m : \mathbb{N} \cdot (n + m) - m = n$.

9.12 Prove that, if $n > m$, then $n - m > \mathsf{Zero}$.

9.13 Prove that subtraction (sometimes) commutes over addition: $\forall l, m, n : \mathbb{N} \cdot l \geq m \Rightarrow (l + n) - m = (l - m) + n$.

9.14 Another approach to defining subtraction is to start by defining the predecessor of a number, and then defining subtraction as taking the predecessor as many times as the second operand of subtraction, as shown below:

$$predecessor(\mathsf{Successor}(n)) \stackrel{\mathrm{df}}{=} n$$

$$n \sim \mathsf{Zero} \stackrel{\mathrm{df}}{=} n$$
$$n \sim \mathsf{Successor}(m) \stackrel{\mathrm{df}}{=} predecessor(n \sim m)$$

Prove that the two operators give the same result:

$$\forall n, m : \mathbb{N} \cdot n \geq m \Rightarrow n - m = n \sim m$$

9.1.2.4 Multiplication

Multiplication corresponds to repeated addition, and $n \times m$ can be seen as shorthand for the addition of n copies of m. The formal definition follows from this intuitive meaning of multiplication and is based on two observations: (i) zero copies of any number m result in zero; and (ii) $\mathsf{Successor}(n)$ copies of a number m is the same as n copies of m, plus an additional m.

$$\mathsf{Zero} \times m \stackrel{\mathrm{df}}{=} \mathsf{Zero}$$
$$\mathsf{Successor}(n) \times m \stackrel{\mathrm{df}}{=} m + n \times m$$

Lemma 9.6 *Multiplication can also be done by adding the first operand for each successor in the second*: (i) $n \times \mathsf{Zero} = \mathsf{Zero}$; *and* (ii) $n \times \mathsf{Successor}(m) = n + (n \times m)$.

Proof We will prove the second law, and leave (i) as an exercise for the reader. The second proposition will be proved by induction on variable n.

Base case: To prove that $\mathsf{Zero} \times \mathsf{Successor}(m) = \mathsf{Zero} + (\mathsf{Zero} \times m)$.

$$\mathsf{Zero} \times \mathsf{Successor}(m)$$
$$= \{\text{line 1 of the definition of multiplication}\}$$
$$\mathsf{Zero}$$
$$= \{\text{line 1 of the definition of addition}\}$$
$$\mathsf{Zero} + \mathsf{Zero}$$
$$= \{\text{line 1 of the definition of multiplication}\}$$
$$\mathsf{Zero} + (\mathsf{Zero} \times m)$$

Inductive case: We assume that the result is true for $n = k$: $\forall m : \mathbb{N} \cdot k \times \mathsf{Successor}(m) = k + (k \times m)$. Based on this we prove that it also holds for $n = \mathsf{Successor}(k)$: $\forall m : \mathbb{N} \cdot \mathsf{Successor}(k) \times \mathsf{Successor}(m) = \mathsf{Successor}(k) + (\mathsf{Successor}(k) \times m)$.

$$\text{Successor}(k) \times \text{Successor}(m)$$
$$= \{\text{line 2 of the definition of multiplication}\}$$
$$\text{Successor}(m) + (k \times \text{Successor}(m))$$
$$= \{\text{using the inductive hypothesis}\}$$
$$\text{Successor}(m) + (k + (k \times m))$$
$$= \{\text{associativity of addition}\}$$
$$(\text{Successor}(m) + k) + (k \times m)$$
$$= \{\text{Lemma 9.3}\}$$
$$\text{Successor}(m + k) + (k \times m)$$
$$= \{\text{commutativity of addition}\}$$
$$\text{Successor}(k + m) + (k \times m)$$
$$= \{\text{Lemma 9.3}\}$$
$$(\text{Successor}(k) + m) + (k \times m)$$
$$= \{\text{associativity of addition}\}$$
$$\text{Successor}(k) + (m + (k \times m))$$
$$= \{\text{line 2 of the definition of multiplication}\}$$
$$\text{Successor}(k) + (\text{Successor}(k) \times m)$$

Using the inductive principle, the result thus follows. ∎

Based on Lemma 9.6, we can now prove commutativity of multiplication.

Theorem 9.7 *Multiplication is commutative*: $\forall n, m : \mathbb{N} \cdot n \times m = m \times n$.

Proof The proof is by induction on variable n.

Base case: For $n = \text{Zero}$:

$$\text{Zero} \times m$$
$$= \{\text{line 1 of the definition of multiplication}\}$$
$$\text{Zero}$$
$$= \{\text{using Lemma 9.6(i)}\}$$
$$m \times \text{Zero}$$

Inductive case: We assume that the property holds for $n = k$: $\forall m : \mathbb{N} \cdot k \times m = m \times k$. We proceed to prove that it thus also holds for $n = \text{Successor}(k)$.

$$\text{Successor}(k) \times m$$
$$= \{\text{line 2 of the definition of multiplication}\}$$
$$m + (k \times m)$$
$$= \{\text{using the inductive hypothesis}\}$$
$$m + (m \times k)$$
$$= \{\text{using Lemma 9.6(ii)}\}$$
$$m \times \text{Successor}(k)$$

The result thus holds by induction. ∎

Exercises

9.15 Prove law (i) of Lemma 9.6: $\forall n : \mathbb{N} \cdot n \times \mathsf{Zero} = \mathsf{Zero}$.

9.16 Prove that multiplication is associative:

$$\forall l, m, n : \mathbb{N} \cdot l \times (m \times n) = (l \times m) \times n$$

9.17 Prove that multiplication on the left distributes over addition:

$$\forall l, m, n : \mathbb{N} \cdot l \times (m + n) = (l \times m) + (l \times n)$$

9.18 Prove that if $n > \mathsf{Zero}$, then $n \times m \geq m$.

9.19 Prove that multiplication on the right distributes over subtraction:

$$\forall l, m, n : \mathbb{N} \cdot n \geq m \Rightarrow (n - m) \times l = (n \times l) - (m \times l)$$

9.20 In this exercise, we will explore the definition of natural number exponentiation. We will write n^m to denote n to the power of m: $n \times n \times \cdots \times n$ (with m copies of n). In the same way that addition is repeated incrementation, and multiplication is repeated addition, one can define powers of natural numbers as repeated multiplication.

   (i) Keeping in mind that a number to the power of zero is equal to Successor(Zero), recursively define n^m using pattern-matching on m.

   (ii) Prove that $\forall l, m, n : \mathbb{N} \cdot (l^m)^n = l^{m \times n}$.

   (iii) Prove that $\forall l, m, n : \mathbb{N} \cdot l^m \times l^n = l^{m+n}$.

9.1.3 Strong Induction

Informally, we find numeric induction to be a reasonable principle by noting that the property holds for Zero (the first proof obligation for induction), which implies that it also holds for Successor(Zero) (using the second proof obligation on the previous statement), which in turns implies that it also holds for Successor(Successor(Zero)), etc., until we reach the desired number. Moving up the number line uses the inductive case: if it holds for value k, then it must also hold for Successor(k).

Now consider another proof principle, with two proof obligations: (i) that the property holds for Zero (just like normal induction), and (ii) that if the property holds for all numbers up to and including k, then it must also hold for Successor(k). Informally, this principle also seems to be a sound one, but note that the inductive hypothesis is not simply that the property holds for k, but that it holds for *all* values up to and including k. In some cases, this stronger assumption enables us to prove properties which would otherwise not be directly provable using normal induction.

Formally, the principle of strong induction is the following:

$$\frac{\pi(\mathsf{Zero}) \qquad \forall k : \mathbb{N} \cdot (\forall k' : \mathbb{N} \cdot k' \leq k \Rightarrow \pi(k')) \Rightarrow \pi(\mathsf{Successor}(k))}{\forall n : \mathbb{N} \cdot \pi(n)}$$

Here is an informal example of the use of strong induction.

Example Every number n can be written as a sum of different powers of 2. For example, 19 can be written as the sum of three powers $2^0 + 2^1 + 2^4$, and 24 can be written as the sum of two powers $2^3 + 2^4$. This statement can be proved by strong induction on n.

Base case: For $n = \mathsf{Zero}$. One can express zero as an empty sum of powers of two.
Inductive case: We will assume that the result holds for all values up to and including k. We now need to prove that it also holds for $\mathsf{Successor}(k)$.
   Consider the value of α such that $2^\alpha \leq \mathsf{Successor}(k) < 2^{\alpha+1}$. Since $\mathsf{Successor}(k) - 2^\alpha < \mathsf{Successor}(k)$, we can use the inductive hypothesis to split $\mathsf{Successor}(k) - 2^\alpha$ into a sum of distinct powers of two. Adding 2^α to the sum, we get a new sum of powers of two which adds up to $\mathsf{Successor}(k)$. Are the powers still distinct? It suffices to show that 2^α could not have appeared in the powers of two adding up to $\mathsf{Successor}(k) - 2^\alpha$.
   Since $2^{\alpha+1} = 2 \times 2^\alpha = 2^\alpha + 2^\alpha$, and $\mathsf{Successor}(k) < 2^{\alpha+1}$, we can conclude that $\mathsf{Successor}(k) < 2^\alpha + 2^\alpha$. This confirms that $\mathsf{Successor}(k) - 2^\alpha < 2^\alpha$, and therefore the powers of two adding up to $\mathsf{Successor}(k) - 2^\alpha$ cannot include the power α.

This completes the proof by strong induction. Note that this result justifies how we encode numbers in binary on a computer.   ◇

   What is particularly interesting is that the proof rule for strong induction can be proved using normal induction. An informal argument of the proof follows.

Theorem 9.8 *Strong induction is a sound principle.*

Proof To prove that strong induction is a sound principle, we assume the antecedents of the rule, and prove that the conclusion is true. Let us assume the following two predicates:

(i) $\pi(\mathsf{Zero})$
(ii) $\forall k : \mathbb{N} \cdot (\forall k' : \mathbb{N} \cdot k' \leq k \Rightarrow \pi(k')) \Rightarrow \pi(\mathsf{Successor}(k))$

Based on this we would like to show that: $\forall n : \mathbb{N} \cdot \pi(n)$, which would suffice to show that strong induction is a sound principle.
   Consider the property π', defined as follows:

$$\pi'(n) \stackrel{\mathrm{df}}{=} \forall n' : \mathbb{N} \cdot n' \leq n \Rightarrow \pi(n')$$

Note that $\pi'(n)$ implies $\pi(n)$. Therefore, if we manage to prove property $\forall n : \mathbb{N} \cdot \pi'(n)$, then we can conclude the desired result that $\forall n : \mathbb{N} \cdot \pi(n)$.

We will prove property $\pi'(n)$ for all values of n using normal induction.

Base case: For $n = \mathsf{Zero}$. We need to prove $\pi'(\mathsf{Zero})$:

$$
\begin{aligned}
&\pi'(\mathsf{Zero}) \\
={}& \{\text{definition of } \pi'\} \\
&\forall n' : \mathbb{N} \cdot n' \leq \mathsf{Zero} \Rightarrow \pi(n') \\
={}& \{\text{only Zero is less than or equal to Zero}\} \\
&\pi(\mathsf{Zero})
\end{aligned}
$$

But by (i), we know that $\pi(\mathsf{Zero})$ holds, hence so does $\pi'(\mathsf{Zero})$.

Inductive case: We assume that $\pi'(k)$ holds: $\forall n' : \mathbb{N} \cdot n' \leq k \Rightarrow \pi(n')$, based on which we would like to prove that $\pi'(\mathsf{Successor}(k))$ also holds: $\forall n' : \mathbb{N} \cdot n' \leq \mathsf{Successor}(k) \Rightarrow \pi(n')$.

$$
\begin{aligned}
&\{\text{inductive hypothesis}\} \\
\Leftrightarrow{}& \forall n' : \mathbb{N} \cdot n' \leq k \Rightarrow \pi(n') \\
\Rightarrow{}& \{\text{by (ii)}\} \\
&(\forall n' : \mathbb{N} \cdot n' \leq k \Rightarrow \pi(n')) \wedge \pi(\mathsf{Successor}(k)) \\
\Leftrightarrow{}& \forall n' : \mathbb{N} \cdot n' \leq \mathsf{Successor}(k) \Rightarrow \pi(n')
\end{aligned}
$$

Hence the inductive case also holds.

By using normal induction, we have thus proved that $\forall n' : \mathbb{N} \cdot \pi'(n)$, which is sufficient to prove the conclusion of strong induction. ∎

We will be seeing other applications of strong induction in the rest of the chapter.

9.1.4 Division and Prime Numbers

The notion of prime numbers, numbers greater than 1 which are exactly divisible only by 1 and themselves, has a long history in mathematics. Prime numbers correspond closely to the notion of atomic numbers—numbers which cannot be constructed as a product of smaller numbers. Although studied since the Ancient Greeks, they have only found practical applications very recently. Nowadays, prime numbers play a crucial role in encryption algorithms, and securely buying from an online store would not be possible without them.

9.1.4.1 Multiples, Divisors and Remainders

Since prime numbers are based on division, we start by formally defining this operator. It is tempting to view division as a single operator which acts as the inverse of multiplication. For example, the value of $12 \div 3$ should be the value of x satisfying $x \times 3 = 12$, making it equal to 4. However, this approach leaves $7 \div 3$ undefined, since there is no natural number which satisfies $x \times 3 = 7$. Without substantial changes, this approach would thus limit us to talk about exact division. Instead, we split the notion of division into two parts: the quotient, or how many times a number fits into another, and the remainder. For instance, $7 \div 3$ would give two results: (i) the quotient part equal to 2, which says that 3 fits at most twice in 7; and (ii) the remainder part equal to 1, which says that removing two threes from the value of 7, leaves 1 unused. We will write these as $7 \div_q 3$ and $7 \div_r 3$, respectively.

Let us start by looking at the remainder part of division.

Definition 9.9 *The remainder of a natural number division can be algorithmically defined in the following manner. To calculate the remainder of n when divided by m, we subtract m from n repeatedly, until $n < m$, in which case we just take n:*

$$n \div_r m \stackrel{\mathrm{df}}{=} \begin{cases} n & \text{if } n < m \\ (n - m) \div_r m & \text{otherwise} \end{cases}$$

Two things to note are that (i) the definition assumes that $m \neq$ Zero, otherwise it would not be well defined, since the parameters of the recursive call are not smaller; and (ii) for the second value to be taken, it must be the case than $n \geq m$, which means that the subtraction is well-defined. ∎

One important observation of this operator is that the remainder should always be smaller than the divisor.

Theorem 9.10 *The remainder of a division operation is always smaller than the divisor: $\forall n, m : \mathbb{N} \cdot n \div_r m < m$.*

Proof This proof uses strong induction on variable n. Keep in mind that $m >$ Zero for the division to be meaningful.

Base case: For $n =$ Zero. We need to prove that $\forall m : \mathbb{N} \cdot$ Zero $\div_r m < m$.

> Zero $\div_r m$
> $=$ {since $m >$ Zero, it follows that Zero $\div_r m =$ Zero}
> Zero
> $<$ {since $m >$ Zero}
> m

Inductive case: Since we are using strong induction, we assume that $\forall m : \mathbb{N} \cdot k' \div_r$ $m < m$ for all $k' \leq k$. Now let us consider the property for $n =$ Successor(k). We consider two cases: (a) Successor$(k) < m$; and (b) Successor$(k) \geq m$.

Case (a): $\mathsf{Successor}(k) < m$:

$$\mathsf{Successor}(k) \div_r m$$
$$= \{\text{by line 1 of the definition of remainder and}$$
$$\qquad\text{the fact that } \mathsf{Successor}(k) < m\}$$
$$\mathsf{Successor}(k)$$
$$< \{\text{since } \mathsf{Successor}(k) < m\}$$
$$m$$

Case (b): $\mathsf{Successor}(k) \geq m$:

$$\mathsf{Successor}(k) \div_r m$$
$$= \{\text{by line 2 of the definition of remainder and}$$
$$\qquad\text{the fact that } \mathsf{Successor}(k) \geq m\}$$
$$(\mathsf{Successor}(k) - m) \div_r m$$
$$< \{\text{using inductive hypothesis (since } m > \mathsf{Zero} \text{ and}$$
$$\qquad\text{thus } \mathsf{Successor}(k) - m < \mathsf{Successor}(k))\}$$
$$m$$

Since the result follows in both cases, which cover all possibilities, we can conclude that the inductive case holds.

This completes the proof using strong induction.   ∎

It is worth noting that some of the lines of the above proof were not fully rigorous, since they used results which we have not proved elsewhere. However, we *do* have the tools to prove these results. We will now start taking larger steps in our proofs, knowing however, that we would be able to prove the additional results if necessary.

We can now also define the quotient part of the division.

Definition 9.11 *The quotient of n divided by m is defined in the following manner: if $n < m$, the result is* Zero, *otherwise calculate the quotient of $n - m$ divided by m adding one to the result.*

$$n \div_q m \stackrel{\mathrm{df}}{=} \begin{cases} \mathsf{Zero} & \text{if } n < m \\ \mathsf{Successor}((n - m) \div_q m) & \text{otherwise} \end{cases}$$ ∎

These operators allow us to reason about natural number division in a complete manner. The following fundamental theorem shows that, given two numbers, there is only one reasonable choice of quotient and remainder.

Theorem 9.12 *For any natural numbers n and m, there exist unique natural numbers q and r, with $r < m$, such that $n = q \times m + r$.*

Proof This proof is split into two parts: we first need to prove that there do exist natural numbers which satisfy the equality, and secondly, we prove that these natural numbers q and r are unique. For the first part, we note that $q = n \div_q m$ and $r = n \div_r m$ are solutions to the equation. We leave it as an exercise to prove that: (i) $n \div_r m < m$, and (ii) $n = (n \div_q m) \times m + (n \div_r m)$.

The proof of uniqueness is a proof *by contradiction*. Assuming a statement, we reach a contradiction, from which we can conclude that the assumed statement was false.

Let us assume that we have two different solutions q_1, r_1 and q_2, r_2, with $q_1 \leq q_2$. We can reason as follows:

$$\{\text{both } q_1, r_1 \text{ and } q_2, r_2 \text{ pairs are solutions of } n = q \times m + r\}$$
$$\Rightarrow n = q_1 \times m + r_1 \wedge n = q_2 \times m + r_2$$
$$\Rightarrow \{\text{transitivity and commutativity of equality}\}$$
$$q_1 \times m + r_1 = q_2 \times m + r_2$$
$$\Rightarrow \{\text{since } a + b \geq a, (a + b) - a \text{ is a valid subtraction}\}$$
$$(q_1 \times m + r_1) - (q_1 \times m) = (q_2 \times m + r_2) - (q_1 \times m)$$
$$\Rightarrow \{\text{subtraction distributes over multiplication and}$$
$$\text{commutes with addition (Exercises 9.19 and 9.13)}\}$$
$$(q_1 \times m - q_1 \times m) + r_1 = (q_2 - q_1) \times m + r_2$$
$$\Rightarrow \{\text{since } n - n = \mathsf{Zero} \text{ and zero is the identity of addition}\}$$
$$r_1 = (q_2 - q_1) \times m + r_2$$

We now split $q_1 \leq q_2$ into two cases: $q_1 = q_2$ and $q_1 < q_2$ and consider them separately:

Case (a): If $q_1 = q_2$, from $r_1 = (q_2 - q_1) \times m + r_2$ we can conclude that $r_1 = r_2$. Therefore, in this case, $q_1 = q_2$ and $r_1 = r_2$—the two solutions are not distinct at all, which contradicts the assumption that the solutions are different.

Case (b): If $q_1 < q_2$ (or equivalently that $q_2 > q_1$), we can reason as follows:

$$q_2 > q_1$$
$$\Rightarrow \{\text{by Exercise 9.12}\}$$
$$q_2 - q_1 > \mathsf{Zero}$$
$$\Rightarrow \{\text{by Exercise 9.18}\}$$
$$(q_2 - q_1) \times m \geq m$$
$$\Rightarrow \{\text{by Exercise 9.8}\}$$
$$(q_2 - q_1) \times m + r_2 \geq m$$
$$\Rightarrow \{\text{since } r_1 = (q_2 - q_1) \times m + r_2\}$$
$$r_1 \geq m$$

This contradicts the original statement that $r_1 < m$.

In both cases, we have discovered a contradiction, implying that the our assumption that there exist distinct solutions for q and r is wrong. Therefore, there must be no more than one solution. ∎

The opposite direction is also the case. Combining a number using a quotient and remainder will return the original quotient and remainder upon division:

Corollary 9.13 *For any natural numbers q, d and r $(r < d)$, $((q \times d) + r) \div_r d = r$ and $((q \times d) + r) \div_q d = q$.*

Proof Define $\alpha = q \times d + r$. But by the first part of the previous theorem, we also know that $\alpha = (\alpha \div_q d) \times d + (\alpha \div_r d)$. Since the solutions for q and r are unique, we can conclude that $((q \times d) + r) \div_r d = r$ and $((q \times d) + r) \div_q d = q$. ∎

Division is considerably more complex than the other operators we have encountered so far. However, the results we proved in this section are crucial for reasoning about prime numbers in the next section.

Exercises

9.21 Prove that $\forall n : \mathbb{N} \cdot n \div_q 1 = n$.

9.22 Prove that $\forall n : \mathbb{N} \cdot n \div_r 1 = \mathsf{Zero}$.

9.23 Prove that $\forall n, m : \mathbb{N} \cdot n \div_r m < m$.

9.24 Prove that $\forall n, m : \mathbb{N} \cdot n = (n \div_q m) \times m + (n \div_r m)$.

9.25 Prove that, if $l \div_r m = \mathsf{Zero}$ and $m \div_r n = \mathsf{Zero}$, then $l \div_r n = \mathsf{Zero}$.

9.1.4.2 Prime Numbers

In studying objects and phenomena, an approach which frequently works is that of decomposing the object into its constituent parts. Starting by understanding the constituent parts, then how they interact and combine together, gives insight into how and why an object behaves as it does. We see this approach everywhere, from chemists studying compounds by looking at the elements which make them up, physicists studying how atoms work by looking at their constituent parts, linguists who study language by splitting sentences into phrases, to system developers, who when faced with a huge program seek to understand it by looking at the constituent parts. One of the fascinating things about this approach is that it can be iterated— we know that matter is made up of molecules, which are made up of atoms, which are made up of a combination of protons, neutrons and electrons, and so on. When to stop breaking down the objects into parts depends on the level of abstraction one would like to reason at, and whether further decomposition gives any further insight. For instance, when trying to understand a program written in a high-level language, it is rarely useful to study the machine code produced by the compiler, or the electrical signals going through your computer when the program is being executed.

The Ancient Greeks were interested in studying how numbers can be decomposed into smaller ones. Using the two main operators of addition and multiplication, they categorised numbers according to how they can be split into parts. For instance, the notion of square numbers corresponds to those which can be decomposed via multiplication into two equal parts, or according to addition as the sum of n copies of a number n—thus allowing a square placement of the items. Similarly, triangular numbers are the ones which can be decomposed as the sum of all the numbers from 1 to some number n—allowing a triangular placement.

For example, 10 is a triangular number, because it is equal to the sum $1 + 2 + 3 + 4$:

Splitting numbers down in terms of multiplication—splitting a number into two divisors, neither of which is 1—turned out to be particularly interesting. Although, a number can be split in different ways—for example, 12 can be split as 2×6 or 3×4 —if one reiterates the process until no further breaking down is possible, it turns out that there is only one way of splitting a number. For example, if 12 is split as 2×6, we note that 2 cannot be split any further, but 6 can be split into 2×3, neither of which can be split any further—thus splitting 12 as $2 \times 2 \times 3$. Another way of splitting 12 is into 3×4—once again, 3 cannot be split any further, but 4 can be split into 2×2, thus breaking down 12 into $3 \times 2 \times 2$. Other than the order of the parts, we note that the two decompositions are equivalent. It turns out that this works for any number, which provides a tool to analyse and reason about numbers. The numbers which cannot be split any further give us a notion of indivisible or atomic numbers.

To be able to define and reason about prime numbers, we start by formally defining the notion of exact divisors, or factors of a number.

Definition 9.14 *We say that a number n exactly divides another number m, written* $n \overset{\cdot}{\to} m$, *if dividing m by n leaves no remainder:* $m \div_r n = \mathsf{Zero}$.

The set of numbers which exactly divide a non-zero number n are called the divisors, *or* factors *of n:*

$$divisors(n) \overset{\mathrm{df}}{=} \{m : \mathbb{N}_1 \mid m \overset{\cdot}{\to} n\}$$

The proper divisors *of a number $n \in \mathbb{N}_1$ are the divisors of n excluding 1 and n:*

$$divisors^+(n) \overset{\mathrm{df}}{=} divisors(n) \setminus \{1, n\} \qquad \blacksquare$$

Although the definition of divisors of a number leaves open the possibility that they may be larger than the number itself, we can prove that this is impossible:

Proposition 9.15 *Divisors of a number are not larger than the number itself:* $\forall n, d :$ $\mathbb{N} \cdot d \in divisors(n) \Rightarrow d \leq n$. *Furthermore, proper divisors are strictly smaller than the number itself.*

Proof The proof is by contradiction. Take $n \in \mathbb{N}_1$, and $d \in divisors(n)$ and let us assume that $d > n$. By the first line of the definition of remainder, we know that $n \div_r d = n$ (since $d > n$). But since d is a divisor of n, $n \div_r d = 0$, which means that $n = 0$, which is impossible. Hence the assumption is wrong: $\neg(d > n)$, or $d \le n$.

Since, by definition, a proper divisor d of n cannot be equal to n, it follows that $d < n$.   ∎

One important corollary of this proposition is that we can write a program to generate all factors of a number n, by checking the finite set of numbers from 1 to n.

We are now ready to define the notion of prime numbers.

Definition 9.16 *A number $n \ge 2$ is said to be* prime *if it has no proper divisors:* $prime(p) \overset{\text{df}}{=} divisors^+(p) = \emptyset$.   ∎

Since, by the previous proposition, the exact divisors of a number are not larger than the number itself, we can write an algorithm that finds all prime numbers up to a maximum M:

```
primes := ∅
foreach n in (2...M) {
    isprime := true
    foreach i in (2...n − 1) {
        if (n ÷r i = 0) then isprime := false
    }
    if (isprime) then primes := primes ∪ {n}
}
```

The program is far from being an efficient generator of prime numbers. However, we can prove properties allowing us to make the program more efficient.[2] Consider the following theorem, which says that it is sufficient to check whether a number has prime factors:

Theorem 9.17 *A number ($n \ge 2$) is prime if and only if it has no proper prime divisors.*

Proof Outline Clearly, if a number is prime it has no proper divisors, and therefore has no proper prime divisors.

On the other hand, we need to prove that a number which has no proper prime divisors has to be prime. We will give an informal sketch of how this can be proved using strong induction.

---

[2]As the complexity of the proofs in the rest of this chapter increases, we will not be giving full rigorous proofs, since these would be extremely long and tedious to follow. Instead, we will be giving proof sketches, which indicate the outline of the structure of a rigorous proof if we were to write one.

For the base case (in this case, we take $n = 2$), the proof is straightforward. The number 2 has no proper prime divisors, and is prime.

For the inductive case, we assume that all number up to and including k satisfy the property. Does Successor(k) also satisfy the property? We need to show that, if Successor(k) has no prime divisors, then it is prime.

Now, either Successor(k) is prime or it is not. In the first case, if Successor(k) is prime, then the result follows trivially. If it is not a prime, then it must have proper divisors. Let d be one of them. If d is prime, then the proof is done (since the condition of the implication is false). If d is not prime, and since $d <$ Successor(k) (since d is a proper divisor of Successor(k)), by the inductive hypothesis it follows that d must have proper prime divisors—let us call one of them d'. Since d' is a proper divisor of d, and d is a proper divisor of Successor(k), it follows from Exercise 9.25 that d' is a divisor of Successor(k). We have thus found a proper prime divisor for Successor(k), which concludes the inductive case. □

How does this theorem allow us to produce prime numbers more efficiently? If we start building a list of the prime numbers as we did before, we need only check for divisors amongst the prime numbers we have already generated. This algorithm was already known to the Ancient Greeks, and is known as the Sieve of Eratosthenes:

$$
\begin{aligned}
&primes := \emptyset \\
&foreach\ n\ in\ (2\ldots M)\ \{ \\
&\quad isprime := true \\
&\quad foreach\ i\ in\ primes\ \{ \\
&\quad\quad if\ (n \div_r i = 0)\ then\ isprime := false \\
&\quad \} \\
&\quad if\ (isprime)\ then\ primes := primes \cup \{n\} \\
&\}
\end{aligned}
$$

The divisors of a number n cannot be greater than n itself, implying it can only have a finite number of divisors. Can the same be said of prime numbers? Is there a largest prime number, implying that there exists a finite list of prime numbers through which we can construct all the natural numbers using multiplication? The following is a proof that no such finite list exists—a result which was already known to the Ancient Greeks.

Theorem 9.18 *There is an infinite number of primes.*

Proof Outline The proof is a proof by contradiction. Let us assume that there is a finite number of prime numbers, which we will call p_1 to p_n. Now consider the number: $\alpha = p_1 \times p_2 \times \cdots \times p_n + 1$. Is p_1 a divisor of α? Using Corollary 9.13 we know that $\alpha \div_r p_1 = 1$, implying that p_1 is not a divisor of α. The same can be said of all the primes. The number α thus has no prime divisors, which, by Theorem 9.17 implies that it is itself prime. But α is larger than all the primes, and was therefore

not included in the original set of primes. This contradicts the original assumption
that there is a finite set of prime numbers.   □

We started off by motivating prime numbers as providing us with a unique way of
decomposing a number into constituent parts. The *Fundamental Theorem of Arithmetic* gives us a proof of this result. Before we prove it, we will require the following
result:

Lemma 9.19 *Every number $n \geq 2$ can be written as a product of primes.*

Proof Outline The proof is by strong induction on n, starting from 2. The result
clearly holds for the base case with $n = 2$.

For the inductive case, let us assume that any number up to (and including) k
can be written as a product of primes. We have to prove that Successor(k) can also
be written in this form. Now Successor(k) is either (i) prime, or (ii) has at least
one proper divisor, and can thus be written as the product of two smaller values
Successor(k) $= \alpha \times \beta$. In case (i), we can write Successor(k) in the desired format
(simply as Successor(k), since it is prime). In case (ii), we know, by the inductive
hypothesis, that both α and β can be written as products of primes:

$$\alpha = p_1 \times p_2 \times \cdots p_i$$
$$\beta = q_1 \times q_2 \times \cdots q_j$$

Since Successor(k) $= \alpha \times \beta$, we can combine the two products together to obtain a
prime factorisation of Successor(k):

$$\text{Successor}(k) = \alpha \times \beta = p_1 \times p_2 \times \cdots p_i \times q_1 \times q_2 \times \cdots q_j$$

The result thus follows by strong induction.   □

With this result in hand we are ready to prove that numbers have a unique prime
factorisation.

Theorem 9.20 *The Fundamental Theorem of Arithmetic (also referred to as the
Unique Prime-Decomposition Theorem): Given a number $\alpha \geq 2$, there is exactly
one way of expressing α as an ordered product of prime numbers.*

Proof Outline Lemma 9.19 tells us that any number can always be expressed as a
product of primes, which can be ordered. What we have to show is thus that such an
ordered product of primes is unique.

The proof is by strong induction on α. If α is a prime number, then the theorem
trivially holds, and therefore it is trivially true for the base case with $\alpha = 2$.

For the inductive case, we will assume that the theorem holds for all values up to
and including k. We have to prove that Successor(k) also permits only one ordered

product of primes. Consider Successor(k) expressed as two ordered products of primes:

$$\text{Successor}(k) = p_1 \times p_2 \times \cdots p_n = q_1 \times q_2 \times \cdots \times q_m$$

Since they are ordered, we assume that, for all i: $p_i \leq p_{i+1}$ and $q_i \leq q_{i+1}$. Without loss of generality, we will assume that $p_n \geq q_m$. We now consider two different cases: (i) $p_n > q_m$, and (ii) $p_n = q_m$.

Case (i): For $p_n > q_m$. Note that p_n is a divisor of Successor(k) and thus also of $q_1 \times q_2 \times \cdots \times q_m$. Furthermore, since p_n is a prime, we note that p_n must be a divisor of some q_i. However $p_n > q_m$, and for every i, $q_m \geq q_i$. It thus follows that p_n is larger than each q_i and cannot be one of its divisors. This case is thus not possible, leaving us with case (ii).

Case (ii): If $p_n = q_m$, the remaining products must also be equal: $p_1 \times \cdots \times p_{n-1} = q_1 \times \cdots \times q_{m-1}$. Let us call this value k'. Since $k < $ Successor(k) (because p_m is a proper factor of Successor(k)), we can apply the inductive hypothesis to k'—any two ordered prime decompositions of k' must be equivalent: meaning that $n = m$ and $p_i = q_i$ for each $1 \leq i \leq n - 1$.

Therefore, by strong induction, any number permits exactly one ordered product decomposition of prime numbers. □

9.1.4.3 Prime Numbers for Fun and Profit

Consider the following puzzle: you are given a finite sequence of non-zero natural numbers, which you want to encode as a single number, in such a way that you can decode it back to the original sequence of numbers. For instance, consider the sequence $\langle 132, 927, 3, 3 \rangle$. We cannot simply join the numbers together into a single number 13292733, since we would not know whether this originally came from, for instance, the sequence $\langle 13, 29, 2733 \rangle$ or the original one. Padding the numbers with zeros to obtain 132927003003 does not work either, since we do not have a limit on the size of the numbers in the sequence (hence the number of digits)—and the number shown can also be decoded to $\langle 132927, 3003 \rangle$. Think about this puzzle, and try to find a solution before reading any further.

There are various solutions to this puzzle. One of the most notable solutions is due to the mathematician Kurt Gödel, and is known as Gödel numbering. Given a sequence of length n: $\langle v_1, v_2, \ldots v_n \rangle$, we take the first n prime numbers: 2, 3, 5, $\ldots p_n$ and calculate the number $2^{v_1} \times 3^{v_2} \times \cdots \times p_n^{v_n}$. For example, the sequence $\langle 132, 927, 3, 3 \rangle$ would be encoded as $2^{132} \times 3^{927} \times 5^3 \times 7^3$. To decode, we simply decompose the number into its constituent prime factors and reconstruct the sequence. For example, given a sequence encoded as 600, we factorise into $600 = 2^3 \times 3^1 \times 5^2$, which means that the sequence was $\langle 3, 1, 2 \rangle$. This encoding works thanks to two results we have proved in this section. Can you identify which ones?

The more obvious result we use is the unique decomposition theorem, which guarantees that factorising will always give us back the original sequence. The other crucial theorem is that there is an infinite number of primes, guaranteeing that we have sufficient prime numbers no matter what length the original sequence has. Do not be misled into thinking that this encoding is a simple way of remembering the whole telephone directory by transforming the sequence of numbers into a single number and memorising that single number. The resulting number would have so many digits, that you would have been better off remembering the original numbers in the first place. Still, this gives us an important mathematical tool, which we will use to reason about computability.

Prime numbers have been studied for over 2,000 years, and yet, there are various seemingly simple questions to which we have no answer. For instance, consider the prime pairs 3 and 5, 11 and 13, 29 and 31, all of which are exactly two apart. We know that there is an infinite number of primes, but can one find an infinite number of prime pairs exactly two apart? Despite the apparent simplicity of the question, no one knows its answer.

How much computing power is required to decompose a number into its factors is another open question, even if it is widely believed that factorising huge numbers is a computationally intensive task. This property is used in cryptography, for instance by the RSA algorithm,[3] in which two huge prime numbers (huge as in having hundreds of digits) are multiplied together to generate an even larger number which is given out to be used to encrypt messages. Messages encrypted in this manner can, however, only be decrypted efficiently if the two original prime numbers are known. This approach can be used, for instance, so that an online store gives me the product to encrypt my credit card number, which no one but the store can decrypt, even if a malicious party were to have listened to the number which was passed on to me by the store. Such algorithms are used to encrypt sensitive data over the Internet, thus enabling secure transactions involving private and financial information.

9.2 Beyond the Natural

We have looked at how the counting numbers can be defined. However, frequently we require more complex numbers, for instance to handle negative values, or to talk about fractions. In this section we will briefly look at ways in which these types can be defined mathematically to enable reasoning about them. It is important to keep in mind that the best approach to reason about these sets of numbers is usually to axiomatise them directly. However, here we will look at ways of formulating them in terms of the concepts we already have.

---

[3]RSA stands for Rivest, Shamir and Adleman, the names of the scientists who published this algorithm.

9.2.1 Integers

The next step after the natural numbers is to look at the integers—the set of whole numbers, including negative ones, usually denoted by $\mathbb{Z}$. One way of encoding the integers is as a structured type using the natural numbers:

$$\mathbb{Z} \quad ::= \quad \mathsf{Pos}\ \mathbb{N}\ |\ \mathsf{Neg}\ \mathbb{N}_1$$

Two constructors are used to encode positive values including zero (Pos) and negative ones (Neg). Needless to say, there are various other ways in which integers could have been encoded, for instance adding another constructor for zero and excluding it from the positive numbers (by changing the type of the parameter of Pos to $\mathbb{N}_1$). Another way could have been to encode them as a pair Boolean $\times\ \mathbb{N}$, with the first denoting the sign (positive or negative) and the second denoting the value. Note that with this encoding, zero would have had two representations (*true*, Zero) and (*false*, Zero), which would have meant that we would have had to do additional work to reason about, for instance, equality of integers.

Addition of integers can now be calculated in terms of addition and subtraction of natural numbers:

$$(\mathsf{Pos}\ n) +_{\mathbb{Z}} (\mathsf{Pos}\ m) \stackrel{\mathrm{df}}{=} \mathsf{Pos}\ (n+m)$$

$$(\mathsf{Neg}\ n) +_{\mathbb{Z}} (\mathsf{Neg}\ m) \stackrel{\mathrm{df}}{=} \mathsf{Neg}\ (n+m)$$

$$(\mathsf{Pos}\ n) +_{\mathbb{Z}} (\mathsf{Neg}\ m) \stackrel{\mathrm{df}}{=} \begin{cases} \mathsf{Pos}\ (n-m) & \text{if } n \geq m \\ \mathsf{Neg}\ (m-n) & \text{otherwise} \end{cases}$$

$$(\mathsf{Neg}\ n) +_{\mathbb{Z}} (\mathsf{Pos}\ m) \stackrel{\mathrm{df}}{=} \begin{cases} \mathsf{Pos}\ (m-n) & \text{if } m \geq n \\ \mathsf{Neg}\ (n-m) & \text{otherwise} \end{cases}$$

Note that, due to the six possible cases (four possibilities for all the combinations of positive and negative values, two of which split further into two), proofs have to handle all the cases separately. For instance proving that integer addition is commutative would involve looking at all the six possibilities.

Proposition 9.21 *Addition of integers is commutative:* $\forall n, m : \mathbb{Z} \cdot n +_{\mathbb{Z}} m = m +_{\mathbb{Z}} n.$

Proof We consider all possible cases of the two parameters.

- When $n = \mathsf{Pos}\ n'$ and $m = \mathsf{Pos}\ m'$

$$n + m$$
$$= \mathsf{Pos}\ n' +_{\mathbb{Z}} \mathsf{Pos}\ m'$$
$$= \{\text{by definition of integer addition}\}$$
$$\mathsf{Pos}\ (n' + m')$$
$$= \{\text{addition of natural numbers is commutative}\}$$
$$\mathsf{Pos}\ (n' + m')$$
$$= \{\text{by definition of integer addition}\}$$
$$\mathsf{Pos}\ m' +_{\mathbb{Z}} \mathsf{Pos}\ n'$$
$$= m + n$$

- When $n = \mathsf{Neg}\ n'$ and $m = \mathsf{Neg}\ m'$ the proof is very similar to the previous case when both n and m are positive.
- When $n = \mathsf{Pos}\ n'$ and $m = \mathsf{Neg}\ m'$.
  This case splits further into: when $n' \geq m'$ and when $n' < m'$.
  - When $n' \geq m'$:

$$n + m$$
$$= \mathsf{Pos}\ n' +_{\mathbb{Z}} \mathsf{Neg}\ m'$$
$$= \{\text{definition of integer addition and } n' \geq m'\}$$
$$\mathsf{Pos}\ (n' - m')$$
$$= \{\text{definition of integer addition and } n' \geq m'\}$$
$$\mathsf{Neg}\ m' +_{\mathbb{Z}} \mathsf{Pos}\ n'$$
$$= m + n$$

  - When $n' < m'$:

$$n + m$$
$$= \mathsf{Pos}\ n' +_{\mathbb{Z}} \mathsf{Neg}\ m'$$
$$= \{\text{definition of integer addition and } n' < m'\}$$
$$\mathsf{Neg}\ (m' - n')$$
$$= \{\text{definition of integer addition and } n' < m'\}$$
$$\mathsf{Neg}\ m' +_{\mathbb{Z}} \mathsf{Pos}\ n'$$
$$= m + n$$

- When $n = \mathsf{Neg}\ n'$ and $m = \mathsf{Pos}\ m'$, the proof is very similar to the previous case. ∎

Other operators can be similarly defined by considering all the cases. With the number of different cases one has to consider, proofs tend to become long and tedious using this way of encoding integers.

Exercises

9.26 Define subtraction over integers, and prove that $\forall n : \mathbb{Z} \cdot n -_{\mathbb{Z}} n = \mathsf{Zero}$.

9.27 Define the comparison operators over integers: $\geq_{\mathbb{Z}}, \leq_{\mathbb{Z}}, >_{\mathbb{Z}}$ and $<_{\mathbb{Z}}$.

9.28 Define multiplication over integers.

9.2.2 The Rational Numbers

A limitation with the integers is that they enable us to reason only about whole values. For instance fairly splitting three cakes between two persons cannot be done if we can only use whole numbers of cakes. Fractional numbers are crucial to allow us to reason about such situations. If we limit ourselves to positive fractions, the set of fractional, or rational, numbers $\mathbb{Q}$ can be encoded as a pair of natural numbers $\mathbb{N} \times \mathbb{N}_1$, with (p, q) corresponding to our notion of the fraction $\frac{p}{q}$. Note that q is not allowed to be zero.

One problem with this encoding is that a single fraction can be expressed in different ways. For instance the notion of a half can be encoded as $(1, 2)$ or as $(2, 4)$ or in a multitude of other ways. To be able to reason about equality of rational numbers we need a notion of the lowest form of a fraction.

A fraction (p, q) is said to be in its lowest form if p and q have no common proper divisors: $divisors^+(p) \cap divisors^+(q) = \emptyset$. The lowest form of a fraction (p, q) is defined to be (p', q') such that, for some value of α, $p = \alpha \times p'$ and $q = \alpha \times q'$, and (p', q') is in its lowest form. It can be proved that any fraction has a unique lowest form. Two fractions can now be defined to be equal if they have the same lowest form.

As in the case of integers, we can define operators by giving how to compute the result. For instance, multiplication of two rational numbers $(p, q) \times_{\mathbb{Q}} (p', q')$ is defined to be $(p \times p', q \times q')$.

Exercises

9.29 Define addition over the rational numbers.

9.30 Define the comparison operators over the rational numbers.

9.31 Prove that multiplication of rational numbers is commutative and associative.

9.32 Modify the encoding of rational numbers to enable reasoning about negative fractions.

9.2.3 The Real Numbers

Using the rational numbers to reason about non-exact values is very convenient, since it reduces all reasoning to the use of two whole numbers. The Ancient Greeks, however, discovered that not all numbers can be expressed as fractions. Consider the diagonal of a square with sides of length 1, which by Pythagoras' theorem is of length $\sqrt{2}$:

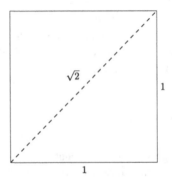

What fraction corresponds to the length of the diagonal? For instance, the fraction $\frac{5}{4}$ is too small to be equal to $\sqrt{2}$, since $\frac{5}{4} \times \frac{5}{4} = \frac{25}{16}$ which is smaller than 2. On the other hand, $\frac{10}{7}$ is too large, since $\frac{10}{7} \times \frac{10}{7} = \frac{100}{49}$, which is larger than 2. No matter how hard you try to identify a fraction which exactly corresponds to $\sqrt{2}$, you are doomed to fail.

Theorem 9.22 *The square root of 2 is not a rational number:* $\sqrt{2} \notin \mathbb{Q}$.

Proof Outline The proof is by contradiction. We assume that $\sqrt{2}$ can be written as a fraction in its lowest form: $\sqrt{2} = \frac{p}{q}$. Squaring both sides of the equation, we get that $2 \times q^2 = p^2$. Since p^2 is even, we can conclude that p must also be even. Let us write p as $2 \times x$.

Rewriting the equation we get $2 \times q^2 = (2 \times x)^2$, from which we can conclude that $q^2 = 2 \times x^2$. Therefore, q^2 is even, which means that so is q.

Since both p and q are even, the fraction $\frac{p}{q}$ was not in its lowest terms, which contradicts our original statement. We can thus conclude that our only assumption, that $\sqrt{2}$ can be written as a fraction, must be false.   □

This theorem confirms that there are numbers which are not rational. Various other numbers have been shown not to be rational, such as π, the ratio between the diameter and the circumference of a circle, as well as the mathematical constant e.

The set of numbers, including ones which are not rational, is called the set of real numbers, and is usually denoted by $\mathbb{R}$. They correspond to numbers with an infinite decimal expansion, such as $3.14159\ldots$, $0.121212\ldots$ and $0.5000\ldots$. If the sequence of digits eventually starts repeating (an individual number or block of numbers), then the number can be represented as a fraction, and is thus rational. The rest of the numbers, such as $\sqrt{2}$ and π, are called the irrational numbers. The real numbers are made up of the rational and irrational numbers.

One way of encoding real numbers is as an infinite sequence of digits, giving the decimal expansion of the number, with an additional symbol to denote the position of the decimal point. Unfortunately, such representations yield complex encodings of real number operators such as addition, and it is usually the case that real numbers are directly axiomatised.

Exercises

9.33 Prove that, for any natural number n, if n^2 is even, then so is n. Hint: Proving the contrapositive of the result is easier.

9.34 Modify the proof to show that $\sqrt{3}$ is also irrational. Which part of the proof would not work if we try to use the same approach to prove that $\sqrt{4}$ is irrational?

9.3 Cardinality

We started off this chapter arguing that the concept of natural numbers stems directly from the everyday activity of counting objects. We have formalised natural numbers and shown how operators can be defined on these numbers. We have even looked into how one could go about reasoning about more complex numeric types. However, despite the fact that the natural numbers seem to correspond to the numbers we use to count in, we have not shown how we can use them to *count* the items in, for instance, a set of objects. In this section we will explore how a counting scheme can be defined, allowing us, for instance, to compare the size of sets.

9.3.1 Counting with Finite Sets

In real-life we usually count finite collections of objects.[4] How many items are there in the set: {Red, Lighten Blue, Yellow}? Does the set $\{7, \ 3\frac{1}{2}\}$ contain more or fewer items? Let us start by comparing a set with another to see whether it contains more, less, or an equal number of items. Based on this, we can then formally define what we mean by the size of a set.

Let us start by examining a simple puzzle, or rather three variants of a puzzle. A boy has two boxes of toys—a green box and a red box. He decides to tie strings from toys in the green box to toys which lie in the red box—all strings can be assumed to be tied at both ends.

Scenario 1: If, upon examination, we find that (i) each item in the green box has at most one string tied to it, and (ii) all toys in the red box are tied to at least one string, not knowing anything else about the toys, which box do you think contains most toys?

Scenario 2: If, instead we discover that (i) each item in the green box has at least one string tied to it, and (ii) no toy in the red box is tied to more than one string, would you change your mind?

---

[4]Even in myths and legends in which someone is assigned a seemingly never-ending task, it usually consists of counting a finite, even if very big, collection—counting the grains of sand on a beach, or the number of drops of water in an ocean, as opposed to counting infinite collections of objects, such as counting all the numbers one can think of.

Scenario 3: Finally, if each item in both boxes is tied to exactly one string, which box would you now choose?

Let us look at the scenarios separately. In the first case, we would reason that, if there were more toys in the red box, then some toys would not have been tied with a string, since the strings (at most one from each toy in the green box) would have run out. This means that the green box has at least as many toys as the red one, and that is the one to choose.

In the second scenario, we realise that each toy in the red box was tied with one or no string. Since every toy in the green box has a string tied to it, there are at least as many toys in the red box as there are in the green one, possibly more. The red box is the one to choose.

There are different ways of reasoning about the last scenario. The most straightforward is that the observation that each item in the red box is tied to exactly one string satisfies both the first and second scenarios, because (i) all the toys in the red box are tied to at least one string, and (ii) no toy in the red box is tied to more than one string. In the first scenario we concluded that the green box had at least as many toys as there were in the red one, while in the second we concluded the opposite. The only way to satisfy both constraints is to have the same number of toys in both boxes.

Fine, we seem to be able to tie toys to compare the number of items in different boxes. But how can we formalise this to be able to use the mathematics we have developed so far to reason about counting? String theory is not the solution.

One way of looking at a situation with toys from the green box connected to ones in the red box using strings is as a relation. If we see the boxes as two sets, and the strings as items related by the relation, the scenarios can be formalised.

The first scenario is when each toy in the green box has at most one string tied to it, meaning that the strings correspond to a function. Furthermore, all the items in the destination set are tied to at least one string—the relation covers all the codomain, a surjective relation. If we thus find a surjective function from a set to another, the former has at least as many different elements as the latter.

The second scenario, is when each toy in the green box has at least one string tied to it, corresponding to a total relation—because each item in the source set has at least one outgoing arrow. The toys in the red box are tied to at most one string—no item in the codomain has two or more incoming arrows, corresponding to injectivity. Therefore, if we find a total injective relation from a set to another, we know that the latter has at least as many elements as the former.

The final setting is when the relation turns out to be total, functional, injective and surjective, allowing us to say that two sets are of the same size if there is a total bijective function between the two.

We will use these notions to define comparison operators between the sizes, or cardinality, of sets.

Definition 9.23 *Given two sets A and B, we say that A has cardinality not less than B, written A $\geq_\#$ B, if there exists a surjective function from A to B. We say that A has cardinality not more than B, written A $\leq_\#$ B, if B $\geq_\#$ A. Finally, we say that they are of the same cardinality, written A $=_\#$ B, if both A $\geq_\#$ B and A $\leq_\#$ B.* ∎

Although we defined $\leq_\#$ and $=_\#$ in terms of $\geq_\#$, they could have been defined directly.

Proposition 9.24 *The cardinality comparison relations can be defined in a different way: (i) A $\leq_\#$ B holds if and only if there exists a total injective relation; (ii) A $=_\#$ B if and only if there is a total bijective function between A and B.*

Proof The proof of (i) is split into two parts. If there exists a total injective relation r between A and B, we know from the results about relations that r^{-1} is a surjective function from B to A. Therefore, $B \geq_\# A$ and hence $A \leq_\# B$. On the other hand, if $A \leq_\# B$, then it follows that $B \geq_\# A$, implying the existence of a surjective function f from B to A. However, based on the proofs about relations, we know that f^{-1} is a total injective relation.

The proof of (ii) follows immediately from (i). ∎

This notion of comparing the cardinality of sets satisfies a number of expected laws.

Proposition 9.25 *Cardinality comparison satisfies a number of basic laws: (i) every set is of the same cardinality as itself A $=_\#$ A; (ii) both $\leq_\#$ and $\geq_\#$ are transitive relations; (iii) cardinality equality ($=_\#$) is an equivalence relation.*

Proof Outline To prove (i) we simply note that the identify function over A is a total bijective function, allowing us to conclude that $A =_\# A$.

For property (ii), we know that a surjective function f exists from A to B, and another such function g exists from B to C. But $f \,\fatsemi\, g$ is a surjective function from A to C (since the composition of two surjective functions is itself a surjective function), and thus $A \geq_\# C$. Similarly, $\leq_\#$ can be proved to be transitive.

Property (iii) follows from the definition of $=_\#$ and results (i) and (ii). □

Example We expect that adding items to a set will never reduce the size of the set. At worst, if we add items already in the set, the size of the set will remain constant. We can confirm this observation by proving that $A \cup B \geq_\# A$.

To show this we need to show that there exists a surjective function from $A \cup B$ to A. Consider the partial function $f \in A \cup B \to A$ defined as $\{(x, x) \mid x \in A\}$. Note that (i) f is functional, since if $(x, y) \in f$ and $(x, z) \in f$, it follows that $x = y$ and $x = z$, and hence $x = y$; and (ii) f is surjective over A, since for any $x \in A$, $f(x) = x$. If thus follows that $A \cup B \geq_\# A$. ◇

We thus have a means of comparing sets together, in terms of size. This still does not provide a way of *counting* the elements in a set. What does it mean to say that set $C = \{\text{Red, Lighten Blue, Yellow}\}$ has three elements? Since we can now compare sets, we can now equate this statement with saying that C has the same cardinality as the set $\{1, 2, 3\}$. In other words, there is a way of enumerating all items in C, starting from 1 till 3. This notion can be generalised for finite sets. We will be using the notation $\mathbb{F}X$ to denote all finite subsets of X—similar to $\mathbb{P}X$, but excluding infinite subsets.

Definition 9.26 *Given a finite set $A : \mathbb{F}X$, we say that the cardinality of A is n, written $\#A = n$, if $A =_\# 1...n$. Note that the set $1...n$ is defined to be $\{i : \mathbb{N} \mid 1 \le i \le n\}$.* ∎

Before we start using this definition, it is essential that we check that it makes sense. Is the size of a finite set uniquely defined? Is there a set for which we can prove that $\#A = n$ and $\#A = m$ for different values of n and m? To prove that this is not the case, we require the following preliminary result:

Lemma 9.27 *There exists a surjective function $f \in 1...n \twoheadrightarrow 1...m$ if and only if $n \ge m$.*

Proof Outline For the backward direction of the proof, we note that, if $n \ge m$, then $\lambda i \in 1...m \cdot i$ can be shown to be a surjective function from $1...n$ to $1...m$.

In the forward direction, the proof follows by induction on n.

For the base case, we take $n = 0$:

$$\exists f \cdot f \in 1...0 \to 1...m \text{ where } f \text{ is a surjective function}$$
$\Rightarrow$ {since $1...0 = \emptyset$ and definition of surjectivity}
$$\exists f \cdot f \in \emptyset \to 1...m \land ran(f) = 1...m$$
$\Rightarrow$ {by definition of relations}
$$\exists f \cdot f \subseteq \emptyset \times 1...m \land ran(f) = 1...m$$
$\Rightarrow$ {since $A \times \emptyset = \emptyset$}
$$\exists f \cdot f \subseteq \emptyset \land ran(f) = 1...m$$
$\Rightarrow$ {only $\emptyset$ is a subset of $\emptyset$}
$$\exists f \cdot f = \emptyset \land ran(f) = 1...m$$
$\Rightarrow$ {using the one-point rule}
$$ran(\emptyset) = 1...m$$
$\Rightarrow$ {the range of the empty relation is the empty set}
$$\emptyset = 1...m$$
$\Rightarrow$ {only $m = 0$ satisfies this equality}
$$m = 0$$
$\Rightarrow$ {natural number inequality}
$$0 \ge m$$

For the inductive case, we assume that, if there exists a surjective function $f \in 1...k \twoheadrightarrow 1...m$, then $k \ge m$, and prove that it also holds for $k + 1$: if there exists a surjective function $f \in 1...k + 1 \twoheadrightarrow 1...m$, then $k + 1 \ge m$.

To prove the implication, we assume that there exists a surjective function $f' \in 1...k+1 \rightarrow 1...m$. We are now required to prove that $k+1 \geq m$. Note that, when $m = 0$, the inequality trivially holds, so we will consider the case with $m > 0$.

We can visualise f' as a collection of arrows from the set $1...k+1$ to the set $1...m$. We identify two particular values α and β such that $f'(\alpha) = m$ and $f'(k+1) = \beta$. Note that α is guaranteed to exist by the surjectivity of f'. For the special case when $f'(k+1)$ is not defined, we note that this would mean that f' is a surjective function from $1...k$ to the set $1...m$, implying that $k \geq m$ by the inductive hypothesis, and thus $k+1 \geq m$. Let us thus turn our attention back to when $f'(k+1)$ is defined:

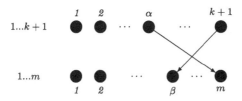

Now consider function f, which removes $k+1$ from the domain and switches the outgoing arrow from α to β, which would be defined as follows:[5]

$$f \in 1...k \rightarrow 1...m-1$$

$$f(i) \stackrel{\text{df}}{=} \begin{cases} \beta & \text{if } i = \alpha \\ f'(i) & \text{otherwise} \end{cases}$$

Note that f is (i) functional (since f' is functional), and (ii) surjective over $1...m-1$ (since f' is surjective). Therefore, by the inductive hypothesis, we can conclude that $k \geq m-1$ and hence $k+1 \geq m$.

Using induction, we can thus conclude that the lemma holds. □

We can now prove that set cardinality is well defined.

Theorem 9.28 *Given a finite set A, if $\#A = n$ and $\#A = m$, then $n = m$.*

Proof Outline Since $\#A = n$ and $\#A = m$, we know that there exist total bijective functions f_n between A and $1...n$, and f_m between A and $1...m$. Using these, we

---

[5]The special case when $f'(k+1) = m$ has to be treated differently, and is left as an exercise.

can define the relation $f = f_n^{-1} \, \mathring{\,}_9 \, f_m$ between $1...n$ and $1...m$. Since total bijective functions are closed under relational inverse and relational composition, we can conclude that f is a total bijective function between $1...n$ and $1...m$. Similarly, f^{-1} is a total bijective function between $1...m$ and $1...n$.

Since f is a surjective function from $1...n$ to $1...m$, we can conclude by Lemma 9.27, that $n \geq m$. Similarly, since f^{-1} is a surjective function from $1...m$ to $1...n$, using the same lemma, we can conclude that $m \geq n$. Hence $n = m$.   □

We can also show that comparing cardinalities is the same as comparing the size of sets.

Theorem 9.29 *If* $\#A = n$ *and* $\#B = m$ *then* (i) $A \geq_\# B$ *if and only if* $n \geq m$, (ii) $A \leq_\# B$ *if and only if* $n \leq m$, *and* (iii) $A =_\# B$ *if and only if* $n = m$.

Proof Outline We start by proving (i). Let $f_A \in A \rightarrow 1...n$ be a total bijective function showing that $\#A = n$, and similarly f_B be the equivalent function for B. We prove the two directions of the bi-implication separately:

Forward implication: If $A \geq_\# B$, then by definition of $\geq_\#$, there exists a surjective function $f \in A \rightarrow B$. Using the results from the chapters about relations, we know that (i) the inverse of a total bijective function is itself a total bijective function, and (ii) the composition of surjective functions is a surjective function. Therefore, $f_A^{-1} \, \mathring{\,}_9 \, f \, \mathring{\,}_9 \, f_B$ is a surjective function from $1...n$ to $1...m$. Hence, by Lemma 9.27, we can conclude that $n \geq m$.

Backward implication: If $n \geq m$, we can conclude from Lemma 9.27, that there exists a surjective function $f \in 1...n \rightarrow 1...m$. Using the same results about the composition of functions, we know that $f_A \, \mathring{\,}_9 \, f \, \mathring{\,}_9 \, f_B^{-1}$ is a surjective function from A to B, and thus $A \geq_\# B$.

Hence $A \geq_\# B$ is equivalent to $n \geq m$.

The proof of (ii) follows directly from the proof of (i):

$$A \leq_\# B$$
$$\Leftrightarrow \{\text{by definition of } \leq_\#\}$$
$$B \geq_\# A$$
$$\Leftrightarrow \{\text{by result (i)}\}$$
$$m \geq n$$
$$\Leftrightarrow \{\text{numeric inequalities}\}$$
$$n \leq m$$

The proof of (iii) similarly follows directly.   □

Proposition 9.30 *For any finite set* A, *and* $x \notin A$, *the size of* $\{x\} \cup A$ *is one more than the cardinality of* A: $\#(\{x\} \cup A) = 1 + \#A$.

Proof Outline Since A is a finite set, $\#A = n$ for some value of n. This means that there exists a total bijective function f from A to $1...n$. Now, let us define

$f' \in \{x\} \cup A \rightarrow 1...n+1$ as follows:

$$f'(y) \overset{\text{df}}{=} \begin{cases} f(y) & \text{if } y \in A \\ n+1 & \text{otherwise} \end{cases}$$

The function f' can be shown to be a total bijection, thus allowing us to conclude that $\#(\{x\} \cup A) = 1 + n$ and therefore $\#(\{x\} \cup A) = 1 + \#A$.   □

Since we now have a relationship between finite sets and their cardinality as a natural number, we can derive a rule of induction for finite sets, which will correspond to induction on the size of the set. The base case is when the cardinality is zero, equivalent to proving the property for the empty set. For the inductive case, we assume the property is true for a set A and prove that it still holds when we add another element to the set:

$$\frac{\pi(\emptyset) \qquad \forall k : X, \ K : \mathbb{F}X \cdot \pi(K) \Rightarrow \pi(\{k\} \cup K)}{\forall A : \mathbb{F}X \cdot \pi(A)}$$

This rule of induction can be proved to be correct using natural number induction.

Example Now that we can measure the size of a set, and not just compare the cardinality between sets, we can refine the previous result which showed that adding elements to a set cannot decrease its size. What is the cardinality of the union of two finite sets $A \cup B$? If we add the cardinality of A and that of B, we would end up with an over-approximation, since elements in both A and B would be counted twice. To obtain the exact value, we would then need to subtract the number of such common elements so that the overall effect is that of counting them once. We can prove that $\#(A \cup B) = \#A + \#B - \#(A \cap B)$.

The proof is by finite set induction on A.

Base case: For the base case, when $A = \emptyset$, we have to prove that $\#(\emptyset \cup B) = \#\emptyset + \#B - \#(\emptyset \cap B)$.

$$\begin{aligned} &\#(\emptyset \cup B) \\ ={} &\{\text{the empty set is the identity of union}\} \\ &\#B \\ ={} &\{\text{zero is the identity of addition}\} \\ &\#\emptyset + \#B \\ ={} &\{\text{law of cancellation}\} \\ &(\#\emptyset - \#\emptyset) + \#B \\ ={} &\{\text{subtraction commutes with addition}\} \\ &\#\emptyset + \#B - \#\emptyset \\ ={} &\{\text{the empty set is the zero of intersection}\} \\ &\#\emptyset + \#B - \#(\emptyset \cap B) \end{aligned}$$

Inductive case: For the inductive case, we will assume that the property holds for a finite set K: $\#(K \cup B) = \#K + \#B - \#(K \cap B)$. We now have to prove it also holds for $\{k\} \cup K$. We consider three distinct cases: (i) when $k \in K$, (ii) when $k \notin K$ and $k \in B$, and (iii) when $k \notin K$ and $k \notin B$. It is not difficult to confirm that the three cases cover all possibilities.

Case (i) $k \in K$.

$$\#(((\{k\} \cup K) \cup B)$$
$$= \{\text{since } k \in K\}$$
$$\#(K \cup B)$$
$$= \{\text{by inductive hypothesis}\}$$
$$\#K + \#B - \#(K \cap B)$$
$$= \{\text{since } k \in K\}$$
$$\#(\{k\} \cup K) + \#B - \#((\{k\} \cup K) \cap B)$$

Case (ii) $k \notin K, k \in B$.

$$\#(((\{k\} \cup K) \cup B)$$
$$= \{\text{commutativity and associativity of union}\}$$
$$\#(K \cup (\{k\} \cup B))$$
$$= \{\text{since } k \in B\}$$
$$\#(K \cup B)$$
$$= \{\text{by inductive hypothesis}\}$$
$$\#K + \#B - \#(K \cap B)$$
$$= \{\text{laws of arithmetic}\}$$
$$(1 + \#K) + \#B - (1 + \#(K \cap B))$$
$$= \{\text{since } k \notin K, \text{ and thus } k \notin K \cap B,$$
$$\qquad \text{we can apply Proposition 9.30}\}$$
$$\#(\{k\} \cup K) + \#B - \#(\{k\} \cup (K \cap B))$$
$$= \{\text{since } k \in B\}$$
$$\#(\{k\} \cup K) + \#B - \#((\{k\} \cup K) \cap B)$$

Case (iii) $k \notin K, k \notin B$.

$$\#(((\{k\} \cup K) \cup B)$$
$$= \{\text{associativity of union}\}$$
$$\#(\{k\} \cup (K \cup B))$$
$$= \{\text{since } k \notin K \cup B\}$$
$$1 + \#(K \cup B)$$
$$= \{\text{by inductive hypothesis}\}$$
$$1 + (\#K + \#B - \#(K \cap B))$$
$$= \{\text{laws of arithmetic}\}$$
$$(1 + \#K) + \#B - \#(K \cap B)$$

$$= \{\text{since } k \notin K\}$$
$$\#(\{k\} \cup K) + \#B - \#(K \cap B)$$
$$= \{\text{since } k \notin B\}$$
$$\#(\{k\} \cup K) + \#B - \#((\{k\} \cup K) \cap B)$$

This completes the inductive case of the proof.

The property thus holds by finite set induction on A.   ◇

The notion of counting gives rise to an important mathematical principle frequently used in mathematics and computer science—that of the *pigeon-hole principle*. The informal idea behind the pigeon-hole principle is that, if one has n pigeons which are all placed in m pigeon-holes, where $n > m$, then there must be one pigeon-hole with at least two pigeons inside it. The principle can be used, for instance, to show that amongst 13 persons, at least two must share the month of their birthday—the persons correspond to the pigeons, the months to the pigeon holes. Formally, we express the notion of placing a set of pigeons A in a set of pigeon-holes B (with $\#A > \#B$ since we have more pigeons than pigeon-holes) as a total function f from A to B. Having at least two different pigeons p_1 and p_2 (with $p_1 \neq p_2$) share a pigeon hole thus corresponds to $f(p_1) = f(p_2)$, implying that f is not injective.

Theorem 9.31 *The pigeon-hole principle: Given two sets A and B such that $\#A > \#B$, any total function $f \in A \to B$ cannot be injective.*

Proof Outline We use a proof-by-contradiction approach. Let us assume that there exists a total injective function $f \in A \to B$. Using the results about relations, we know that f^{-1} is (i) surjective (since f is total), and (ii) functional (since f is injective). From this, we can conclude that there exists a surjective function from B to A, implying that $B \geq_{\#} A$. From Theorem 9.29 we can conclude that $\#B \geq \#A$ which contradicts the fact that $\#A > \#B$. Hence, no total injective function from A to B exists.   □

Example You are given 10 different whole numbers between 1 and 100. What are the chances of being able to pick two disjoint sets of numbers (of the given 10) such that both sets have the same sum?

For instance, given the numbers $\{19, 23, 24, 39, 42, 50, 71, 75, 81, 97\}$ we can choose the sets $\{39, 97\}$ and $\{23, 42, 71\}$ which are disjoint and both add up to 136.

It turns out that two such sets exist no matter which 10 numbers you start off with. Here is an informal proof.

Given a set N of 10 numbers ranging over 1 to 100, there are 2^{10} or 1024 different subsets we can choose. Each one of these has a sum which at the very least is 0 (for the empty set), and at the very most $91 + 92 + \cdots + 100 = 955$ (when we choose the numbers to be all the ones from 91 to 100). Consider the total function $s \in \mathbb{P}N \to 0...955$, which calculates the sum of the set of numbers chosen from N. By the pigeon-hole principle, since $\#\mathbb{P}N > \#(0...955)$, it follows that s is not injective, and thus there are two subsets of N, let us call them N_1 and N_2, with the same sum ($s(N_1) = s(N_2)$). By removing any common elements from both N_1 and N_2 we can thus obtain two disjoint sets with the same sum.

This proof is interesting not only because it shows that we can choose sets with the same sum, but also because it is not a constructive proof—*it does not show how we can compute the subsets*. We know that two such subsets exist, but we still have no clue how to find them.   ◇

Here is another informal example showing an application of the pigeon-hole principle to graphs.

Example Consider a graph $G = (V, L, E)$ with a finite set of vertices V. Any path in G of length $\#V + 1$ or longer must contain a loop. This can be shown using the pigeon-hole principle. Consider a path of length $\#V + 1$ or longer, and the function f mapping each position in such a path to the vertex it goes through. Since the size of the set of vertices is $\#V$, and hence less than the number of positions in the path, f is not injective, and thus, two different positions must map to the same vertex. This means that the path loops through that repeated node.

This result has various important implications. For instance, we can show that, if there is no path in G starting from vertex v and finishing at vertex v' with length less than $\#V$, then v' cannot be reached from v in any number of steps. How? If the shortest path from v to v' is longer than $\#V$, then it must contain a loop, meaning that it can be shortened (by removing the loop). Therefore, the shortest path can never by longer than $\#V$. This means that, to check whether a vertex is reachable from another, it suffices to analyse paths of length up to $\#V$ in the graph.   ◇

We have thus managed to define the notion of the number of elements in a set and use it to draw conclusions about sets based on their size. What is particularly interesting about this approach, is that the definition of the size of a finite set is defined in terms of set cardinality comparison, which in turn is defined in terms of the existence of particular forms of relations between two sets. As we shall see in the coming sections, this approach allows us, not only to count items in a finite set, but also to compare infinite sets.

Exercises

9.35 Show that the empty set is not larger than any other set: $A \geq_\# \emptyset$.

9.36 The cardinality of the empty set is zero. Prove it.

9.37 Prove that, if $A \subseteq B$, then $A \leq_\# B$.

9.38 Give a proof sketch to show that, if the cardinality of a set A is the same as that of B, then $A =_\# B$.

9.39 Sketch a proof to show that, if $\#A \geq \#B$, then $A \geq_\# B$.

9.40 In the proof of Lemma 9.27, we left out the analysis for when $f'(k + 1) = m$. Show that a function f can also be constructed in such a case, thus completing the proof.

9.41 Given a drawer with 10 red and 10 blue socks, how many socks must I blindly take out of the drawer to guarantee that I have a matching pair. Show how the pigeon-hole principle can be used to justify your answer.

9.42 The *influence factor* of a person p in a group on a social networking site is defined to be the number of members of the group to which p is connected. So John would have an influence factor of 7 in a group where he is connected to 7 of the members. Assuming that the connected-to relation is symmetric (p is connected to p' if and only if p' is connected to p) and not reflexive (no person is connected to him- or herself), use the pigeon-hole principle to show that, in any group with two or more members, there must be at least two persons with the same influence factor. Hint: In a group with n members, there cannot be both a person with an influence factor of 0 and another with an influence factor of $n - 1$.

9.3.2 Extending Cardinality to Infinite Sets

Although the cardinality of a set only works with finite-sized sets, the underlying tool of set-size comparisons using classes of relations can also be applied to infinite sets. We will be informally visiting some of the results, originally proposed by Georg Cantor and David Hilbert in the late 19th and early 20th century. Although it may seem simpler to limit our concept of sets to finite collections, we cannot do without infinite ones. For instance, to be able to count the number of items in a finite set, we need the concept of the natural numbers—which is an infinite collection. Similarly, without the notion of infinite sets, we cannot formulate prime numbers, which are crucial in number theory.

Although the set of natural numbers is an infinite set, it is important to realise that all the numbers in the set are finite ones. One can use induction to prove this—zero is a finite number, and for any number k, if k is finite, then so is its successor. The cardinality of a set can thus only be applied to finite sets, and the cardinality of the natural numbers #$\mathbb{N}$ or the prime numbers #Primes are ill-defined concepts, since there is no natural number n such that the set $1...n$ can be put in a one-to-one correspondence with the set in question. The impossibility to give the size of a set is what makes it infinite.

Definition 9.32 *A set A is said to be infinite if there is no natural number n such that* #$A = n$. ∎

For instance, when we proved that the prime numbers are infinite, we assumed that such a value n exists, and then showed that there is at least one additional prime number which we did not include. Hence, no such value can exist.

Despite the fact that we cannot talk about the size of an infinite set, we can still use the previous notions of set cardinality comparison ($\geq_\#$, $\leq_\#$ and $=_\#$) for infinite sets. On one hand, common sense may suggest that all infinite sets, being infinite, are of the same size. Similarly, common sense tells us that, since the set of prime numbers is a proper subset of the natural numbers, the set of prime numbers must be smaller than the set of natural numbers. As we will see, both these common-sense arguments turn out to be wrong.

The arguments will be illustrated using *Hilbert's Hotel*, which contains an infinite number of rooms. This imaginary hotel is named after David Hilbert, a German mathematician who was responsible for many of the concepts presented here.

Example Tourism has been steadily increasing for a number of years, and Hilbert's Hotel was built precisely to ensure that, no matter how many tourists visit the city, they will always find a place to stay. The hotel's brochure boasts an infinite number of rooms on a single floor, numbered sequentially starting from room 0.

On the busiest day of the year, Hilbert's Hotel was completely full. A few minutes after the receptionist handed over the keys to the last free room, a regular guest arrived at the hotel asking for a room. The receptionist explained that the hotel was already full and that he would have to find a room elsewhere. The new arrival made such a fuss that the manager was called over. A former mathematician, the hotel manager proposed a solution to give the visitor an empty room without sending out any of the current guests or having any of them share a room (and without building another room in the hotel). How did he manage to do so?

The solution: The manager made an announcement on the hotel PA system: "Can all guests kindly move their belongings to the next room down the corridor, thus moving to a room whose number is one higher than the one they currently occupy?" The person in room 0 would move to room 1, whose current occupant would be moving to room 2, etc. Although the guests were not so happy to have to move rooms, they all found an empty room (since the person in that room was also moving), and at the end of the process, room 0 was not occupied. The visitor was handed the key to this empty room.

Wasn't the hotel already full? How could he add another guest into a full hotel?   ◇

Although seemingly counter-intuitive at first, the moral of the story is that one can always add an object to an infinite set without changing its cardinality. This approach, of moving all persons from room n to room $n+1$, can be used to show that the natural numbers and the natural numbers less zero are of the same cardinality.

Proposition 9.33 $\mathbb{N}$ *and* $\mathbb{N}_1$ *have the same cardinality:* $\mathbb{N} =_\# \mathbb{N}_1$.

Proof Outline To prove that the two sets are of the same cardinality, we need to find a total bijective function from the first to the second. The function we desire is the one which simply increments the given value by one: $f(n) \stackrel{df}{=} n+1$. It is not difficult to prove that $f \in \mathbb{N} \to \mathbb{N}_1$ is: (i) total on $\mathbb{N}$, (ii) a function, (iii) surjective on $\mathbb{N}_1$, and (iv) injective.   □

In fact, using this process, the manager can host any finite number of new guests arriving at the hotel. If k new guests arrive, the current guests are asked to move k

rooms down the corridor, from room n to room $n + k$. This leaves k rooms empty for the new guests.

Example The following morning, with Hilbert's Hotel still full, saw an infinite bus full of tourists stopping at its front door. Since the hotel was the only infinite one in town, the guide explained that Hilbert's Hotel was the only place where they could be hosted. The receptionist realised that there was no way he could ask the current guests to move sufficiently down the hallway to make space for an infinite number of new guests. The manager was called in once again, and came up with an even smarter solution than the previous time to ensure that everyone was hosted in the hotel. What was his solution?

The solution: This time round, the manager announces that each current guest is to move to the room whose number is double the one he or she is currently in: the person in room n moves to room $2 \times n$, thus leaving the odd rooms empty. The tourists on the bus were then told to move into the hotel, such that the person sitting at seat n takes room $2 \times n + 1$ (assuming that the seats on the bus are also numbered starting from 0). Not only did everyone get a room, but the regular guest who had just arrived the previous night, and was in room 0, did not have to change rooms.

Weirder still than what went on the previous night, a full hotel managed to find space for an infinite number of new guests. Can it get any weirder than this? (Yes, it can.)   ◇

Even more counter-intuitive than the previous example, this example shows that adding two infinite sets together does not change the cardinality. For instance, if we take the even numbers (an infinite set) and add the odd numbers (another infinite set) we end up with the natural numbers. Does the set of natural numbers contain more elements than the even numbers? Not only is the set of even numbers a proper subset of the natural numbers, but it has infinitely fewer elements (it lacks all the odd numbers). We can show that, in fact, their cardinalities are the same.

Proposition 9.34 *The natural numbers* $\mathbb{N}$ *and the set of even numbers Even are of the same cardinality*: $\mathbb{N} =_{\#} Even$.

Proof Outline To prove that the two sets have the same cardinality, we will give a total bijective function $f \in \mathbb{N} \to Even$. The function which doubles a number $f(n) \overset{\mathrm{df}}{=} 2 \times n$ does the trick. It can be proved that (i) f is total on $\mathbb{N}$; (ii) it is a function; (iii) it is surjective on $Even$; and (iv) it is an injective function.   □

This approach can be used to show that the cardinality of $\mathbb{N}$ is also the same as that of the square numbers (by mapping n to n^2) and the prime numbers (by mapping n to p_n, the nth prime number), despite the fact that in both these sets we progressively drop more elements from the natural numbers as we progress. It can also be used to show that there are as many natural numbers as integers.

Proposition 9.35 *The natural numbers* $\mathbb{N}$ *and the integers* $\mathbb{Z}$ *are of the same cardinality*: $\mathbb{N} =_{\#} \mathbb{Z}$.

Proof Outline The total bijective function which does the trick is slightly more complicated than before, but corresponds very closely to what the manager did. Consider function $f \in \mathbb{Z} \to \mathbb{N}$ defined as follows:

$$f(n) \stackrel{\text{df}}{=} \begin{cases} 2 \times n & \text{if } n \geq 0 \\ 2 \times (-n) - 1 & \text{otherwise} \end{cases}$$

Note that the non-negative numbers are mapped from n to $2 \times n$ (just like the persons already in the hotel), while the negative numbers are mapped from $-n$ to $2 \times n - 1$ (just like the persons on the bus, except that in this case no one is sitting on seat 0).

It can be proved that (i) f is total on $\mathbb{Z}$; (ii) it is a function; (iii) it is surjective on $\mathbb{N}$; and (iv) it is an injective function. Hence the natural numbers and the integers are of the same cardinality.   □

We will look at one final example of an unexpected result about infinite sets.

Example Hilbert's Hotel became so popular that the owners decided to build further stories above the (infinite) ground-level corridor. Having made so much money from that last busload of tourists, they decided to build not one, not two, but an infinite number of stories. Rooms were now numbered as a pair e.g., *L17-R41*, indicating room 41 on level 17. The first room on the ground floor was *L0-R0*. The demand for rooms in the hotel did not diminish, and the hotel was completely full when a safety inspector turned up. The lack of sufficient fire escapes from the new floors led to an immediate request to evacuate all the rooms on the new floors due to safety concerns. The manager, as usual, found a solution, and all the guests ended up hosted in rooms on the ground floor. How did he manage to do so?
The solution: Consider the layout of the hotel as shown in the figure below:

$L0$-$R3$	$L1$-$R3$	$L2$-$R3$	$L3$-$R3$	$\cdots$
$L0$-$R2$	$L1$-$R2$	$L2$-$R2$	$L3$-$R2$	$\cdots$
$L0$-$R1$	$L1$-$R1$	$L2$-$R1$	$L3$-$R1$	$\cdots$
$L0$-$R0$	$L1$-$R0$	$L2$-$R0$	$L3$-$R0$	$\cdots$

The manager assigned each room to a number by going around in a spiral-like movement as shown below:

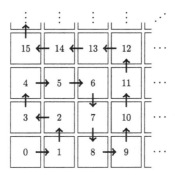

This can be written as a function from the level and room number to a single number so that every guest knows which room to move to. Note that, in this way, everyone still has a room, although they will all be housed on the ground floor. ◇

Using the manager's intuition, we can show that there are as many rational numbers as natural numbers.

Theorem 9.36 *The cardinality of the rational numbers is the same as that of the naturals:* $\mathbb{Q} =_{\#} \mathbb{N}$.

Proof Outline Every rational number $\frac{p}{q}$ can be seen as a room in Hilbert's Hotel *Lp-Rq*. Using the manager's redistribution function, we have a way of mapping the rationals to the natural numbers such that the mapping is a total bijective function. Hence $\mathbb{Q} =_{\#} \mathbb{N}$. □

This result runs very much counter to what one would expect. Intuitively, there seem to be so many more rational numbers than natural ones. Mathematically, however, it turns out that this is not the case. With the definition of equality of cardinality of infinite sets as equivalent to the existence of a one-to-one relationship between the sets, the two sets turn out to be of equivalent cardinality.

The cardinality of the natural numbers, the even numbers, the integers and the rationals are thus all the same. Such sets which have cardinality equal to that of the natural numbers are called *denumerable* sets or *countably infinite* sets. Given that all these sets of numbers have been shown to be denumerable, the question which naturally arises is whether there are sets whose cardinality is larger than that of the natural numbers.

9.3.2.1 Beyond Denumerability

Are there infinite sets which are not denumerable? We will show that the real numbers, which as we have already seen, contain more items than the rationals, are not denumerable. There are more real numbers than natural ones.

Theorem 9.37 *The real numbers are not denumerable.*

Proof Outline To prove this, we show that the real numbers between 0 and 1 are not denumerable, and then apply Exercise 9.37 to conclude the result. The proof is by contradiction—we assume that the real numbers between 0 and 1 are denumerable, and then proceed to show that this leads to a contradiction.

Let us assume that the real numbers in the interval $(0, 1)$, are denumerable. This means that there is a total bijective function f between the natural numbers and this set. In other words, we can enumerate all the reals in $(0, 1)$ listing them by applying f to the natural numbers in order:

$$f(0) = 0.a_1 a_2 a_3 \ldots$$
$$f(1) = 0.b_1 b_2 b_3 \ldots$$
$$f(2) = 0.c_1 c_2 c_3 \ldots$$

$$\vdots$$

Now, let us define an operator $\bar{d}$, which takes a digit and changes it into 1 or 2, such that d is different from $\bar{d}$:

$$\bar{d} \stackrel{\mathrm{df}}{=} \begin{cases} 2 & \text{if } d = 1 \\ 1 & \text{otherwise} \end{cases}$$

Now consider the real number: $n = 0.\bar{a}_1 \bar{b}_2 \bar{c}_3 \ldots$—is this number listed in the table we showed earlier? The number does not match the first line since $a_1 \neq \bar{a}_1$, neither does it match the second number since $b_2 \neq \bar{b}_2$, etc. If the ith number was $0.x_1 x_2 x_3 \ldots$, this does not match n since $x_i \neq \bar{x}_i$. Since n is not listed in the table, it means that f is not surjective, and thus not a total bijective function. This contradiction means that our original assumption that the numbers between 0 and 1 are denumerable is false. It thus follows that the real numbers are not denumerable. $\square$

The way we have constructed a counterexample in this proof uses a technique called diagonalisation. We take the table of values and change the ones lying on the diagonal so as to construct a row which is not in the table. The approach is used in other contradiction proofs and, as we shall see in the next chapter, allows us to prove that there is a limit on the power of computers.

We have thus shown that there are sets which are not denumerable. Apart from the real numbers, one can show, for instance that the power set of the natural numbers is also not denumerable. Are there sets with cardinality even larger than that of the real numbers? One can show that the power set of real numbers has cardinality even higher than that of the reals. Furthermore, taking the power set of an infinite set always results in a new set with an even higher cardinality, giving us a glimpse of an infinite chain of infinite set cardinalities. Cantor's great achievement was to show that the notion of infinite can be categorised into different cardinalities allowing us to reason more effectively about infinite sets.

9.4 Summary

In this chapter we have looked at how we can formalise and reason about numbers. The natural numbers turned out to be a simple instance of structured types, and the range and complexity of the operators and concepts defined are evidence of how powerful structured types can be. Although we have briefly looked at integers, the rationals and the real numbers, to reason formally about these classes of numbers, one would ideally formalise them directly, rather than encode them using the natural numbers.

What is usually taken to be a basic notion in mathematics—numbers and counting—turns out to involve interesting twists, especially when it comes to reasoning about infinite sets. What may appear to be merely an intellectual exercise in comparing the sizes of infinite sets will, however, turn out to have important implications in computer science. We will be seeing this in the next chapter.

Chapter 10
Reasoning About Programs

We have looked at various topics in mathematics, and although examples from computer science have been used throughout, it may not seem obvious how the mathematical tools we have developed can be concretely and practically applied. In this chapter we will look at three applications:

Correctness of algorithms: The first application is to show how one can prove that an algorithm really satisfies a specification. Frequently, writing a specification, explaining *what* a program should do, is simpler than specifying *how* it should be done. A typical example is sorting a list of numbers. Explaining what the outcome of calling a sorting routine should be ('return a list of numbers containing the same items as the original list, but such that each item is smaller or equal to all the items following it') is simpler than explaining how Bubblesort, or Quicksort works. One can then prove that Quicksort does satisfy the requirements to be a correct sorting algorithm.

As another example, consider the specification: 'No user may be allowed to log on to the system more three times in any half hour period.' There are various ways the system developer may choose to satisfy this requirement. One solution may be to disable a user account for 10 minutes after every logout, while another solution would be to disable a user from logging in for 10 minutes from the moment that he or she logs in. If one formalises the specification and solutions and tries to prove that the solutions really work, it will be discovered that the first solution fails to work if a user is allowed to have multiple open logins at the same time—by logging in three times in a row without logging out, the specification is violated. The second would, however, work. We have already seen proofs of how mathematically specified algorithms of operators such as subtraction and multiplication satisfy certain properties. In Sect. 10.1, we will look at other examples of such proofs.

Giving meaning to programs: Once an algorithmic solution to a specification is identified, one would still have to implement it as a concrete computer program in a programming language. Since these programs are the artefacts that will be executed, giving them a mathematical meaning allows us to reason directly about system behaviour: Is one program equivalent to or better than another? Does a

G.J. Pace, *Mathematics of Discrete Structures for Computer Science*,
DOI 10.1007/978-3-642-29840-0_10, © Springer-Verlag Berlin Heidelberg 2012

program exhibit certain behaviours and not others? We will show how this approach, of giving programs a formal meaning, or semantics, also enables us to prove properties of program transformations. For example, a compiler typically performs many optimising changes to a program before an executable file is produced. Clearly, it is crucial that these transformations do not change the behaviour of the original program, since that could introduce bugs in programs which would otherwise have been correct. We will see how one can reason about programs in Sect. 10.2.

The limits of computing: Can any specification be implemented as a computer program, or are there tasks which computers cannot perform? Note that we are not asking whether there are specifications which we still do not know how to implement on a computer, but whether there are ones for which no implementation even exists. Surprisingly, we can prove that such specifications do exist. For instance, we can prove that there exist languages for which no parser, which reads a sentence from the input and confirms whether or not it is syntactically correct, can ever be written. Another example of an impossible task is to write a program which identifies whether or not a given program with a given input terminates. You will never be able to buy a compiler which always tells you whether the program you are compiling terminates or not. We will see that it is not simply that until this day no one has discovered how to do it, but that such a program does not exist. We will see a proof of this in Sect. 10.3.

The applications of mathematics to computer science go well beyond these examples. The aim is not to be exhaustive, but rather to illustrate such applications.

10.1 Correctness of Algorithms

To start off, we will be looking at how one can prove the correctness of algorithms. We have already seen instances of this earlier. For instance consider the concept of natural number subtraction. On one hand, we gave an algorithmic definition of how one can compute the result of subtracting one number from another—by progressively discarding successor constructors from both operands, until the second operator reduces to zero. On the other hand, we have also seen properties we expect from this result—the most important one is that subtraction negates addition: $(n + m) - m = n$. The first description tells us *how* to calculate the result, while the latter tells us *what* the result should satisfy.

There are two ways to look at algorithm correctness. In the first, one sees the algorithm as the main description, with properties one expects to hold specified separately. This was the approach we adopted when talking about natural number subtraction. We use this approach when we know how to compute something and want to deduce properties which hold for the result of the algorithm. The other approach is to consider the specification to be the golden standard telling us what the result should be. Typically, mathematicians prefer to define subtraction as an operator which acts as an inverse of addition. The way to compute a result which satisfies this specification is a secondary question. If one requires an algorithmic way of computing the specification, one then develops an algorithm which does obey the specification. This approach is particularly useful when one wants to leave different options open for the implementation. For example, consider the specification stating: 'The items in the input and output lists should be identical, although possibly in a different order. The largest item appearing in the input list should be at the head of the output list.' An algorithm which sorts the whole list, or one which just moves the largest item to the front, or even one which moves the largest item to the front while changing the order of the rest are all possible ways of implementing the specification. The context in which the algorithm is to be used dictates which one to adopt.[1]

Whichever way one sees it, it remains the case that, given an algorithm and a property, one would like to be able to prove that the *how* and *what* match.

10.1.1 Euclid's Algorithm

Consider a 'what' specification you have probably encountered: *Given two natural numbers, their greatest common divisor (GCD), sometimes also called the highest common factor (HCF), is the largest natural number which divides both numbers without leaving any remainder.* For instance, the GCD of 24 and 32 is 8 (since no other number larger than 8 exactly divides both 24 and 32), and the GCD of 78 and 52 is 26. Although it may appear to be a theoretical concept, it turns out to be very useful in practice—from reducing fractions to their lowest form (which enables easier arithmetic with rational numbers), to encryption.

Definition 10.1 *Given two natural numbers n and m (at least one of which is not zero), the greatest common divisor (GCD) is the largest number which is a divisor of both n and m*:

$$GCD(n, m) \stackrel{\text{df}}{=} max(divisors(n) \cap divisors(m)) \qquad ■$$

Since the divisors of a non-zero number cannot exceed the number itself, the sets of divisors are finite, which in turn means that we can always take the maximum of their intersection. If the intersection were infinite, the maximum would not have been defined; e.g. the maximum of the set of prime numbers is not a well defined value. We thus know that the GCD of any two natural numbers is well defined.[2]

---

[1]If, whenever we use this algorithm, we always also require a sorted version of the list, then we might as well sort the list in the first place. If, however, we just require the largest element, sorting the sequence would be unnecessarily wasteful.

[2]Actually, the argument is slightly more involved since one of the operands of GCD can be zero. However, the intersection of a finite set with another set is always finite, and thus, given that at least one of the two operands is non-zero guarantees that GCD is well defined.

Furthermore, also note that the commutativity of the GCD operator follows immediately from the commutativity of intersection.

Proposition 10.2 *Given two natural numbers n and m (at least one of which is not zero), GCD(n, m) is well defined, and is equal to GCD(m, n).*

One case for which the GCD of two numbers can be computed easily is when one of the operands is zero: $GCD(n, \mathsf{Zero}) = n$. Otherwise, although it is conceptually easy to understand and define what the result of the GCD should be, how to compute it is not so obvious. One method, which closely follows the definition, is to compute all factors of the two numbers and find the largest common one. A variant of this method, sometimes taught in schools, is to decompose the two numbers into their prime factors and select the common prime factors. However, these approaches require the decomposition of a number into its divisors, which is a computationally expensive operation. While the method works when applied to small numbers, it fails to scale up for larger numbers. Luckily, over 2,000 years ago, Euclid devised a simple and efficient algorithm to compute the GCD of two numbers. Euclid's algorithm is based on a property of the GCD of two numbers: $GCD(n, m) = GCD(n \div_r m, m)$. Let us first see how to prove this property before using it to obtain the algorithm.

Recall that calculating the remainder when dividing a number n by another number m can be computed by repeatedly subtracting m until the result is less than m. We start by proving that subtracting m from n will not change the GCD, based on which, we will be able to prove Euclid's equivalence.

Lemma 10.3 *Given two numbers n and m, with $n \geq m$:*

$$GCD(n, m) = GCD(n - m, m)$$

Proof We prove that $divisors(n) \cap divisors(m) = divisors(n - m) \cap divisors(m)$, from which we can conclude the result directly from the definition of GCD.

To prove equality of the two sets, we will separately prove set inclusion in both directions:

$$i \in divisors(n) \cap divisors(m)$$
$$\Rightarrow \{\text{theorem about intersection}\}$$
$$i \in divisors(n) \wedge i \in divisors(m)$$
$$\Rightarrow \{\text{if } i \xrightarrow{\div} x \text{ and } i \xrightarrow{\div} y, \text{ where } x \geq y, \text{ then } i \xrightarrow{\div} x - y\}$$
$$i \in divisors(n - m) \wedge i \in divisors(m)$$
$$\Rightarrow \{\text{theorem about intersection}\}$$
$$i \in divisors(n - m) \cap divisors(m)$$

To prove set inclusion in the other direction, we use the following argument:

$$
\begin{aligned}
&i \in divisors(n - m) \cap divisors(m) \\
\Rightarrow\ &\{\text{theorem about intersection}\} \\
&i \in divisors(n - m) \wedge i \in divisors(m) \\
\Rightarrow\ &\{\text{if } i \overset{\div}{\to} x \text{ and } i \overset{\div}{\to} y \text{ then } i \overset{\div}{\to} x + y\} \\
&i \in divisors(n) \wedge i \in divisors(m) \\
\Rightarrow\ &\{\text{theorem about intersection}\} \\
&i \in divisors(n) \cap divisors(m)
\end{aligned}
$$

This means that the two sets are equal, and therefore, so is their maximum. ∎

We can now prove Euclid's law about the GCD operator:

Theorem 10.4 *Given natural numbers n and m (not both zero), $GCD(n, m) = GCD(n \div_r m, m)$.*

Proof Outline The proof is by strong induction on n. For the base case, when $n = $ Zero, we need to prove that $GCD(\text{Zero}, m) = GCD(\text{Zero} \div_r m, m)$, which is true since $\text{Zero} \div_r m = \text{Zero}$.

For the inductive case, assume that the property holds for all values less than n, and we prove that the result for n. Note that, if $n < m$, then $n \div_r m = n$ and it thus immediately follows that $GCD(n, m) = GCD(n \div_r m, m)$. Let us thus focus on when $n \geq m$.

$$
\begin{aligned}
&GCD(n, m) \\
=\ &\{\text{using the previous lemma, and } n \geq m\} \\
&GCD(n - m, m) \\
=\ &\{\text{using the inductive hypothesis (since } n - m < n)\} \\
&GCD((n - m) \div_r m, m) \\
=\ &\{\text{since } n \div_r m = (n + m) \div_r m\} \\
&GCD(n \div_r m, m)
\end{aligned}
$$

This completes the inductive case.   □

So how does this theorem help us calculate the GCD of two numbers. Consider the following recursive function definition:

$$
GCD_0(n, m) \overset{df}{=} \begin{cases} n & \text{if } m = \text{Zero} \\ GCD_0(m, n) & \text{if } n < m \\ GCD_0(m, n \div_r m) & \text{otherwise (if } n \geq m > \text{Zero)} \end{cases}
$$

What is this function doing? It is split into three cases (i) whenever the second operand is zero, it returns the first operand, which we know is correct thanks to Exercise 10.2; (ii) whenever the first operand is smaller than the second, it performs a recursive call with the operands swapped, which we know is correct thanks to commutativity of GCD; and (iii) in the remaining case, when $n \geq m$ and $m \neq \mathsf{Zero}$, it performs a recursive call with m and $n \div_r m$, which we also know is correct according to the theorem we have just proved together with the commutativity of GCD.

Do we have a correct implementation of GCD? One last thing we should check is whether the definition is a valid one. Recall that we allowed recursive definitions provided that the parameters somehow structurally decrease in size. Otherwise, the recursive definition may never terminate. Consider the second parameter of GCD_0. In the first recursive case, when $n < m$, the second parameter goes from m to n, which we know is smaller. In the second recursive case, we go from m to $n \div_r m$, which we know to be smaller than m. Since the function is also defined for all values (the conditions cover all the different cases), the function is total and well defined.

We thus have a recursive definition of the GCD of two numbers which is guaranteed to be correct. Consider an application of GCD_0 to the numbers 78 and 52:

$$
\begin{aligned}
& GCD_0(78, 52) \\
={} & GCD_0(52, 78 \div_r 52) \\
={} & GCD_0(52, 26) \\
={} & GCD_0(26, 52 \div_r 26) \\
={} & GCD_0(26, 0) \\
={} & 26
\end{aligned}
$$

In just three applications of the definition, the result is obtained. In general, it can be shown that the algorithm computes the result in at worst on the order of $\log(n)$ recursive calls, where n is the smaller operand. This is a huge gain over the original algorithm which started by finding all divisors.

We have thus shown that, from a mathematical definition of an operator, we can prove results which allow us to derive an implementation which is guaranteed to be correct.

Exercises

10.1 Prove that, if $i \overset{.}{\to} n$ and $i \overset{.}{\to} m$, then $i \overset{.}{\to} n + m$.

10.2 Using just the definition of GCD, prove that $GCD(n, \mathsf{Zero}) = n$.

10.3 The definition of GCD we started off with already gave us a way of computing the result. However, we may have started from an even more property-based definition:

> Given two natural numbers n and m (at least one of which is non-zero), the GCD of n and m, written $GCD'(n, m)$ is the number satisfying:
>
> (i) it divides both n and m: $GCD'(n, m) \overset{.}{\to} n$ and $GCD'(n, m) \overset{.}{\to} m$;
>
> (ii) it is the largest such divisor: for any l such that $l \overset{.}{\to} n$ and $l \overset{.}{\to} m$, $l \leq GCD'(n, m)$.

Prove that GCD' and GCD are equivalent definitions.

10.1.2 Sorted Binary Trees

In Chap. 8, we have already seen how we can encode binary trees using algebraic types, and how to prove properties of functions defined over the type. One important application is *sorted binary trees*—in which, starting from any node, the nodes reachable from the left branch contain a value smaller or equal to the value at that node, while the nodes reachable from the right branch contain a value larger than that of the value at the node. For instance, the binary tree depicted below is a sorted one:

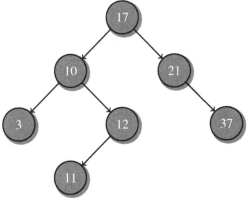

Sorted binary trees make looking for items more efficient than if we were to search in an unordered list. For example, consider the process to check whether 20 appears in the tree shown above. Since the topmost node, has value 17, it is sufficient to look at its right subtree, whose root is 21, meaning that it is sufficient to look at its left subtree, which is empty. We can thus be sure that 20 does not appear in the tree, through just two comparisons, even though the tree contains seven items.

The type of binary trees was defined as follows:

$$\mathsf{Tree}\,\alpha ::= \mathsf{Nil} \mid \mathsf{Node}(\alpha \times \mathsf{Tree}\,\alpha \times \mathsf{Tree}\,\alpha)$$

Functions which take trees as parameters can be defined using pattern-matching, separating the definition for Nil from that of Node values. For instance, the function *objects* takes a tree and returns the set of all values appearing in the nodes of the tree:

$$objects \in \mathsf{Tree}\,\alpha \to \mathbb{P}\,\alpha$$

$$objects(\mathsf{Nil}) \stackrel{\mathrm{df}}{=} \emptyset$$

$$objects(\mathsf{Node}(x, lt, rt)) \stackrel{\mathrm{df}}{=} \{x\} \cup objects(lt) \cup objects(rt)$$

Unless a tree t is sorted, we would have to go through $objects(t)$ to check whether an item occurs in the tree. However, if it is sorted, checking for inclusion can be done more efficiently:

$$in(x, \mathsf{Nil}) \stackrel{\mathrm{df}}{=} false$$

$$in(x, \text{Node}(x', lt, rt)) \stackrel{\text{df}}{=} \begin{cases} true & \text{if } x = x' \\ in(x, lt) & \text{if } x < x' \\ in(x, rt) & \text{otherwise} \end{cases}$$

The definition of *in* takes advantage of the fact that the tree is sorted, and would not work if it were not so. We will thus formally define what it means for a binary tree to be sorted, to enable us to prove the correctness of *in* when applied to a sorted binary tree.

Definition 10.5 *A binary tree $t \in \text{Tree}\,\alpha$ is said to be sorted (or ordered) if either it is empty or if* (i) *all the objects in the left subtree of the root are smaller or equal to the value of the root;* (ii) *all the objects in the right subtree are larger than the value at the root; and* (iii) *both the left and right subtree are sorted binary trees:*

$$sorted(\text{Nil}) \stackrel{\text{df}}{=} true$$

$$sorted(\text{Node}(x, lt, rt)) \stackrel{\text{df}}{=} sorted(lt) \wedge sorted(rt)$$

$$\wedge\ \forall y : \alpha \cdot y \in objects(lt) \Rightarrow y \le x$$

$$\wedge\ \forall y : \alpha \cdot y \in objects(rt) \Rightarrow y > x \qquad \blacksquare$$

We can now prove that checking for an element in an ordered tree using the function *in* gives the same result as checking whether the element appears in the *objects* of the tree:

Theorem 10.6 *Given a sorted tree, checking whether a value appears in the tree using in is the same as checking whether it appears in the set of values in the nodes:*

$$\forall t : \text{Tree}\ \alpha \cdot sorted(t) \Rightarrow \forall x : \alpha \cdot (x \in objects(t) \Leftrightarrow in(x, t))$$

Proof The proof uses structural induction on t.

Base case: Starting with the base case, when $t = \text{Nil}$, we need to prove that: $sorted(\text{Nil}) \Rightarrow \forall x : \alpha \cdot (x \in objects(\text{Nil}) \Leftrightarrow in(x, \text{Nil}))$. Since $sorted(\text{Nil})$ is true, we need to prove the bi-implication:

$$x \in objects(\text{Nil})$$
$$\Leftrightarrow \{\text{by definition of } objects\}$$
$$x \in \emptyset$$
$$\Leftrightarrow \{\text{by definition of } \emptyset\}$$
$$false$$
$$\Leftrightarrow \{\text{by definition of } in\}$$
$$in(x, \text{Nil})$$

Inductive case: For the inductive case, we start by assuming that the statement holds for two trees lt and rt:

$$sorted(lt) \Rightarrow \forall x : \alpha \cdot (x \in objects(lt) \Leftrightarrow in(x, lt))$$
$$sorted(rt) \Rightarrow \forall x : \alpha \cdot (x \in objects(rt) \Leftrightarrow in(x, rt))$$

We now prove that the property holds for $\mathsf{Node}(v, lt, rt)$:

$$sorted(\mathsf{Node}(v, lt, rt)) \Rightarrow$$
$$\forall x : \alpha \cdot (x \in objects(\mathsf{Node}(v, lt, rt)) \Leftrightarrow in(x, \mathsf{Node}(v, lt, rt)))$$

To prove the implication, we assume that $sorted(\mathsf{Node}(v, lt, rt))$, and prove the bi-implication. Note that, since the combined tree is sorted, by definition of $sorted$, we can conclude that (i) both subtrees lt and rt are also sorted and that (ii) if $x \in objects(lt)$, then $x \leq v$, and if $x \in objects(rt)$, then $x > v$.

$\quad x \in objects(\mathsf{Node}(v, lt, rt)$
$\Leftrightarrow$ {by definition of $objects$}
$\quad x \in \{v\} \cup objects(lt) \cup objects(rt)$
$\Leftrightarrow$ {set union}
$\quad x \in \{v\} \vee x \in objects(lt) \vee x \in objects(rt)$
$\Leftrightarrow$ {$x \in \{v\}$ is equivalent to $x = v$}
$\quad x = v \vee x \in objects(lt) \vee x \in objects(rt)$
$\Leftrightarrow$ {from knowledge of statement (ii)}
$\quad x = v \vee (x \leq v \wedge x \in objects(lt)) \vee (x > v \wedge x \in objects(rt))$
$\Leftrightarrow$ {using the inductive hypotheses and
$\qquad\quad$ the fact that both subtrees are sorted (i)}
$\quad x = v \vee (x \leq v \wedge in(x, lt)) \vee (x > v \wedge in(x, rt))$
$\Leftrightarrow$ {by definition of in}
$\quad in(x, \mathsf{Node}(v, lt, rt)))$ ∎

This guarantees that, if we have a sorted binary tree, then we can check whether an item appears in its nodes in an efficient manner. The problem remains of how to ensure that the tree is sorted, since checking every time whether a tree is sorted would defeat the whole purpose. The solution usually adopted is to have a way of adding new values to an already sorted tree in such a way that the tree remains sorted. In this manner, if we start with an empty tree (which is sorted by definition) or a tree which we know to be sorted, no matter how many items we add, the tree is guaranteed to remain sorted, thus relieving us from having to confirm this every time we check for inclusion. We claim that the following function achieves this:

$$insert(v, \mathsf{Nil}) \stackrel{\mathrm{df}}{=} \mathsf{Node}(v, \mathsf{Nil}, \mathsf{Nil})$$

$$insert(v, \mathsf{Node}(x, lt, rt)) \stackrel{\mathrm{df}}{=} \begin{cases} \mathsf{Node}(x, insert(v, lt), rt) & \text{if } v \leq x \\ \mathsf{Node}(x, lt, insert(x, rt)) & \text{otherwise} \end{cases}$$

Does *insert* really satisfy our requirements? Its correctness can be split into two: (i) *insert* really adds the item to the tree; and (ii) if we start with a sorted binary tree and add an item, the resulting tree is still sorted. It is worth noting that, if we omit to prove the first property, the function which throws away the new item to be inserted $insert'(x, t) \overset{df}{=} t$ could be mistaken for a correct implementation. We will prove the second property, and leave the first one as an exercise.

Theorem 10.7 *Inserting an item into a sorted tree returns a sorted tree*:

$$\forall x : \alpha, \ t : \mathsf{Tree}\ \alpha \cdot sorted(t) \Rightarrow sorted(insert(x, t))$$

Proof The proof is by induction on tree t. Recall that, to prove a property over trees, we have to prove that (i) the property holds for Nil and (ii) if the property holds for two trees lt and rt, it must also hold for $\mathsf{Node}(x, lt, rt)$ for any value of x.

Base case: To prove the property for $t = \mathsf{Nil}$, we have to prove that:

$$sorted(\mathsf{Nil}) \Rightarrow sorted(insert(x, \mathsf{Nil}))$$

Since Nil is sorted, we have to prove that $sorted(insert(x, \mathsf{Nil}))$:

$$
\begin{aligned}
& sorted(insert(x, \mathsf{Nil})) \\
\Leftrightarrow\ & \{\text{definition of } insert\} \\
& sorted(\mathsf{Node}(x, \mathsf{Nil}, \mathsf{Nil})) \\
\Leftrightarrow\ & \{\text{definition of } sorted\} \\
& sorted(\mathsf{Nil}) \wedge sorted(\mathsf{Nil}) \wedge \\
& (\forall y : \alpha \cdot y \in objects(\mathsf{Nil}) \Rightarrow x \geq y) \wedge \\
& (\forall y : \alpha \cdot y \in objects(\mathsf{Nil}) \Rightarrow x < y) \\
\Leftrightarrow\ & \{\text{definition of } sorted \text{ and } objects\} \\
& (\forall y : \alpha \cdot y \in \emptyset \Rightarrow x \geq y) \wedge \\
& (\forall y : \alpha \cdot y \in \emptyset \Rightarrow x < y) \\
\Leftrightarrow\ & \{x \in \emptyset \text{ is always false}\} \\
& true
\end{aligned}
$$

Inductive case: For the inductive case, we have two inductive hypotheses, which we will assume to hold:

$$
\begin{aligned}
sorted(lt) &\Rightarrow sorted(insert(x, lt)) \quad &\text{(IH1)} \\
sorted(rt) &\Rightarrow sorted(insert(x, rt)) \quad &\text{(IH2)}
\end{aligned}
$$

Based on these, we have to prove that any tree built up from these two subtrees lt and rt still satisfies the property:

$$sorted(\mathsf{Node}(x, lt, rt)) \Rightarrow sorted(insert(v, \mathsf{Node}(x, lt, rt))).$$

To prove this implication, we assume the left hand part of the implication, $sorted(\mathsf{Node}(x, lt, rt))$ and prove the right hand side. By definition of *sorted* we are thus assuming that:

$$\left.\begin{array}{l} sorted(lt) \wedge sorted(rt) \wedge \\ (\forall y : \alpha \cdot y \in objects(lt) \Rightarrow x \geq y) \wedge \\ (\forall y : \alpha \cdot y \in objects(rt) \Rightarrow x < y) \end{array}\right\} \quad \text{(Assumption)}$$

We prove this by considering two cases: (i) $x \geq v$; (ii) $x > v$. The proof of the first case follows.

$\qquad sorted(insert(v, \mathsf{Node}(x, lt, rt)))$
$\Leftrightarrow$ {definition of *insert* and the fact that $v \leq x$}
$\qquad sorted(\mathsf{Node}(x, insert(v, lt), rt))$
$\Leftrightarrow$ {definition of *sorted*}
$\qquad sorted(insert(v, lt)) \wedge sorted(rt) \wedge$
$\qquad (\forall y : \alpha \cdot y \in objects(insert(v, lt)) \Rightarrow x \geq y) \wedge$
$\qquad (\forall y : \alpha \cdot y \in objects(rt) \Rightarrow x \leq y)$
$\Leftrightarrow$ {from (Assumption), we know that rt is sorted and
$\qquad\qquad$ all objects in rt are greater than x}
$\qquad sorted(insert(v, lt)) \wedge$
$\qquad \forall y : \alpha \cdot y \in objects(insert(v, lt)) \Rightarrow x \geq y$
$\Leftrightarrow$ {Exercise 10.4}
$\qquad sorted(insert(v, lt)) \wedge$
$\qquad \forall y : \alpha \cdot y \in \{v\} \cup objects(lt) \Rightarrow x \geq y$
$\Leftrightarrow$ {we are considering the case when $x \geq v$}
$\qquad sorted(insert(v, lt))$
$\Leftrightarrow$ {by the inductive hypothesis (IH1) and
$\qquad\qquad$ the fact that lt is sorted by (Assumption)}
$\qquad true$

The other case for $x > v$ follows similarly.

The theorem thus holds by structural induction on trees.   ■

In computer science, one encounters various data structures and algorithms to manipulate them. In this section, we have thus shown how one can prove that these algorithms work as expected. The algorithms are written using mathematical notation, but can be implemented in a straightforward manner.

Exercises
10.4 Prove that the set of objects after inserting an item x into tree t is the same as the objects in tree t with the addition of x:

$$\forall x : \alpha, \ t : \mathsf{Tree}\ \alpha \cdot objects(insert(x, t)) = objects(t) \cup \{x\}$$

10.5 Define a function *subtrees*, which given a binary tree t, returns the set of all subtrees appearing in t.

10.6 Formally state the property that, if a tree is sorted, then so are all its subtrees. Prove this property.

10.7 We will say that a node is *sandwiched* if its left child (the node immediately below it to the left, if it has one) is smaller or equal to its value, while its right child (similarly, the immediate node below it to its right if it has one) is larger. Define a function which, given a tree, returns whether the root is sandwiched.

10.8 An alternative way of defining sorted binary trees is to say that all nodes in the tree are sandwiched. Formally write this alternative definition.

10.9 Prove that the two definitions of sorted binary trees are equivalent.

10.2 Assigning Meaning to Programs

In the previous section we have seen how one can build a constructive mathematical function from a set of properties, or conversely to prove properties of a mathematical definition. The mathematical definitions can then be implemented as a program, with the guarantee that the algorithm works.

Conversely, sometimes we are given a program which we need to analyse mathematically, so as to be able to prove things about its behaviour. In this section we will look at means of giving a mathematical meaning to programs. Instead of starting with the mathematical function definitions, we start with actual programs, which are transformed into mathematical notions. We look at two different ways in which this can be done.

10.2.1 Numeric Expressions

Consider expressions in computer programs, such as `(6+count)*cost`. These expressions may refer to variables (such as `count` and `cost`), constant natural numbers, and numeric operators. We will look into how we can give a formal semantics to such expressions.

Before we can give a meaning, we need to define the type of valid expressions about which we will make formal statements. Structured types give us a clean way of defining this type:[3]

$$\text{Exp} ::= \text{Val } \mathbb{N}$$
$$| \quad \text{Var Identifier}$$
$$| \quad \text{Add (Exp} \times \text{Exp)}$$
$$| \quad \text{Mul (Exp} \times \text{Exp)}$$

Expressions can be either a numeric value (using constructor Val), a variable with a name over type Identifier and about which we will avoid giving any further information (using constructor Var), or the addition or multiplication of two expressions (us-

---

[3]For simplicity, we limit our operators to addition and multiplication.

ing the Add and Mul constructors). For example, the expression `(6+count)*cost` would be written as:

$$\text{Mul(Add(Val 6, Var count), Var cost)}$$

The inductive principle for numeric expressions is:

$$
\begin{array}{l}
\forall n : \mathbb{N} \cdot \pi(\text{Val } n) \\
\forall v : \text{Identifier} \cdot \pi(\text{Var } v) \\
\forall e_1, e_2 : \text{Exp} \cdot \pi(e_1) \wedge \pi(e_2) \Rightarrow \pi(\text{Add}(e_1, e_2)) \\
\forall e_1, e_2 : \text{Exp} \cdot \pi(e_1) \wedge \pi(e_2) \Rightarrow \pi(\text{Mul}(e_1, e_2)) \\
\hline
\forall e : \text{Exp} \cdot \pi(e)
\end{array}
$$

The type Exp of expressions gives us a way of describing the syntax of expressions. But what about their meaning? We would expect, for instance, that Add(Val 1, Val 2) is semantically equal to Add(Val 2, Val 1), even though they are syntactically different.

What is the semantic meaning of a numeric expression? A numeric value seems to be a reasonable type. However, we get stuck when we try to evaluate variables—what is the value of Add(Val 6, Var count)? Well, it depends. It depends on the value of count at the time of execution. So as to be able to evaluate, or give a meaning to, an expression, we need to know the values of the variables at the point in the program where it will be evaluated. We can represent the state of the variables as a function from variables to numeric values, which we call a *valuation*:

$$\text{Valuation} \overset{\text{df}}{=} \text{Identifier} \rightarrow \mathbb{N}$$

We are now ready to give a meaning to a numeric expression. Given a valuation and an expression, we can evaluate its value in the following manner:

$$
\begin{array}{l}
evaluate \in \text{Valuation} \times \text{Exp} \rightarrow \mathbb{N} \\
evaluate(\sigma, \text{Val } n) \overset{\text{df}}{=} n \\
evaluate(\sigma, \text{Var } v) \overset{\text{df}}{=} \sigma(v) \\
evaluate(\sigma, \text{Add}(e_1, e_2)) \overset{\text{df}}{=} evaluate(\sigma, e_1) + evaluate(\sigma, e_2) \\
evaluate(\sigma, \text{Mul}(e_1, e_2)) \overset{\text{df}}{=} evaluate(\sigma, e_1) \times evaluate(\sigma, e_2)
\end{array}
$$

For example, evaluating Mul(Add(Val 6, Var count), Var cost) when count is 8 and cost is 10, we get:

$evaluate(\{(\mathsf{count}, 8), (\mathsf{cost}, 10)\},$
$\quad\quad \mathsf{Mul}(\mathsf{Add}(\mathsf{Val}\,6,\,\mathsf{Var}\,\mathsf{count}),\,\mathsf{Var}\,\mathsf{cost}))$
$= \{\text{using line 4 of the definition of } evaluate\}$
$\quad evaluate(\{(\mathsf{count}, 8), (\mathsf{cost}, 10)\},\,\mathsf{Add}(\mathsf{Val}\,6,\,\mathsf{Var}\,\mathsf{count}))\times$
$\quad\quad evaluate(\{(\mathsf{count}, 8), (\mathsf{cost}, 10)\},\,\mathsf{Var}\,\mathsf{cost}))$
$= \{\text{using line 3 of the definition of } evaluate\}$
$\quad (evaluate(\{(\mathsf{count}, 8), (\mathsf{cost}, 10)\},\,\mathsf{Val}\,6)+$
$\quad\quad evaluate(\{(\mathsf{count}, 8), (\mathsf{cost}, 10)\},\,\mathsf{Var}\,\mathsf{count})\times$
$\quad\quad\quad evaluate(\{(\mathsf{count}, 8), (\mathsf{cost}, 10)\},\,\mathsf{Var}\,\mathsf{cost}))$
$= \{\text{using lines 1 and 2 of the definition of } evaluate\}$
$\quad (6 + \{(\mathsf{count}, 8), (\mathsf{cost}, 10)\}(\mathsf{count}))\times$
$\quad\quad \{(\mathsf{count}, 8), (\mathsf{cost}, 10)\}(\mathsf{cost})$
$= \{\text{function application}\}$
$\quad (6 + 8) \times 10$
$= \{\text{basic arithmetic}\}$
$\quad 140$

Although tedious, the calculation is easy to perform. We can now indeed verify that, with any variable valuation, the expression $\mathsf{Add}(\mathsf{Val}\,1,\,\mathsf{Val}\,2)$ is semantically equivalent to $\mathsf{Add}(\mathsf{Val}\,2,\,\mathsf{Val}\,1)$:

$evaluate(\sigma,\,\mathsf{Add}(\mathsf{Val}\,1,\,\mathsf{Val}\,2))$
$= \{\text{using line 3 of the definition of } evaluate\}$
$\quad evaluate(\sigma,\,\mathsf{Val}\,1) + evaluate(\sigma,\,\mathsf{Val}\,2)$
$= \{\text{addition of natural numbers is commutative}\}$
$\quad evaluate(\sigma,\,\mathsf{Val}\,2) + evaluate(\sigma,\,\mathsf{Val}\,1)$
$= \{\text{using line 3 of the definition of } evaluate\}$
$\quad evaluate(\sigma,\,\mathsf{Add}(\mathsf{Val}\,2,\,\mathsf{Val}\,1))$

We can show that addition of numeric expressions—the constructor Add—is commutative under evaluation:

Proposition 10.8 *Addition of two expressions is commutative under evaluation*:

$$\forall \sigma : \mathsf{Valuation} \cdot evaluate(\sigma, \mathsf{Add}(e_1, e_2)) = evaluate(\sigma, \mathsf{Add}(e_2, e_1))$$

Proof The proof is a straightforward application of the definition of *evaluate* and commutativity of natural number addition:

$evaluate(\sigma,\,\mathsf{Add}(e_1,\,e_2))$
$= \{\text{using line 3 of the definition of } evaluate\}$
$\quad evaluate(\sigma,\,e_1) + evaluate(\sigma,\,e_2)$
$= \{\text{addition of natural numbers is commutative}\}$
$\quad evaluate(\sigma,\,e_2) + evaluate(\sigma,\,e_1)$
$= \{\text{using line 3 of the definition of } evaluate\}$
$\quad evaluate(\sigma,\,\mathsf{Add}(e_2,\,e_1)) \quad\quad\quad\quad\quad\quad\quad\quad\quad \blacksquare$

Note that this proposition tells us that addition at the topmost operator of an expression is commutative. To prove that any addition subexpression can be commuted would require a stronger result.

Let us consider a slightly more involved example. A compiler writer decides that for efficiency reasons, any subexpression of the form $0 \times e$ or $e \times 0$ will be simplified to 0. Does this change the value of an expression? Let us look at the compiler writer's simplification algorithm:

$$simplify(\mathsf{Val}\ n) \overset{\mathrm{df}}{=} \mathsf{Val}\ n$$

$$simplify(\mathsf{Var}\ v) \overset{\mathrm{df}}{=} \mathsf{Var}\ v$$

$$simplify(\mathsf{Add}(e_1, e2)) \overset{\mathrm{df}}{=} \mathsf{Add}(simplify(e_1),\ simplify(e_2))$$

$$simplify(\mathsf{Mul}(e_1, e_2)) \overset{\mathrm{df}}{=} \begin{cases} \mathsf{Val}\ 0 \\ \quad \text{if } simplify(e_1) = \mathsf{Val}\ 0 \text{ or } simplify(e_2) = \mathsf{Val}\ 0 \\ \mathsf{Mul}(simplify(e_1),\ simplify(e_2)) \\ \quad \text{otherwise} \end{cases}$$

We can now prove that the compiler writer's modification has no effect on the final result, and is thus a correct optimisation of an expression.

Theorem 10.9 *Simplification does not change the value of an expression*:

$$\forall \sigma : \mathsf{Valuation},\ e : \mathsf{Exp} \cdot evaluate(\sigma, e) = evaluate(\sigma, simplify(e))$$

Proof The proof will be by structural induction on expression e. Note that we have two base cases: for values and variables, and two inductive cases: for addition and multiplication:

Base case for values: $e = \mathsf{Val}\ n$. We have to prove that, for any valuation σ and number n: $evaluate(\sigma, \mathsf{Val}\ n) = evaluate(\sigma, simplify(\mathsf{Val}\ n))$. The result holds immediately, since from its definition, we know that *simplify* leaves values intact.

Base case for variables: $e = \mathsf{Var}\ v$. Exactly as in the previous case, the result follows immediately, since *simplify* leaves variables intact.

Inductive case for addition: We assume that the property holds for two expressions e_1 and e_2:

$$\forall \sigma : \mathsf{Valuation} \cdot evaluate(\sigma, e_1) = evaluate(\sigma, simplify(e_1)) \quad \text{(IH1)}$$
$$\forall \sigma : \mathsf{Valuation} \cdot evaluate(\sigma, e_2) = evaluate(\sigma, simplify(e_2)) \quad \text{(IH2)}$$

Based on these, we have to prove that it also holds for $\mathsf{Add}(e_1, e_2)$:

$$\forall \sigma : \mathsf{Valuation} \cdot$$
$$evaluate(\sigma, \mathsf{Add}(e_1, e_2)) = evaluate(\sigma, simplify(\mathsf{Add}(e_1, e_2)))$$

The proof is rather straightforward:

$$evaluate(\sigma, \mathsf{Add}(e_1, e_2))$$
$= \{\text{by line 3 of the definition of } evaluate\}$
$$evaluate(\sigma, e_1) + evaluate(\sigma, e_2)$$
$= \{\text{using the two inductive hypotheses}\}$
$$evaluate(\sigma, simplify(e_1)) + evaluate(\sigma, simplify(e_2))$$
$= \{\text{by line 3 of the definition of } evaluate\}$
$$evaluate(\sigma, \mathsf{Add}(simplify(e_1), simplify(e_2)))$$
$= \{\text{by line 3 of the definition of } simplify\}$
$$evaluate(\sigma, simplify(\mathsf{Add}(e_1, e_2)))$$

Inductive case for multiplication: As in the case of addition, we have two inductive hypotheses for subexpressions e_1 and e_2, from which we have to prove that it also holds for $\mathsf{Mul}(e_1, e_2)$:

$\forall \sigma : \mathsf{Valuation}\cdot$
$$evaluate(\sigma, \mathsf{Mul}(e_1, e_2)) = evaluate(\sigma, simplify(\mathsf{Mul}(e_1, e_2)))$$

We split this inductive case into two subcases: (i) when at least one of $simplify(e_1)$ or $simplify(e_2)$ is equal to $\mathsf{Val}\ 0$; (ii) when neither of the two simplified subexpressions is $\mathsf{Val}\ 0$.
Case (ii) is almost identical to the case of addition, so we will only prove case (i). We will also restrict ourselves to a proof for when $simplify(e_1) = \mathsf{Val}\ 0$, since the other case is similar. The proof proceeds as follows:

$$evaluate(\sigma, \mathsf{Mul}(e_1, e_2))$$
$= \{\text{by definition of } evaluate\}$
$$evaluate(\sigma, simplify(e_1)) \times evaluate(\sigma, simplify(e_2))$$
$= \{\text{since we know that } simplify(e_1) = \mathsf{Val}\ 0\}$
$$evaluate(\sigma, \mathsf{Val}\ 0) \times evaluate(\sigma, simplify(e_2))$$
$= \{\text{using the definition of } evaluate\}$
$$0 \times evaluate(\sigma, simplify(e_2))$$
$= \{\text{definition of multiplication}\}$
$$0$$
$= \{\text{by definition of } evaluate\}$
$$evaluate(\sigma, \mathsf{Val}\ 0)$$
$= \{\text{by definition of } simplify\}$
$$evaluate(\sigma, simplify(\mathsf{Mul}(e_1, e_2)))$$

The result thus holds by structural induction.   ∎

The first natural reaction to a proof that addition in numeric expressions is commutative or that simplifying $0 \times n$ to 0 is correct, is "So what?" Since natural number addition is commutative, and evaluating an expression transforms Add into $+$, the result may seem too obvious to require proof. Similarly, from the definition of natural number multiplication, we know that multiplying by zero returns zero. Surprisingly, however, in most imperative programming languages, such as Java, C, C♯ and C++, addition *is not* commutative and blindly optimising subexpressions multiplied by zero may change the meaning of an expression. The main reason for this is the presence of *side-effects*. In mathematics, evaluating an expression returns its value, but does not change anything—evaluating an expression once or a hundred times will not make any difference (except that we take longer to evaluate it a hundred times). However, in imperative programming languages, evaluating an expression may change the state of the machine as a side-effect, and evaluating it a second time may return a different result if the expression depends on the machine state. Two operators you may have encountered from programming are the two increment operators: v++ and ++v. Both increment the value of variable v, with the first returning the incremented value of v, while the second returns the value before being incremented. Thus, for example, if the value of v is 10, (v++)*3 would return 33 while (++v)*3 would return 30. In both cases, the value of v after evaluation of the expression will be 11. It should already be clear that evaluating either expression a second time will give a different result. But what does this have to do with commutativity?

Let us add the ++v operator to our numeric expressions. It will be encoded using the Inc constructor:

$$\text{Exp} ::= \text{Val } \mathbb{N}$$
$$\mid \quad \text{Var Identifier}$$
$$\mid \quad \text{Add } (\text{Exp} \times \text{Exp})$$
$$\mid \quad \text{Mul } (\text{Exp} \times \text{Exp})$$
$$\mid \quad \text{Inc Identifier}$$

Consider the expression Add(Inc v, Var v). Evaluating it should (i) increment variable v; (ii) take the new value of v; and finally (iii) add them together. Note that the variable valuation may change as one evaluates an expression. The type of *evaluate* has to be changed to be able to handle this. The new type will take a valuation and an expression and return, not just the numeric value of the expression, but also the updated valuation:

$$evaluate \in (\text{Valuation} \times \text{Exp}) \to (\text{Valuation} \times \mathbb{N})$$

Defining evaluation of the enriched expressions is straightforward for basic values and variables, for which the valuation does not change:

$$evaluate(\sigma, \text{Val } n) \stackrel{\text{df}}{=} (\sigma, n)$$
$$evaluate(\sigma, \text{Var } v) \stackrel{\text{df}}{=} (\sigma, \sigma(v))$$

Evaluating Inc v updates the valuation for the value of v, and returns the old value of v:

$$evaluate(\sigma, \text{Inc } v) \stackrel{\text{df}}{=} (\sigma \oplus \{(v, \sigma(v) + 1)\}, \sigma(v))$$

Finally, we consider addition and multiplication. Whereas before we used the initial valuation to calculate the result, now, evaluating the first subexpression may change the valuation—and the updated valuation is the one we should use to evaluate the second subexpression:

$$evaluate(\sigma, \text{Add}(e_1, e_2)) \stackrel{\text{df}}{=} (\sigma'', n_1 + n_2) \text{ where}$$
$$(\sigma', n_1) = evaluate(\sigma, e_1)$$
$$(\sigma'', n_2) = evaluate(\sigma', e_2)$$
$$evaluate(\sigma, \text{Mul}(e_1, e_2)) \stackrel{\text{df}}{=} (\sigma'', n_1 \times n_2) \text{ where}$$
$$(\sigma', n_1) = evaluate(\sigma, e_1)$$
$$(\sigma'', n_2) = evaluate(\sigma', e_2)$$

Using this updated definition, let us calculate the value of Add(Inc v, Var v) when v starts off with value 0:

$$evaluate(\{(v, 0)\}, \text{Add}(\text{Inc } v, \text{Var } v)$$
$$= (\sigma'', n_1 + n_2) \text{ where}$$
$$(\sigma', n_1) = evaluate(\sigma, \text{Inc } v)$$
$$(\sigma'', n_2) = evaluate(\sigma', \text{Var } v)$$
$$= (\sigma'', n_1 + n_2) \text{ where}$$
$$(\sigma', n_1) = (\{(v, 0)\} \oplus \{(v, \sigma(v) + 1)\}, \{(v, 0)\}(v))$$
$$(\sigma'', n_2) = (\{(v, 0)\} \oplus \{(v, \{(v, 0)\}(v)) + 1)\})(v)$$
$$= (\sigma'', n_1 + n_2) \text{ where}$$
$$(\sigma', n_1) = (\{(v, 1)\}, 0)$$
$$(\sigma'', n_2) = (\{(v, 1)\}, 1)$$
$$= (\{(v, 1)\}, 1)$$

The expression thus updates v to value 1 and returns the value 1. Now let us calculate the value of Add(Var v, Inc v) also when v starts off with value 0:

$$evaluate(\{(v, 0)\}, \text{Add}(\text{Var } v, \text{Inc } v)$$
$$= (\sigma'', n_1 + n_2) \text{ where}$$
$$(\sigma', n_1) = evaluate(\sigma, \text{Var } v)$$
$$(\sigma'', n_2) = evaluate(\sigma', \text{Inc } v)$$
$$= (\sigma'', n_1 + n_2) \text{ where}$$
$$(\sigma', n_1) = (\{(v, 0)\}, \{(v, 0)\}(v))$$
$$(\sigma'', n_2) = (\{(v, \{(v, 0)\}(v) + 1)\}, \{(v, 0)\}(v))$$
$$= (\sigma'', n_1 + n_2) \text{ where}$$
$$(\sigma', n_1) = (\{(v, 0)\}, 0)$$
$$(\sigma'', n_2) = (\{(v, 1)\}, 0)$$
$$= (\{(v, 1)\}, 0)$$

Variable v has, once again, been updated to 1, but the expression now returned 0. Note that the two expressions are identical except that the two operands of addition have been switched over. With the incrementation operator, addition is thus no longer commutative.

Side-effects also lead to other issues which you may never have asked about a programming language. For instance, our evaluation function started by evaluating the left operand, followed by the right operand. However, if we started with the right operand, some expressions would give different results. What evaluation order does your favourite imperative language use? Maybe it is time to read the language reference manual to check. It is worth noting that some pure functional programming languages such as Haskell avoid side-effects altogether, and thus one can reason about the correctness of such programs much more easily, although at the cost of using a completely different paradigm from the imperative one.

In this section, we have seen ways of reasoning about concrete programs—in this case a simple numeric expression language—but also how one can prove things about the programming language as a whole. For instance, if the language has no side-effect operators (the first case we saw), addition is commutative, and a compiler writer is free to switch them over. In the next section, we will look at an alternative approach to reasoning about programs.

Exercises

10.10 Enrich the (side-effect-free) expression type with integer division and re-mainder operators and add the necessary definitions to *evaluate*.

10.11 We can extend the simplification function to also cover incrementation: $simplify(\text{Inc } e) \stackrel{\text{df}}{=} \text{Inc}(simplify(e))$. Identify a valuation σ and expression e for which the new simplification procedure yields an incorrect result: $evaluate(\sigma, e) \neq evaluate(\sigma, simplify(e))$.

10.12 Define $evaluate_r$, which evaluates expressions but starting with the right operand in the case of addition and multiplication. Calculate $evaluate_r(\{(v, 0)\}, \text{Add}(\text{Var } v, \text{Inc } v))$ to see how it differs from the worked example we have seen in this section.

10.2.2 Program Semantics

We have just seen how one can reason about programs, or program fragments, by transforming them to calculate their outcome. In this section we will see another way of expressing the semantics of a more complex programming language. Let us look at a typical program we would like to be able to reason about. Look at the following program, and understand what it does when initially n and m are non-negative, and not both zero, with $n \geq m$:

```
while (m != 0)
   begin
      t = n;
      n = m;
      m = t mod m;
   end
```

This is an iterative implementation of the GCD algorithm, leaving the result $GCD(n, m)$ (where n and m are the initial values of variables n and m) in variable n. How can one prove that this algorithm is correct? We could try to extend the function *evaluate* to work on more complex programs than the simple expressions we have seen. What would applying *evaluate* on a while-loop look like? There are two cases to consider—when the valuation tells us that the loop condition is false, and when it is true. When it is true, it will evaluate the loop body and then the whole loop again. Recall that this type of recursion in mathematical functions is not allowed, since the recursive parameter is not becoming strictly smaller. The problem is that, if given an infinite loop to check, *evaluate* will be undefined. How can we get around this problem?

One common solution is to express mathematically what it means to simulate a program along 'one' step—this can be expressed as a relation over valuations and programs, telling us that with a particular valuation, a particular program ends up with a new valuation and new program left to be executed. For instance: $(\{(n, 4), (m, 2)\}, \; n = m + 1; m = 1)$ would be related to $(\{(n, 3), (m, 2)\}, \; m = 1)$. Note that the valuation is necessary, since otherwise there would be no way to evaluate $m + 1$.

Let us start by building a type to be able to express imperative programs:

$$
\begin{aligned}
\mathsf{Prg} ::= \;&\mathsf{Done} \\
| \;&\mathsf{Assign} \, (\mathsf{Identifier} \times \mathsf{Exp}) \\
| \;&\mathsf{Seq} \, (\mathsf{Prg} \times \mathsf{Prg}) \\
| \;&\mathsf{IfThenElse} \, (\mathsf{Exp} \times \mathsf{Prg} \times \mathsf{Prg}) \\
| \;&\mathsf{While} \, (\mathsf{Exp} \times \mathsf{Prg})
\end{aligned}
$$

The constructor Assign takes a variable and an expression, corresponding to variable assignment. We will be using the type of expressions we have seen in the previous section, adding division operators, but leaving out side-effects for simplicity. Seq corresponds to sequential composition—executing the two given programs in sequence, starting from the first parameter. IfThenElse and While both take an expression as a condition. Our programming language will interpret expressions which reduce to 0 to be false and everything else to true. Note that the conditional takes two further programs as parameters (the *then* and the *else* branches) while the loop construct takes one further parameter—the loop body. The GCD algorithm we saw earlier would thus be expressed as:

$$
\begin{aligned}
GCD_{Prg} &\overset{\mathrm{df}}{=} \\
&\mathsf{While} \, (\mathsf{Var} \, m, \\
&\quad \mathsf{Seq} \, (\\
&\qquad \mathsf{Assign}(t, \mathsf{Var} \, n), \\
&\qquad \mathsf{Seq} \, (\\
&\qquad\quad \mathsf{Assign}(n, \mathsf{Var} \, m), \\
&\qquad\quad \mathsf{Assign}(n, \mathsf{Mod}(\mathsf{Var} \, t, \mathsf{Var} \, m)) \\
&\qquad) \\
&\quad) \\
&)
\end{aligned}
$$

What about the Done constructor? We will be using this constructor to represent a terminated program, and it will not appear in programs which the user writes, but will be produced as part of our relation to reason about the execution of a program. This will become clearer as we start looking at the programming language semantics.

We will define the semantics of the programming language Prg as a homogeneous relation $step$: (Valuation, Prg) $\leftrightarrow$ (Valuation, Prg). If running program p with memory σ for one step (instruction), changes the state to σ' and leaves program p' to be executed, then $((\sigma, p), (\sigma', p')) \in step$.

Let us consider the assignment instruction. Performing one step along a program consisting of just a single assignment will update the value of the variable being assigned (to the value of the expression according to the current valuation) and terminate. The following set comprehension identifies the pairs which should appear in $step$:

$$\{\sigma : \text{Valuation}, \ v : \text{Identifier}, \ e : \text{Exp} \bullet$$
$$((\sigma, \text{Assign}(v, e)), (\sigma \oplus \{v, evaluate(\sigma, e)\}, \text{Done}))\}$$

Similarly, for instance, one can define the meaning of a conditional statement. The following set of pairs representing when the *then* branch is taken will be part of the $step$ relation:

$$\{\sigma : \text{Valuation}, \ p_1, p_2 : \text{Prg}, \ e : \text{Exp} \ |$$
$$evaluate(\sigma, e) > \text{Zero} \bullet ((\sigma, \text{IfThenElse}(e, p_1, p_2)), (\sigma, \ p_1))\}$$

Similarly, the following set represents when the *else* branch is taken:

$$\{\sigma : \text{Valuation}, \ p_1, p_2 : \text{Prg}, \ e : \text{Exp} \ |$$
$$evaluate(\sigma, e) = \text{Zero} \bullet ((\sigma, \text{IfThenElse}(e, p_1, p_2)), (\sigma, \ p_2))\}$$

In this manner, all the language constructs can be given a meaning, with $step$ being the union of all these sets of pairs. The main drawback with this approach is that, as can be seen from the above expressions, the semantics of a language can become rather difficult to follow. To overcome this, in most computer science texts, one finds the semantics expressed as axioms and rules of inference. We will show the semantics of Prg expressed in such a manner. We will write $(\sigma, p) \mapsto (\sigma', p')$ as shorthand for $((\sigma, p), (\sigma', p')) \in step$.

Assignment: The axiom to be able to reason about assignment is the following— the program consisting of just a single assignment will update the value of the variable and terminate immediately:

$$(\sigma, \text{Assign}(v, e)) \mapsto (\sigma \oplus \{v, evaluate(\sigma, e)\}, \text{Done})$$

Conditionals: The conditional statement requires two axioms, corresponding to the two sets of pairs we saw before:

$$\frac{}{(\sigma, \ \mathsf{IfThenElse}(e, p_1, p_2)) \mapsto (\sigma, \ p_1)} \ evaluate(\sigma, e) > \mathsf{Zero}$$

$$\frac{}{(\sigma, \ \mathsf{IfThenElse}(e, p_1, p_2)) \mapsto (\sigma, \ p_2)} \ evaluate(\sigma, e) = \mathsf{Zero}$$

Loops: While-loops are surprisingly straightforward to give a semantics to. If the expression resolves to zero, the program terminates immediately:

$$\frac{}{(\sigma, \ \mathsf{While}(e, p)) \mapsto (\sigma, \ \mathsf{Done})} \ evaluate(\sigma, e) = \mathsf{Zero}$$

When the expression is not zero, $\mathsf{While}(e, p)$ will be transformed into the program which first executes p, and then executes $\mathsf{While}(e, p)$ again:

$$\frac{}{(\sigma, \ \mathsf{While}(e, p)) \mapsto (\sigma, \ \mathsf{Seq}(p, \mathsf{While}(e, p)))} \ evaluate(\sigma, e) > \mathsf{Zero}$$

Sequential composition: This leaves just sequential composition. How does the program $\mathsf{Seq}(p_1, p_2)$ behave? It will take one step along p_1—after which we have to consider two cases: (i) when p_1 terminates immediately, and (ii) when it does not. The first rule of inference tells us that, if we can show that p_1 terminates after one step, then $\mathsf{Seq}(p_1, p_2)$ can perform the same step to reach p_2:

$$\frac{(\sigma, \ p_1) \mapsto (\sigma', \ \mathsf{Done})}{(\sigma, \ \mathsf{Seq}(p_1, p_2)) \mapsto (\sigma', p_2)}$$

The second rule says that, if p_1 goes to p_1' (which is not Done) after one step, then $\mathsf{Seq}(p_1, p_2)$ can also perform the same step, reaching $\mathsf{Seq}(p_1', p_2)$:

$$\frac{(\sigma, \ p_1) \mapsto (\sigma', \ p_1')}{(\sigma, \ \mathsf{Seq}(p_1, p_2)) \mapsto (\sigma', \mathsf{Seq}(p_1', p_2))} \ p_1' \neq \mathsf{Done}$$

With these seven axioms and rules of inference, we have just characterised the semantics of the programming language. For instance we can formally prove how the GCD program evolves step by step. Although the proofs involve long expressions (representing the programs), they are straightforward to calculate using a machine. The fact that $\mapsto$ is expressed as a relation also allows us to be able to talk about a program evolving along a number of steps to another state using the transitive closure $\mapsto^*$. For instance, one can formally prove that:

$$(\{n = 70, \ m = 28, \ t = 0\}, \ GCD_{Prg}) \mapsto^* (\{n = 14, \ m = 0, \ t = 28\}, \ \mathsf{Done})$$

Using the step relation, we can also write new predicates to express properties of programs. For instance, the property that, starting from a particular valuation, a program may never terminate can be expressed by saying that there is an infinite sequence of valuations and programs along which a program may evolve:

$$\infty(\sigma, p) \overset{df}{=} \exists ss : seq_\infty \mathsf{Valuation}, \; ps : seq_\infty \mathsf{Prg} \cdot$$
$$\forall i : \mathbb{N} \cdot (ss(i), ps(i)) \mapsto (ss(i+1), ps(i+1))$$

We can now specify exactly what we expect of the GCD algorithm written in our imperative language:

For any $\sigma \in \mathsf{Valuation}$ satisfying $\sigma(\mathrm{n}) \geq \sigma(\mathrm{m})$, not both zero:

(i) The program GCD_{Prg} always terminates: $\neg\infty(\sigma, \; GCD_{Prg})$.
(ii) Upon termination, the value of variable n is equal to the GCD of the initial values of n and m: $\forall \sigma' : \mathsf{Valuation} \cdot (\sigma, GCD_{Prg}) \mapsto^*$
$(\sigma', \mathsf{Done}) \Rightarrow \sigma'(\mathrm{n}) = GCD(\sigma(\mathrm{n}), \sigma(\mathrm{m}))$

In our programming language, the step relation is, in fact, a function. Given a valuation and a program, there is only one way of stepping forward. However, for languages with richer operators, this is not always so. For instance, if we add a parallel composition operator: $\mathsf{Par}(p_1, p_2)$, which runs p_1 and p_2 concurrently, we can define the semantics of the new operator as stepping along either of the two programs:

$$\frac{(\sigma, \; p_1) \mapsto (\sigma', \; p_1')}{(\sigma, \; \mathsf{Par}(p_1, p_2)) \mapsto (\sigma', \mathsf{Par}(p_1', p_2))}$$

$$\frac{(\sigma, \; p_2) \mapsto (\sigma', \; p_2')}{(\sigma, \; \mathsf{Par}(p_1, p_2)) \mapsto (\sigma', \mathsf{Par}(p_1, p_2'))}$$

Note that, now, the step relation is no longer functional, since in the case of parallel composition, it has two different steps it may take. Using this approach of specifying the semantics of a programming language makes it rather easily to add new operators to the language and give their semantics. Although proving things about the programs can be long and tedious, one can easily automate certain checks. In fact, given such semantics of a language, it is straightforward to write an interpreter or emulator of the language. The only problem is that, as we will see in the next section, not all checks can be automated.

Exercises

10.13 Add simple branched conditionals (conditionals without an *else* clause) to the language as a new constructor, and extend the operational semantics to handle them.

10.14 Modify the rules to handle expressions with side-effects.

10.15 The parallel composition rules given do not cater for when one of the programs terminates. Modify them, and add new rules as necessary to handle when the program we step along terminates after a single step.

10.3 The Uncomputable

We have seen how programs can be derived or proved to be correct with respect to a specification, and how one can give a precise, mathematical meaning to programs. One question that arises is whether any specification we can write using mathematics can be correctly implemented as a computer program. The short answer is 'No, there are problems which we know a computer program can never solve.' What is probably most surprising about the results we will show in this section is that the problems we will show not to be computable by a machine are precise mathematical ones. We are not saying, for instance, that 'Computers will never be able to process emotions,' or a similar problem which uses vague, non-precise terms. We will be identifying concrete mathematical specifications of problems, which we will show not to be computable by a machine. Also, we will prove that *no computer program can ever* solve these problems—it is not simply that we do not yet know how to solve them today. And as you will see, these problems are not esoteric mathematical ones, but ones which we encounter every day in computer science.

10.3.1 Counting Computer Programs

Before we prove that certain problems cannot be solved by a computer program, we will start by proving that the set of all programs (for any programming language you may prefer) is denumerable—a result we will use in the following sections.

Consider the set of all computer programs written in some particular computer language X. X can be any language of your choice—Java, C, Haskell, Prolog, Pascal, etc. The important thing to note common to all these languages is that they are expressed as strings of symbols. If we encode all the symbols one may use in a table as, for instance, is done using ASCII or Unicode, we can encode each computer program as a finite sequence of natural numbers. You may recall, from Sect. 9.1.4.3, that using Gödel numbering, we can encode a list of natural numbers as a single natural number. Therefore, we can transform programs into numbers, such that no two programs have the same number. Using this, we can order the set of all programs written in language X (each of which is a string) as follows:

$$p_0 \stackrel{\text{df}}{=} \text{the program in } X \text{ with the smallest Gödel number}$$

$$p_1 \stackrel{\text{df}}{=} \text{the program in } X \text{ with the second smallest Gödel number}$$

$$\vdots$$

This gives us a total bijective function between the natural numbers and programs written in language X: $f(i) \overset{\text{df}}{=} p_i$. It is (i) total, since the set of different programs is infinite and each one has a different Gödel number; (ii) surjective, since any program has a finite Gödel number n, and must therefore appear somewhere between p_0 and p_n; (iii) injective, since a program has a unique Gödel number, meaning that it will appear only once in the sequence; and (iv) functional, since no two programs have the same Gödel number. We can thus conclude that the cardinality of the set of all programs written in X is the same as that of the natural numbers.

Exercises

10.16 Another way of proving that programs written in X are denumerable is to sort the set of programs in lexicographic order: x is lexicographically smaller than y if either (i) x is shorter than y; or (ii) x is the same length as y, but alphabetically comes before. The program p_i would now be the program in i^{th} position when sorted lexicographically. Argue how this also gives us a total bijection between $\mathbb{N}$ and programs in X.

10.3.2 Sets of Numbers

Consider the set of prime numbers P. We can write a program which, given a number n, tells us whether $n \in P$. We simply check the remainder obtained when dividing n by numbers from 2 till $n - 1$. If all of them leave a non-zero remainder, then n must be prime. Otherwise, it is not. This may not be the most efficient way of checking for primality, but the point is that a program to compute whether a given number is in P exists. Similarly, we can write a program to check whether a number is in the set of even numbers, or the set of exact square numbers. We say that a program p decides a set S if, when program p is given input n, it always correctly returns whether $n \in S$. The question is whether all sets of natural numbers S are decidable. Can we always write a program which checks whether a given input n is in a set S?

We start by noting that a program which recognises sets of numbers can be characterised as an infinite sequence of zeros and ones. The value at position i of the infinite sequence will tell us whether the program numbered i will report i to be in the set or not. Look at the following table of examples:

	0	1	2	3	4	5	6	7	8	9	10	11	12	13	14	15	16	17	18	19	20	21	22	23	...
evens	1	0	1	0	1	0	1	0	1	0	1	0	1	0	1	0	1	0	1	0	1	0	1	0	...
odds	0	1	0	1	0	1	0	1	0	1	0	1	0	1	0	1	0	1	0	1	0	1	0	1	...
squares	1	1	0	0	1	0	0	0	0	1	0	0	0	0	0	0	1	0	0	0	0	0	0	0	...
primes	0	0	1	1	0	1	0	1	0	0	0	1	0	1	0	0	0	1	0	1	0	0	0	1	...

Each row (apart from the top one) represents the set named in the first column. Note that, for instance, the value in column 19 of primes is 1, since 19 is a prime, while column 19 of evens is 0, since 19 is not an even number. We will use this representation to show that there exist sets of numbers for which we cannot write a program to check inclusion.

Theorem 10.10 *There are sets of natural numbers which are not decidable.*

Proof Outline Since the set of all programs is denumerable, restricting the set to programs which decide a set of numbers still leaves a denumerable set. Let us call these programs d_0, d_1, etc. Now let us draw a table, with an infinite row for each program d_i, each row being the characteristic sequence of zeros and ones of the set which program d_i decides.

	0	1	2	3	4	5	6	$\cdots$
d_0	1	0	1	0	1	0	1	
d_1	0	1	1	1	0	1	0	$\cdots$
d_2	0	0	0	1	0	1	1	
d_3	1	1	1	0	0	0	0	
$\vdots$				$\vdots$				$\ddots$

Now consider the set S with the characteristic sequence being the diagonal of the table, but with the values flipped from 0 to 1 and vice versa. A number i will thus be in set S if and only if calling program d_i with input i returns false. In the example table, this set would correspond to the list $\langle 0, 0, 1, 1, \ldots \rangle$. Now, which program d_n decides this set? It is not d_0, since giving input 0 to d_0, one obtains 0 when a 1 is expected. Neither can it be d_1, since it does not match with input 1. Similarly, it cannot match any d_i, since it will not match upon input i.

We have thus just constructed a set of numbers which is not recognised by any set-decision program.   □

What is fascinating about this proof is that we make no assumption about the programming language, except that the programs must be strings of symbols. The conclusion is that there are sets of numbers which no program can decide, no matter how smart the programmer may be.

10.3.3 Parsing Languages

We have thus shown that some problems cannot be solved by a program. Some might argue 'Fine, but how many times do we write a program which checks for inclusion of a number in a set? Are there interesting computer science problems which a program may not solve?'

Let us consider parsing of a language. A language can be represented as a set of strings which are part of the language. For instance, English can be seen as the set of all valid English sentences, while Java can be seen as the set of all valid Java programs. A parser of a language L takes a string s as input, and returns whether that string is in the language: $s \in L$. Compilers start by parsing the input; user input validation on web forms uses parsers; tools which handle natural language use parsers—parsing is a central problem in computer science. A valid question would be whether any given language can be parsed by a computer program. The answer is 'No.'

Theorem 10.11 *There are languages which no program can parse.*

Proof Outline The proof is almost identical to the proof of sets of numbers, but with a short prelude. Recall from the discussion about programs that the set of strings over a number of symbols is denumerable. In other words, we can order all possible strings as s_0, s_1, s_2, etc. We now proceed as we did with the number decision programs, but limit ourselves to programs which parse strings. We will enumerate these programs and call them r_0, r_1, r_2 etc., based on which, we can draw a table just as before:

	s_0	s_1	s_2	s_3	s_4	s_5	s_6	$\cdots$
r_0	1	0	1	0	1	0	1	
r_1	0	1	1	1	0	1	0	$\cdots$
r_2	0	0	0	1	0	1	1	
r_3	1	1	1	0	0	0	0	

In the above table, parser r_1, for example, would accept strings s_1, s_2 and s_3, but not s_0 or s_4. Just as we did before, we construct a new language L which includes (i) s_0 if and only if r_0 does not accept s_0; (ii) s_1 if and only if r_1 does not accept s_1; etc. Therefore, $s_i \in L$ if and only if r_i does not accept s_i. Which parser accepts L? Just as before, it is not r_0 (since L and the language accepted by r_0 disagree on input s_0); it is not r_1 since they disagree on s_1; and in general, it is not r_i, since L and the language accepted by r_i do not agree on s_i. Therefore, no parser accepts L.   □

That sounds like bad news. Whenever we design a new programming language, maybe we should be worried about whether a parser can be written for that language. Luckily, computer scientists have developed techniques for describing subclasses of languages for which parsers can always be written. Still, the fact remains that there are languages for which no parser exists.

10.3.4 The Halting Problem

We will now look at yet another incomputable problem. When we were giving formal semantics to the programming language Prg, we defined a predicate which says whether or not a program starting from a particular valuation terminates. This may sound like a useful feature to have in a compiler—knowing whether a program can end up in an infinite loop would be a great debugging tool. The naïve approach of simulating the program, and upon termination, reporting to the user does not work, because when applied to non-terminating programs, this approach never gives an answer. But maybe there is a smarter solution which does not involve simulating the code. Can it be done? Why does no compiler ever tell you whether your program will ever enter an infinite loop? The sad truth is that, in general, it is impossible.

Theorem 10.12 *It is impossible to write a program halt, which given a program p and its input i, calculates whether p with input i terminates.*

Proof Outline The proof of the impossibility of solving the *halting problem* (as it is usually called) can be presented using a proof similar to the previous two we have seen. However, we will look at a slightly different proof which works by contradiction.

Let us start by assuming that *halt* does exist. This program will take two string parameters p and i, respectively, representing a program and its input. The program *halt* will always terminate and report (correctly) whether p terminates with input i.

Let us now define the following program:

```
weird(p) {
    if (halt(p, p)) {
        loop forever;
    } else {
        terminate;
    }
}
```

This program takes one string parameter, and passes it on to *halt* as both parameters. Since the program, like the input, is considered a string, this is a valid step. Now consider the program *weird(weird)*. Either it terminates or it does not—let us consider both cases separately:

- **It terminates:** If *weird(weird)*, then the condition inside the definition of *weird* must have been false. Therefore, *halt(weird, weird)* must have been false, which leads to a contradiction, since in this case, we consider the cases when calling *weird* with itself as a parameter terminates.
- **It does not terminate:** If calling *weird* with *weird* as a parameter does not terminate, then the condition must have been satisfied, since that is the only way in which it can loop forever. Therefore, *halt(weird, weird)* must be true, which contradicts the fact that *weird(weird)* does not terminate.

Either way we have reached a contradiction. Therefore *weird(weird)* can neither terminate, nor not terminate, which is impossible due to the law of the excluded middle. Therefore, our only assumption, that program *halt* exists, is false.   □

This result shows that no compiler which always tells the user whether or not a program terminates may be written. Obviously, for some programs, this is possible to do—for example, if a program has no loops and no recursion, then it certainly terminates. However, in general it is impossible to always give a correct answer.

This third example is certainly something central to computer science. Termination of programs is something which any computer scientist or programmer is interested in. In fact, what is more worrying about this result is that, we have written the specification of the problem mathematically. This means that sometimes, specifications may turn out to be not implementable due to the inherent limitations of computing.

10.4 Summary

We have seen different ways of bridging the gap between mathematical specifications and the implementations. Since programs are written to be executed, the degree of abstraction is limited—for instance, a program cannot stop at describing what is the correct result, but must go into detail of how it is to be computed. On the other hand, mathematical specifications can leave out such unnecessary detail. For example, in the GCD specification it was sufficient to say that the result should be a factor of the two inputs such that there is no larger such factor—without specifying how to find such a number.

The gap between the expressivity of the two became even more evident as we discovered that there are problems which we can specify mathematically, yet for which there is no implementation them. What is surprisingly strong about the arguments we used to identify the limits of computing is that no assumptions had to be made about the programming language, except that programs were to be written as strings of symbols. The real limitation is that the set of all programs in any programming language is a denumerable set, while out there, one can find problems with cardinality higher than that of the natural numbers. Unless this inherent limitation of computing machines can somehow be surpassed, many problems will remain forever undecidable by machine.

Index